Morphogenesis

Paul Bourgine · Annick Lesne
Editors

Morphogenesis

Origins of Patterns and Shapes

Springer

Editors
Dr. Paul Bourgine
CREA
École Polytechnique
Rue Descartes 1
75005 Paris
France
paul.bourgine@polytechnique.edu

Dr. Annick Lesne
CNRS
Laboratoire de Physique Théorique de la Matière Condensée (LPTMC)
Université Paris VI
place Jussieu 4
75252 Paris Cedex 05
France
lesne@lptmc.jussieu.fr

Translation from the French language edition of *Morphogenése* © 2006 Editions Belin, France

ISBN 978-3-642-13173-8 e-ISBN 978-3-642-13174-5
DOI 10.1007/978-3-642-13174-5
Springer Heidelberg Dordrecht London New York

Library of Congress Control Number: 2010936323

Cover design: eStudio Calamar S.L., Heidelberg

Printed on acid-free paper

Springer is part of Springer Science+Business Media (www.springer.com)

Preface

This book has been first published in French by Belin in 2006. It originates in an interdisciplinary Spring School devoted to morphogenesis and held in Berder Island (Morbihan, France) in March 2002. Although the topics were not directly focused on morphogenesis, we should also mention the ensuing annual sessions of this Berder CNRS Thematic School, organized by one of the editor of the book (A. L.), where most of the authors of the book met and interacted. This initial event has been followed by two 1-day meetings, "Journées Complexité" that we organized in Paris at the Institut Henri Poincaré in November 2003 and November 2004, and again specially devoted to morphogenesis and the specific interdisciplinary approaches required to reach a full understanding. These meetings have been among the launching events founding the Institut des Systèmes Complexes de Paris-Ile-de-France. Several other events, mostly organized by ISC-PIF, reinforced the links between the authors and the motivation to put on the paper and share with readers all the benefits and challenges of an interdisciplinary approach of morphogenesis and pattern formation. The collective enterprise that led to the present book is exemplary of the spirit of this institute and its activities.

Morphogenesis has been specially focused among the main issues arising in the science of complex systems, due to the intrinsic interdisciplinarity of the topic. The reader will discover in this book the range of objects where similar questions about the formation and persistence of their shape arise, and the wealth of complementary concepts and methods involved in their study. All the different facets have been considered together to really grasp what is morphogenesis. The ambition of this book is to offer such a multiple account. It aims to present a collective work rather than a compilation of independent papers, in which authors interacted and mutually influenced each other, and shared a similar vision about morphogenesis although substantiated in (very) different instances and contexts. To complete the cohesion and scope of the book, we propose an overview of the central questions raised by morphogenesis and a presentation of the contents in the introduction (Chap. 1).

We acknowledge the support of the European Community for the translation of this book, as being one of the outcome of BioEmergences, a *NEST-Measuring the impossible* project coordinated by one of us (P.B.) and of Embryomics, a *NEST-Adventure* project coordinated by Nadine Peyriéras, author of Chap. 9. We are much grateful to the wonderful job done by Richard Crabtree, who has to face the challenge of translating 18 chapters with different styles and technical terms from as many different disciplines. It was a pleasure to work with him. We hope that both the authors and the readers will appreciate the final result!

Paris, July 2010

Paul Bourgine
Annick Lesne

Contents

About the Authors

Jean-Pierre Aubin is emeritus professor at the University of Paris-Dauphine, researcher at the CREA and the LASTRE (Applied Controlled Tychastic Systems Laboratory), and a specialist in viability theory and its applications.
aubin.jp@gmail.com

Jean-Claude Bacri is professor at the University of Paris-Diderot. His research lies in the field of soft matter: instabilities and structures, flows in porous media, and the physical properties of ferrofluids, particularly their use in the biomedical domain.
jean-claude.bacri@univ-paris-diderot.fr

Arndt Benecke is an experimenter and theorist working on genomic plasticity. He is principal investigator of Systems Epigenomics Group at the Institut de Recherche Interdisciplinaire and the Institut des Hautes Études Scientifiques.
arndt@ihes.fr

Steffen Bohn is a CNRS researcher at the Complex Matter and Systems Laboratory (UMR 7057 CNRS–Paris-Diderot) of the University of Paris-Diderot. His work in experimental and theoretical physics is centred on plant growth and more generally on the formation of structures.
steffen.bohn@univ-paris-diderot.fr

Yves Bouligand directed one of the teams at the Centre de Cytologie Expérimentale of the CNRS in Ivry-sur-Seine. He is now emeritus director at the École Pratique des Hautes Études. He has researched into tissues and cells displaying liquid crystalline phases, including chromosomes and membranes, and stabilised analogues of a composite nature, components of skeletal formations.
yves.bouligand@univ-angers.fr

Paul Bourgine founded the Complex Systems Institute of Paris Ile-de-France. He is director of the CREA at the École Polytechnique.
paul.bourgine@polytechnique.edu

Hugues Chaté is a physicist at CEA-Saclay. He specialises in collective behaviour in non-equilibrium systems.
Hugues.Chate@cea.fr

Stéphane Douady is CNRS research director at the Complex Matter and Systems Laboratory (UMR 7057 CNRS–Paris-Diderot). His research is focused on dynamic systems, granular media, and morphogenesis, particularly in biology and geology.
douady@lps.ens.fr

Florence Elias is a lecturer at the University of Paris 6. Her research is focused on the physical properties of complex fluids: morphologies of systems containing interacting particles, and the macroscopic properties of liquid foams in relation to their internal structure.
florence.elias@univ-paris-diderot.fr

Sara Franceschelli is an epistemologist and historian of science at Ecole Nationale Supérieure de Lyon. Her research is focused on the history of dynamic and complex systems, and on the figure of landscape in these domains.
Sara.Franceschelli@ens-lsh.fr

Jean-Louis Giavitto is a computer science researcher at the CNRS and director of the IBISC (Computer Science, Integrative Biology and Complex Systems) Laboratory at the University of Évry. His work centres on new computational models (especially those inspired by biological processes): the representation of time and space and the use of concepts of combinatorial topology in programming languages.
giavitto@ibisc.univ-evry.fr

Guillaume Grégoire is a lecturer at the University of Paris-Diderot. He is a researcher in the Complex Matter and Systems Laboratory (UMR 7057 CNRS–Paris-Diderot), where he uses statistical physics to study non-equilibrium physical systems.
guillaume.gregoire@univ-paris-direrot.fr

Pascal Hersen is a CNRS researcher in the Complex Matter and Systems Laboratory (UMR 7057 CNRS–Paris-Diderot). His research interests include dune morphogenesis, the dynamics of signaling pathways of eukaryotic cells and animal locomotion.
pascal.hersen@univ-paris-diderot.fr

François Képès, CNRS research director, co-founder and scientific director of the Epigenomics Project of Genopole® (Évry), and former professor of biology at the École Polytechnique, is a cell and systems biologist.
Francois.Kepes@epigenomique.genopole.fr

Annick Lesne is a CNRS researcher at the Theoretical Physics of Condensed Matter Laboratory (UMR 7600 CNRS-Paris 6) and the Institut des Hautes Études Scientifiques. Her research speciality is the mathematical and physical modelling of the regulation mechanisms of living systems.
lesne@ihes.fr

Didier Marchand is a paleontologist at the Biogeosciences Laboratory (UMR 5561 CNRS-University of Burgundy). He specialises in the relationship between morphological evolution and embryogenesis and hence in the formation of bauplans.
didoux.marchand@gmail.com

Andras Paldi is a research director at the École Pratique des Hautes Études and a researcher at Généthon. His main field of research is the epigenetic mechanisms regulating gene expression and epigenetic phenomena.
paldi@genethon.fr

Jean Petitot is a research director at the École des Hautes Études en Sciences Sociales and former director of the CREA (École Polytechnique). His research is focused on mathematical models in cognitive neuroscience.
jean.petitot@polytechnique.edu

Nadine Peyriéras is a CNRS researcher at the DEPSN laboratory of the Institut Alfred Fessard in Gif-sur-Yvette. Her chief interest is the embryogenesis and development of the *Zebrafish*.
nadine.peyrieras@inaf.cnrs-gif.fr

Denise Pumain is a professor at the University of Paris I, member of the Institut Universitaire de France, geographer, founder member of the Géographie-cités laboratory and director of the Cybergeo electronic journal, European Journal of Geography.
pumain@parisgeo.cnrs.fr

Caroline Smet-Nocca is an assistant professor at the University of Lille 1. She has worked with Arndt Benecke at the Interdisciplinary Research Institute on the decoupling of epigenetic signaling and transcriptional regulation in genetic diseases and myeloid leukaemia. Her research interests are focused on the regulation of protein structure and function by posttranslational modifications.
caroline.smet@iri.univ-lille1.fr

Antoine Spicher is a computer science assistant professor at the University of Paris XII – Val de Marne and is a member of the LACL (Laboratory of Algorithmic, Complexity and Logic). His research focuses on the use of concepts of combinatorial topology in programming languages for the modeling and the simulation of dynamical systems.
antoine.spicher@univ-paris12.fr

James Tabony is research director at the Commissariat à l'Énergie Atomique, Direction des Sciences du Vivant, Département Réponse et Dynamique Cellulaires, CEA Grenoble, where he is researching the problem of biological self-organisation.
james.tabony@cea.fr

Chapter 1
Introduction

Annick Lesne and Paul Bourgine

This collective book is devoted to a fundamental issue in natural sciences: *morphogenesis*, that is, the ensemble of mechanisms underlying the reproducible formation of patterns and structures and controlling their shape. An important side issue is to understand the functional character, if any, of these patterns and structures and to elucidate whether it is the cause or the consequence (or both) of their emergence.

This book proposes an overview, unusual in the number of different disciplines involved and the range of questions addressed. Can one distinguish classes of shapes? Or families of morphogenetic processes to be found in different systems and at different scales? What answers have D'Arcy Thompson, Thom or Turing given to such questions? What are the relations between the shapes of a system of cities, a school of fish and a cloud of physical particles? And what events must occur within a living cell for it to participate in the formation of one of our fingers?

1.1 Fundamental Issues

1.1.1 The Notion of Shape

The notion of *shape* or *form* is intuitive and ubiquitous, originating from observation and belonging to ordinary language. It nevertheless brings out several issues, that we shall briefly present in this introduction. The aim of our book is to offer an overview, within the reach of non-specialists, of both the conceptual and technical answers today available in the different disciplines.

Speaking of a form (take a circle, for example) and considering it as an autonomous entity, endowed with a precise mathematical definition, is a first step towards the abstraction and simplification that are necessary to make the reality intelligible. The notions of *invariants* and *symmetries* allow to bring out simple shapes, appearing as the shared features of an *ensemble* of systems and, in the

A. Lesne (✉)
CNRS, Laboratoire de Physique Théorique de la Matière Condensée (LPTMC),
Université Paris VI, place Jussieu 4, 75252 Paris Cedex 05, France
e-mail: lesne@lptmc.jussieu.fr

P. Bourgine, A. Lesne (eds.), *Morphogenesis*, DOI 10.1007/978-3-642-13174-5_1,

above example, to recognise a geometric circle where there are only approximately circular rings. Philosophy and mathematics thus offer a pathway to bridge abstract and material shapes. Physics, with its current use of effective modelling at a given scale encapsulating all the details of lower scales, offers another pathway. Finally, in the many situations where the reality cannot be simplified enough just by looking from a distance (for instance if observations are made at low resolution, or are very sensitive to initial conditions, or are perturbed by uncontrolled external influences), shapes will be described in probabilistic terms, and their relevant features will be statistical: our simple description will then be of the *average shape*, or the *typical shape*.

We shall see that understanding a shape, more precisely its *morphology*, almost always requires us to understand the process of its formation, namely its *morphogenesis*. Pattern formation often implies a *selection* mechanism, whose criteria must be determined. Moreover, as regards biological or artificial shapes, their *function* also enters the discussion.

Finally, it is worth noticing the place taken by shapes outside scientific fields: morphogenesis is present at the origin of all cosmogonies, from the most ancient to the most recent ones. Also interesting is to wonder about the success of some questionable variants: the hidden meaning of shapes, invoked in soothsaying and clairvoyance (tea leaves) or pseudo-sciences like phrenology. Throughout this book, we have replaced such a wish for immediate interpretation by that of a deep scientific understanding, without excluding our wonder for the natural shapes surrounding us – quite the contrary!

1.1.2 Some Paths to Explore the World of Shapes

Several guidelines can be followed to travel across the vast scientific domain concerned with morphogenesis, shapes and pattern formation:

- according to the shape types and contexts, either physical and physico-chemical, biological (intra-cellular level, organisms, populations) or artificial. This inevitably leads to the following question: is it relevant to distinguish living shapes from inanimate ones?
- according to the mechanisms at work in pattern formation and persistence. Three classes arise: shapes *at equilibrium*, transient shapes that are *out of equilibrium* insofar as they have not yet reached equilibrium and are still evolving, and stationary shapes *far from equilibrium*, requiring the injection of energy or matter from outside to endure.
- according to the formalisms adopted to describe the shapes and their origin. We shall meet thermodynamics and its variational formulation of equilibria, classical and fractal geometry, dynamical systems, partial differential equations (hydrodynamics or reaction-diffusion systems), set-valued functions, cellular automata, or the L-systems approach combining computer languages and combinatorics. A classification can possibly be made according to the discrete or continuous nature

of the space, time and phase space variables, to the deterministic or stochastic character of the dynamics, or to the scale of description.
- following the chronology, the history of concepts and their "inventors": Geoffroy Saint Hilaire and his contemporary Goethe (who coined the term "morphogenesis"), D'Arcy Thompson [1] and his forerunner Leduc, Rayleigh, Bénard, Turing [12], Prigogine [2, 9], Waddington [13] and Thom [11]. There are as many celebrated books or seminal papers, and the different chapters of the present book will comment their originality and impact. We shall see how the initially naturalistic approach evolved, integrating both geometrical (symmetries, scale invariance, fractal geometry, frustration) and dynamical aspects (self-organisation, dissipative structures).

1.1.3 Shapes and Their Causes

Understanding morphogenesis means understanding the causes of the shapes we observe. This is a delicate issue. For instance, is it relevant to distinguish between the causes leading to the appearance of a structure and those determining its shape (as we might distinguish, in loaves of bread, between the incisions made before baking and the dilating process that reveals the pattern)? Or does the process generating the structure also impose the geometrical laws governing its shape, as it does in the case of phyllotaxis, liquid crystals, fractures and sand dunes? These examples favour the second explanation, but what about embryogenesis? What is the role of genes and the genetic programme: are they a modern version of the *homunculus* (a miniaturised organism already present in germinal cells) or an archive of a successful combination of factors achieving suitable values for the kinetic rates in morphodynamic models? This latter question joins that of the relation between genotype and phenotype, and between ontogeny and phylogeny. We shall see the lessons drawn from paleontology and the importance of the functional role of the shapes considered.

1.1.4 Modelling Morphogenesis

Two traps are to be avoided in modelling morphogenesis:

- the analogy trap: a model can faithfully reproduce a shape and even its movements and deformations, without necessarily giving a faithful account of the mechanisms actually at work. An example is provided by 3D animation, where the shapes and their motion are achieved following rules that totally differ from the actual mechanical or biological ones. Another example is the band structure in drosophila embryo, similar to the patterns globally generated by the coupling of reaction and diffusion processes (Turing structures) although each band is individually controlled by a specific combination of morphogens [7];

- the exhaustiveness trap: a model is not more convincing if it accounts for more details about the system. On the contrary, the possible flaws are over-interpretation and the lack of robustness of predicted behaviours, if they sensitively depend on the details taken into account and those ignored. The *explanatory power* of a model essentially comes from its *minimal character*: it should integrate the ingredients and mechanisms we believe to be at the origin of the considered shape and show that they are both necessary and sufficient. Agreement between the predicted structures and those observed will support the working hypotheses involved in the model design. Only in a second step, for prediction purposes, can the minimal model thus validated be refined to obtain a better quantitative match with experimental data.

1.2 Morpho-Genesis

1.2.1 Shape-Generating Mechanisms

The investigation of patterns brings to light different situations (although they are often intermingled in real systems):

- shapes prescribed by constraints and boundary conditions;
- shapes generated by an external field or force;
- shapes emerging from static interactions between the elements;
- shapes emerging from dynamics and controlled by its parameters; the dynamics could be described at the level of the elements and their interactions or, at a higher scale, in terms of concentrations.

Morphogenesis is a phenomenon for which *several levels of explanation coexist*: local mechanistic schemes, dynamical stability arguments (that is, a criterion of robustness with respect to perturbations of initial conditions) and finally criteria of functional optimisation (natural selection in the biological context, optimality of fluxes or dissipation in the physico-chemical context, social or economic optimality in the case of cities). One task of the theorist is to articulate these different levels within an integrated model. The challenge is to reconcile identification of *individual, local and instantaneous* mechanisms, with stabilising or amplifying *global* mechanisms, involving couplings and feedbacks between the different levels of organisation.

1.2.2 Equilibrium, Out-of-Equilibrium and Far-from-Equilibrium Shapes

Distinguishing between equilibrium, out-of-equilibrium and far-from-equilibrium shapes, as mentioned in Sect. 1.1.2, is a delicate task. Firstly, the concept of equilibrium is *very relative*, since it is defined with reference to scales in time, space and phase space: speaking of equilibrium makes sense within a given description of the system, specified by a set of state variables (the *observables*) and at a given time scale, when observables no longer evolve at this time scale. By contrast, their evolution could become perceptible on a shorter term, and in a similar way the evolution of more microscopic features could become noticeable at the considered time scale.

Equilibrium can be *mechanical*, *thermal*, or *chemical*, according to the nature of the observables; a system could have reached mechanical equilibrium but not thermal equilibrium, or on the contrary, move under the influence of an external field while being at thermal equilibrium.

In the case of shapes arising in a system of particles, we have to differentiate between interactions between non-reactive species, and dissipative and reactive interactions between chemical species. In this latter case, the structures that arise are either transient, until a least one reactive species is totally consumed, or stationary in open systems with continuous injection of reactive species. In the former case, one speaks of *interactions at equilibrium*, meaning that they occur in a closed system, usually with uniform temperature, and that they do not produce flows, in contrast to the emergence of convection cells or sand dunes. However, pattern formation will involve a transient regime, during which elements settle one with respect to the other. This transient regime can in some cases be very slow and last a long time – long enough to be observed – and is called *out of equilibrium* (meaning "not yet at equilibrium"). Once equilibrium is reached, elements can still be renewed or exchanged.

1.2.3 Irreversibility

The irreversible character of morphogenetic processes appears to be two-fold:

- in a structure at equilibrium, irreversibility originates from the *spontaneous evolution* of the system toward a free energy minimum. Once an organised structure has arisen, it requires energy (for instance heat) to modify or suppress it. This is the case for crystals or viral capsides.
- in a system far from equilibrium, the irreversible character is *enforced* by the coupling between the system and a current or an external field (for instance coupling between dunes and wind, or between reactive chemical species). It disappears as soon as the forcing is suppressed or inverted.

1.2.4 Self-Assembly and Self-Organisation

Among the main types of morphogenetic processes, two apparently close terms are encountered: *self-assembly* and *self-organisation*. We speak, for instance, of the self-assembly of a viral capside or the self-organisation of a mitotic spindle. In both cases, the construction corresponds to the integration of isolated elements into an emerging functional structure. We can nevertheless highlight two differences:

- elements and their properties are not affected by their integration into a self-assembled structure, whereas self-organisation modifies their behaviour, for instance their reactive kinetics;
- once formed, a self-organised structure still depends on the external input of matter or energy, whereas the self-assembled object is stable and autonomous: it is possible to extract and manipulate a viral capside but it is impossible to separate the mitotic spindle from its intra-cellular surroundings and the dynamic processes that take place within it.

To put it briefly and simply, self-assembly refers to processes leading to equilibrium structures, while self-organisation refers to far-from-equilibrium pattern formation.

1.3 Instabilities, Phase Transitions and Symmetry Breaking

1.3.1 Phase Transitions, Bifurcations and Instabilities

Formation of spatial patterns from an homogeneous initial situation represents a qualitative change in the collective behaviour of the constituent elements of the system, what is called a *phase transition*. In their most widely-used meaning, illustrated by liquid-vapour transitions or ferromagnetic transitions, phase transitions are observed in the thermodynamic limit (i.e. the limit when the number of particles tends to infinity at constant density), and correspond to a discontinuity in thermodynamic quantities (first-order transitions) or in their derivatives (second-order transitions). The very existence of a transition can be explained qualitatively as a change in the trade-off between the order resulting from interactions and the disorder induced by thermal fluctuations. A first-order transition is accompanied by a partition of the real space into domains, each occupied by one of the two phases that coexist at the transition point; this partition settles in a way that minimises the energetic cost of boundaries between the domains. By contrast, one of the signatures of a second-order transition is the appearance of fractal structures, for instance in density fluctuations, at the transition point, see Sect. 1.3.4.

The appearance of a rhythm, starting from a stationary situation, is the exact temporal analogue of a phase transition; it is termed a *bifurcation* of the dynamics, that is, a qualitative change in the asymptotic regime in the phase space, leading for instance from a situation where all trajectories converge to a fixed point to one where they wrap closer and closer around a periodic orbit. Bifurcations can also lead to more complicated regimes than a fixed-point or limit cycle, for instance quasi-periodic or chaotic regimes. The phenomenology is even richer in spatially extended systems, typically described by partial differential equations. The transition from one regime to another is then termed an *instability*, and it is most often accompanied by the appearance of regular and reproducible spatial (or spatio-temporal) structures, for instance the convection cells observed in a layer of liquid heated from below or the Turing structures observed in an appropriate reaction mixture.

1.3.2 Symmetry Breaking

A notion closely related to shapes and pattern formation is that of *symmetry breaking*. This statement might be surprising at first sight, since numerous shapes exhibit striking symmetries: the hexagonal symmetry of cells in a beehive, the periodicity of Turing structures (regularly spaced stripes or spots, according to the system geometry) and convection cells, the bilateral symmetry of vertebrate organisms, the five-fold symmetry of echinoderms. And yet symmetry-breaking has occurred in all of these examples, compared to the full symmetry of an homogeneous state (invariant upon any translation, rotation, reflection, inversion or rescaling). More precisely, we speak of symmetry breaking whenever the solution of the equilibrium or evolution equation exhibits *less symmetries than the equation itself.* For instance, in the case of Turing structures, the evolution equations (a system of two coupled partial differential equations) are invariant upon *any* translation $\mathbf{r} \rightarrow \mathbf{r}+\mathbf{r}_0$, whereas the emerging pattern is invariant only upon translations adapted to its periodicity, shifting space points by a distance equal to an integral number of periods.

Shapes can be classified according to the symmetries they exhibit. These symmetries mirror the underlying morphogenetic process and give useful clues for understanding it. Symmetry changes observed in a given system correspond to its phase transitions (at equilibrium) or bifurcations (in a dynamical system). Explaining symmetry breaking thus addresses the central issue of morphogenesis: to account for the reproducible formation of patterns that nothing in the initial or boundary conditions can foreshadow or control. It is in fact the scientific variant of the primordial question of all cosmogonies: to explain how the order of the world and the very existence of structures have emerged from a formless primeval situation ...

1.3.3 Emergence

The appearance of spatial or spatio-temporal structures in an initially homogeneous system of identical elements is widely termed an *emergent property*. This term underlines the spontaneous character of such structures and the impossibility of predicting them on the basis of one sole element. A counterexample is provided by crystals, where knowledge of the elementary crystal cell is sufficient to predict the macroscopic shape, or a building whose construction follows a blue-print. In the case of an emergent structure, by contrast, the shape and its appearance originate in the *collective phenomenon* resulting from the interactions between the elements. Emergent structures typically result from the interplay between local processes[1]

---

[1] Constitutive global mechanisms like long-range interactions, or mechanisms imposed from outside, like boundary conditions, special geometry or an external field, can be superimposed, but such influences, trivially inducing a coherent macroscopic effect, rather confuse the understanding of emergent properties; for this reason, they are ignored in minimal models of self-organisation, phase transitions and collective phenomena (see Chap. 12).

(e.g. chemical reactions) and couplings between neighbouring regions (e.g. through molecular diffusion).

An emergent entity, behaving as a whole, is able to exert a feedback on its elementary causes. It is thus essential to consider (at least) two levels, that of elements and that of collective phenomena, to get a relevant model. In this regard, morphogenesis belongs to the domain of complex systems science, with all the associated difficulties and challenges to obtaining integrated descriptions and explanations.

Another kind of morphogenetic emergence is that resulting from *global optimisation* and *conservation relations* (e.g. mass, energy, or topological invariants). This can be used to account for the formation of loops in an elastic rod (plectonemes), as exemplified by old telephone wires or DNA molecules, or the growth of fractures in a drying film. Pattern formation generally arises in frustrated situations where constraints are strong and conflicting enough to involve a non trivial trade-off (in particular, the homogeneous and formless state that follows from a local optimisation is not a global solution), yielding one or more solutions with remarkable structural features.

1.3.4 Fractal Shapes

One domain that will not be specifically exposed in the present book is that of fractal structures and their morphogenesis, due to the already large number of existing books on this topic, beginning with the seminal publications by Mandelbrot [4, 5]. We shall simply underline a few points specific to fractal shapes in this introduction.

The same mechanisms can give rise either to regular shapes or to fractal structures, according to their regimes (i.e. according to the values of their control parameters). A typical example is encountered in hydrodynamics, where the transition from a laminar regime to a turbulent one, for instance at increasing flow velocity, is accompanied by the appearance of a *fractal hierarchy* of nested eddies (they ensure the transfer of energy from the macroscopic scale at which it is injected into the fluid down to the molecular scale at which the viscous dissipation becomes effective). The fractal structure here allows to reconcile the self-similarity of the hydrodynamic equations with the constraints on the ingoing and outgoing energy fluxes.

More generally, fractal morphogenesis can follow from the *reiteration of local processes* (branchings, subdivisions, hierarchical aggregations). It leads to self-similar structures allowing to reconcile in a flexible and adaptable way (by varying the number of levels in the structure) independent or even competing constraints or requirements, e.g. maximisation of the surface within a given volume.

Fractal structures can also be seen as a *signature of critical phenomena*, either spatial (critical opalescence), temporal (onset of chaos) or spatio-temporal (random walks). In these particular regimes, the competition between interactions and sources of disorder (thermal motion, external noise, boundary conditions) does not favour one over another. On the contrary, it leads to a marginal situation, characterised by long-range spatial and temporal correlations. This statistical *catastrophe* finds expression in the appearance and persistence of fluctuations at all scales, typically self-similar and satisfying scaling laws. Their structural counterparts are fractal features (density inhomogeneities, strange attractors, Brownian paths). We here recover an emergent property, since nothing in the features of the elements and their interactions allows to predict a critical transition, nor its location in the control parameter space.

1.4 Inanimate or Living Shapes

The comparison between physical or physico-chemical morphogenesis and examples from biology or social sciences raises a fundamental issue: are there morphogenetic processes specific to living systems? In particular, what about embryogenesis and the processes that take place during development of the organism? This issue links up with the keen debate opposing the champions of a genetic programme giving instructions in the image of a computer programme, and those who explain the appearance of living structures in terms of self-organisation processes.

1.4.1 Some Questions

By comparing the different examples dealt with in this book, we can raise a series of questions representative of the delicate articulation between physical and biological forms.

- Is the similarity between band patterns observed in ferro-fluids and zebra stripes (Chap. 2) due to common mechanisms? And what about the similarities between a drying gel and leaf veins (Chap. 3)?
- Does the self-assembly of crab shells and other living materials follow the same physico-chemical principles as that of inorganic liquid crystals (Chap. 4)?
- The in vitro self-organisation of a solution of microtubules (biological macromolecules) does not differ from other reaction-diffusion processes (Chap. 5); but does the same hold true in vivo?
- To what extent are sand dunes, which are capable of being born, of reproducing, and of moving without losing their shape, living systems (Chap. 6)?
- Do the mechanisms behind intra-cellular organisation (Chap. 7), cell differentiation (Chap. 8) and embryogenesis (Chap. 9) involve purely physico-chemical principles of pattern formation, or are they, on the contrary, programmed and controlled at the genome level?
- And what can we say about the coherent structures – herds or colonies – that living organisms constitute (Chap. 12)? And about those they build, like hives or cities (Chap. 13)?

Answers will be proposed and discussed in the subsequent chapters, but let us already bring some elements, in particular historical ones, to the debate.

1.4.2 Are Living Shapes Special?

Schrödinger underlined, in 1944, that one specificity of living systems is their ability to behave as a whole, in other words to exist as *organised forms at the macroscopic scale* [10]. This statement raises two questions:

- How can such organised forms appear not only in a reproducible but also in a functional and adaptable way?
- Is this feature specific to living systems? In other words, does a system exhibiting such a feature deserve to be qualified as a "living system"?

To understand the nature of a structure (whether it is living or artificial, for example), Monod recommended, in 1972, that one should analyse the process of its formation [8]. According to Monod, living structures can be differentiated from inanimate ones by the fact that they result from *internal forces*. External forces and elements can intervene as obstacles or constraints (a confining constraint or a field, for instance), or as parameters in the morphogenetic processes (the temperature or the nutrient concentration, for instance), but not as a primary and direct cause indispensable to the emergence of structure.

This point was revisited with the concept of *autopoiesis* introduced by Maturana and Varela: an autopoietic system is autonomous, self-limited and capable of self-reproduction [6]. But it is worth underlining the difficulty of making such formal and generalised distinctions: according to Monod's criteria, crystals belong to the class of living systems, distinguished only by a far smaller degree of complexity, and the lasting persistence of an autopoietic system is impossible without an input of energy or matter to pay the cost of the negative entropy associated with the establishment and maintenance of its organisation.

1.4.3 Functional Shapes

The issue of the relation *between a structure and its function* underlies investigations into biological forms. In particular, it has been amply studied in the relatively simple and well-defined context of *protein folding*, that is, the transition from the linear sequence of amino-acids that constitutes the protein to its shape in space, known as its conformation, or more precisely to a conformation of reference called its native conformation. Moreover, the protein activity (either enzymatic, chemical, or structural) relies on a modification of its conformation. The clearest example is that of allosteric enzymes, where the binding of a first ligand to the enzyme modifies its conformation and triggers its activity at the level of another site.

In this example, we return to the far more general fact, already mentioned in Sect. 1.3, that understanding forms, their emergence and their possible function is largely dependent on the identification of *shape changes*, and then of the mechanisms that induce and control them. In parallel, we have to determine whether the appearance of the form is a mere pattern formation without direct functional consequences (for instance, patterns on some animal coats) or rather a major constraint that must be achieved as a primary step conditioning the following steps in the functional process. The issue of the specificity of living systems can thus be reformulated as follows: are the factors that trigger and determine shapes *of a different nature* in physico-chemical systems, biological systems and artificial systems? To answer this question, we need to analyse the respective roles of the genome and of natural selection, and identify their analogues (if any) in non-biological systems, attaching central importance to the functional character of the emergent form. Living forms are to be understood through the function they perform, established over the course of evolution, but these forms are themselves constrained by the criteria of robustness, adaptability and functional optimisation.

Monod introduced the concept of *teleonomy* to emphasise the fact that the functional character of living structures, like the purposeful design of artificial objects, is of a different nature to the physico-chemical mechanisms at work, which are somehow superimposed. This functional character lies in a different causal scheme, involving feedback from the features of the shape as a whole onto the features and even the potentialities of the underlying elements (a scheme that can be summarised by the term "*downward causation*"). In the case of artificial objects, this feedback is that of the maker who, seeing the shape he has produced, updates the building process or directly acts on the outcome to improve it. In living systems, natural selection plays *a posteriori* the role of the maker.

1.4.4 Genetic Programme, Self-Organisation and Epigenomics

Schrödinger, as early as 1944, highlighted the idea that the heredity substrate (DNA had not yet been discovered) was *at the same time the programme and the means of implementing the programme*. He also insisted on the necessity of understanding the physico-chemical mechanisms of the transition from genotype to phenotype and the inheritance of genetic characteristics. Today, the notion of *epigenetics* (see Chap. 8) reconciles the two extreme and incomplete viewpoints of the "fully genetic" or "fully self-organised" living world. This novel concept, still developing, describes how the coupling between the genetic information substrate (the DNA) and all the chemical and physical processes at work in the cell achieves the control of gene expression. The genome appears as a collection, with a quasi infinite number of combinations, of potential scenarios, together with an archive of the successful ones. More broadly, epigenetic mechanisms enable the appearance of different, complementary cell types over the course of embryogenesis. The development of multicellular organisms is thus based on the control, at the chromatin level, of cell identity, and in particular the control of the parameters of all the intra-cellular reactive subsystems, which in turn control the ensuing forms. This gives rise to a whole paradigm for understanding the morphogenesis of living organisms, that can be summed up in the term of *epigenomics*.

1.4.5 The Robustness and Variability of Living Shapes

One of the specificities of biological morphogenesis is its remarkable reproducibility from one organism to another, or within the same organism over time, despite the complexity of these organisms (far greater than that of physical systems at the same scale). This observation points at the robustness of morphogenesis with respect to external perturbations. Three different kinds of response to perturbations have to be distinguished:

- Shape modification due to the influence of new external conditions on the endogenous morphogenetic processes (assumed to be unchanged); this modification is permanent or transient depending on the *stability*[2] of the system state.
- The modification of mechanisms (in particular their genetic components) in response to external changes, in order to preserve the functional characteristics of the shape (here we speak of *robustness*).
- In complex physical systems (for instance sand dunes) or living systems, a third kind of response is observed: this is the *joint* modification of the elementary mechanisms and the ensuing form, to improve its functional adaptation to its new surroundings (here we speak of *adaptability*). This response is multiscale and involves both the elementary ingredients and the emergent features of the system. It is one of the motors of evolution.

Living systems have to achieve a delicate trade-off between two essential but contradictory properties: robustness on the one hand, and the adaptability of functional forms and processes to changes in the environment on the other. An intuitive idea is that the borderline between robustness and adaptation is controlled by the duration of these environmental changes (robustness with respect to fast or transient changes, adaptation to slow and durable ones) [3].

1.5 Book Overview

The tour starts with model physical systems (ferro-fluids, in Chap. 2, liquid crystals, in Chap. 4), and then visits reactive systems (chemical mixtures, in Chap. 5), and complex and multiscale systems (fractures in Chap. 3, sand dunes in Chap. 6), leading us naturally and almost continuously to the issue of morphogenesis in biology.

Morphogenesis arises at very different levels, from the self-organisation of cellular components (Chap. 7), cell differentiation (Chap. 8), phyllotaxis (Chap. 10), embryogenesis (Chap. 9) and phylogeny, with the viewpoint of a paleontologist

2 *Stability* with respect to perturbations of the system state must be distinguished from *structural stability* with respect to perturbations of the dynamics itself; stability breakdown corresponds respectively to a transition from one basin of attraction to another and sensitivity to initial conditions, or to a bifurcation (an instability in the case of a spatially extended system) or more generally a qualitative change in the phase portrait.

(Chap. 11), up to the organisation of herds and swarms (Chap. 12) and human or animal societies (Chap. 13).

The theoretical analysis and modelling of morphogenetic processes are as inter-disciplinary as the objects investigated, since they resort to non-linear or statistical physics and mathematics (reaction-diffusion equations in Chap. 5, dynamical systems and catastrophe theory in Chap. 15, shape derivatives and other set-valued extensions of functional analysis in Chap. 17), as well as computer science with L-systems and cellular automata in Chap. 18. We cannot fail to cite the founding fathers: D'Arcy Thompson (Chap. 14), Turing (Chap. 5), Waddington (Chap. 16), Thom (Chaps. 15 and 16) and Prigogine (Chap. 5), to mention only the most famous ones, and we shall describe the impact of their work, in most cases still resonant today.

We end this introductory overview by observing that the evolution of the views and explanations of morphogenesis over the ages (and particularly during the twentieth century, when it speeded up) parallels the evolution of leading paradigms in the natural sciences. It also reflects the enthusiasm (sometimes excessive and lacking discernment in some of their applications) for novel and striking theories. So over the course of this book, we shall meet the initial naturalistic approach, hydrodynamics and the notion of instability, self-organisation, dynamical systems theory – often (wrongly) reduced to chaos theory and catastrophe theory – fractal geometry and scale invariance, genetics, complex systems, and the necessary inter-disciplinarity intrinsically required for the study of morphogenesis.

References

1. D'Arcy Wenworth Thompson (1917) *Form and Growth*, Re-edited by Dover (New York), 1992.
2. Glansdorff P. and Prigogine I. (1971) *Thermodynamic theory of structure, stability and fluctuations*, Wiley-Interscience (London).
3. Lesne A. (2008) Robustness: confronting lessons from physics and biology, *Biol. Rev.* **83**, 509–532.
4. Mandelbrot B.M. (1977) *Fractals: form, chance and dimension*, Freeman (San Franscisco).
5. Mandelbrot B.M. (1982) *The fractal geometry of Nature*, Freeman (San Franscisco).
6. Maturana H. and Varela F.J. (1987) *The tree of knowledge*, Shambhala (Boston).
7. Maynard Smith J. (1998) *Shaping life*, Weidenfeld & Nicholson (London).
8. Monod J. (1972) *Chance and necessity*, Vintage Books (New York).
9. Nicolis G. and Prigogine I. (1977) *Self-organization in nonequilibrium systems: from dissipative structures to order through fluctuations*, Wiley-Interscience (London).
10. Schrödinger E. (1946) *What is life*, McMillan (New York).
11. Thom R. (1972) *Stabilité structurelle et Morphogenèse. Essai d'une théorie générale des modèles*, Benjamin (Reading MA), Édisciences (Paris); 2nd edition 1977, InterEditions (Paris), in French. English translation: *Structural stability and morphogenesis* (1975), Benjamin (Reading MA).
12. Turing A.M. (1952) The chemical basis of morphogenesis. *Phil. Trans. Roy. Soc.* **237**, 37–72.
13. Waddington C. (1957) *The strategy of the genes*, Allens & Unwin (London).

Chapter 2
Ferrofluids: A Model System of Self-Organised Equilibrium

Jean-Claude Bacri and Florence Elias

2.1 Introduction: Situation with Regard to the Other Chapters

In this chapter, we shall limit ourselves to studying the morphologies present in an equilibrium state in physical systems: once formed, the structures no longer evolve over time; they remain stable without the need for any further energy to be injected into the system. This means that we can model them by computing the energy balance of the interactions present; the structure obtained and its characteristic length-scale correspond to a minimisation of the system's energy.

We have therefore chosen not to consider dynamic (non-equilibrium) systems, although they can be the site of self-organisation leading to similar periodical patterns. In these dynamic systems, structure formation is generally the result of either dissipative instabilities (such as Rayleigh-Bénard convective instability or Turing instability), or front instabilities (such as Saffman-Taylor instability, or that of Mullins-Sekerka, which leads to the growth of fingers at an interface). Those cases are dealt with in other chapters of the book (see Chaps. 1, 5 and 6).

2.2 Physical Systems in Self-Organised Equilibrium

There is an incredible regularity in the patterns on the coats of leopards, giraffes and zebras. The same patterns can be found throughout the animal kingdom, in seashells, fish, etc. These universal morphologies can be classified according to their elementary design, making up patterns of stripes or spots [1, 5]. Physics also abounds in systems displaying an analogous organisation in equilibrium. They include magnetic films (video and audio tapes), concentrated solutions of surfactants, and many others. These morphologies, in equidistant stripes of equal width or bubbles of the same size spread over a triangular network, can adopt a large-scale architecture and take the form of a labyrinth, for example, or a spiral, a stack of concentric layers

F. Elias (✉)
University of Paris-Diderot, Paris, France
e-mail: florence.elias@univ-paris-diderot.fr

P. Bourgine, A. Lesne (eds.), *Morphogenesis*, DOI 10.1007/978-3-642-13174-5_2,

(onions), or intersecting arches (see Chap. 4). Over the course of this chapter, we shall see that although the systems that display such forms of internal architecture are very diverse, and although the nature of the physical parameter used to describe the pattern varies greatly from one system to another, yet the self-organisation itself derives from the simultaneous presence, within a system, of a small number of physical ingredients: a very short-range repulsive interaction, a medium-range attractive interaction and a long-range repulsive interaction are quite sufficient to produce this multitude of equilibrium conformations.

Before abstracting the few physical ingredients necessary to the formation of structures, it is worth analysing the physical systems that display internal morphologies. Figure 2.1 illustrates a number of examples (without claiming to be exhaustive). Below, we shall briefly describe each of these systems and then examine what they have in common and what is specific to each one.

2.2.1 *Examples of Self-Organised Physical Systems*

2.2.1.1 Magnetic Garnet Films

The first line of the table in Fig. 2.1 presents the example of *magnetic garnet films*. These are thin films made of a material that is ferromagnetic at room temperature. In a ferromagnetic material, the magnetic moments (or spins) of the atoms making up the film tend to align parallel to each other and perpendicular to the plane of the film. This means that the magnetic moments have two possible orientations. When the film is horizontal, these are "upwards" and "downwards". In the absence of any external magnetic field to break the symmetry of the system in relation to the plane of the film, there are an equal number of magnetic moments in each direction, so that the vector sum of the moments for the whole film is zero. But inverting a spin with regard to its neighbour has an energy cost, because this configuration does not respect the ferromagnetic order. The magnetic garnet film therefore has to minimise the number of pairs of opposite moments. Spins having the same orientation are organised into domains. Domains with opposing magnetisations are separated by walls (known as *Bloch walls*) that contain the energy required to inverse the magnetic moments. The domains themselves are then structured into patterns of monodisperse and equidistant stripes or bubbles. Thanks to the regularity of the bubble pattern, which can be as regular as a perfect crystalline network, information can be stored on magnetic tapes. All it needs is the local application of a magnetic field, to inverse a controlled number of magnetic moments on the film. This creates a defect in the structure, a spatial modulation of the magnetisation, that allows information to be encoded at the desired spot on the magnetic garnet film.

2.2.1.2 Superconductors

A second example of a physical system that self-organises into domains in equilibrium is provided by *type I superconductors*. A superconductor is a material whose electrical resistivity is zero for a temperature T lower than its critical temperature T_c

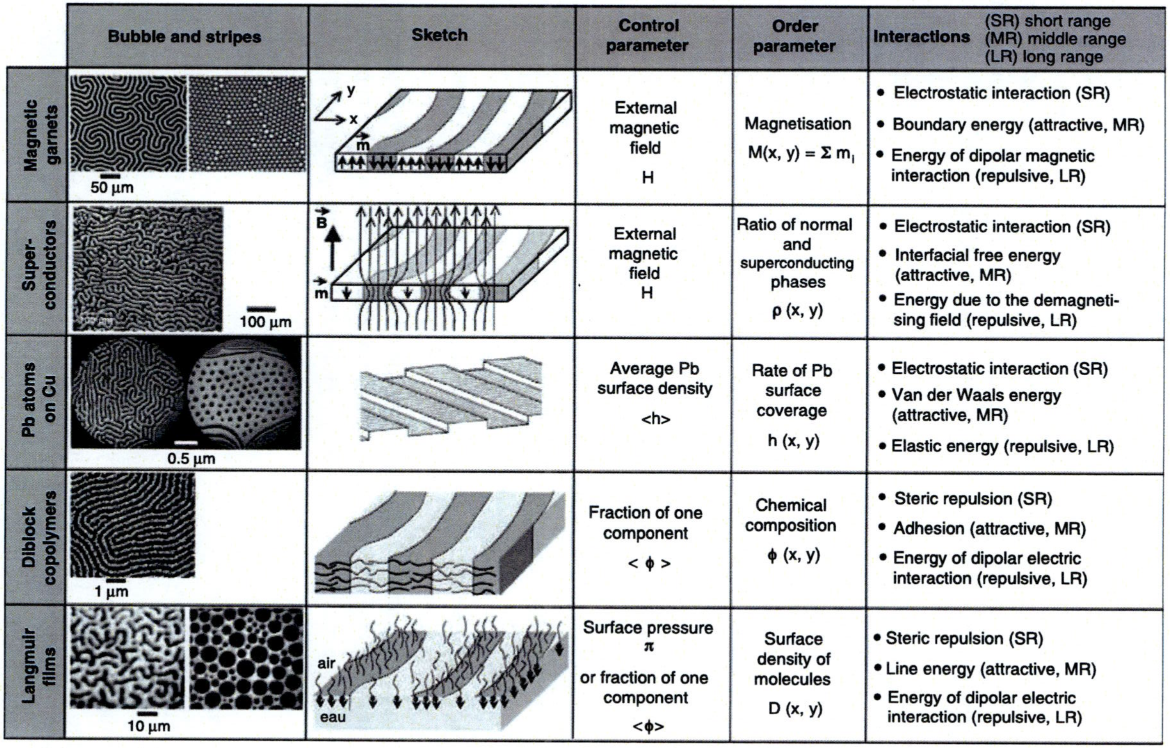

Fig. 2.1 Examples of morphologies in physical systems. Images: courtesy of P. Molho (magnetic garnets), V. Jeudy (superconductors), G. L. Kellogg (Pb atoms on Cu surfaces), H. Jaeger (diblock copolymers) and S. Akamatsu (Langmuir films)

(of the order of between a few degrees Kelvin and a few tens of degrees Kelvin). For a type I superconductor, the phase transition between the normal phase (for $T > T_c$) and the superconducting phase (for $T < T_c$) is of the first order, meaning that there is a range of temperatures below T_c in which domains of superconducting phase coexist with domains of normal phase. The absence of electrical resistivity gives the superconducting phase magnetic properties (diamagnetic, to be more precise). The magnetic field applied is therefore another parameter that, in addition to the temperature, controls the transition between the normal and superconducting phases. When the sample takes the form of a thin film, the application of a magnetic field perpendicular to the plane of the film has the effect of structuring the domains: as in the case of magnetic garnet films, patterns of equidistant stripes or bubbles appear.

2.2.1.3 Lead Atoms on a Copper Surface

Self-organised domains are also formed when a constant flow of lead atoms is deposited on a copper crystal surface, cut along a crystal plane. The interaction between atoms favours the formation of islands of lead on the copper. However, the equilibrium distance between lead atoms is different from the interatomic equilibrium distance of the copper. The interface between the copper and the lead is therefore the site of tensions and compressions, the effect of which is to break up the islands of lead so as to relax the mechanical constraints. These domains are organised into networks of bubbles or stripes separated by about ten nanometres.

2.2.1.4 Diblock Copolymers

Soft matter physics also contains many cases of self-organisation in equilibrium. One example is that of *diblock copolymers*. These are made up of two antagonistic blocks, for example one hydrophobic and the other hydrophilic, joined by a chemical bond. These polymers tend to self-assemble to form a structure aggregating all the blocks with the same affinity, and we can observe the spontaneous appearance of structures whose characteristic thickness is the length of one polymer, about ten nanometres.

2.2.1.5 Langmuir Films

Our last example, also drawn from soft matter physics, is that of *Langmuir films* on the surface of water. These films are composed of molecules of surfactants, which are themselves made up of hydrophilic polar head groups and hydrophobic aliphatic tails. Because of this dual nature, they are also called *amphiphilic molecules*. When they are introduced into water, they tend to adsorb spontaneously to the interfaces, with the heads plunging into the water and the tails pointing out, forming films of monomolecular thickness on the water surface. Depending on the surface density of the amphiphilic molecules present in a Langmuir film, several different phases can be distinguished in these films, analogous to the well-known solid, liquid and gaseous phases in three dimensions. As in the three-dimensional case, different phases can coexist for certain values of the average surface density of the molecules. It is under these conditions of coexistence that we can observe the presence of domains of one of the two phases within the other. And here, once again, the domains are self-organised into patterns of stripes or bubbles.

2.2.2 The Origin of Order

Why do the same patterns emerge in such different systems? These morphologies are obtained *in equilibrium*: once all the external parameters that can influence the system are fixed, the state of the system no longer evolves. Now, a system in equilibrium is a system that minimises its energy. Our search for the origin of these ordered architectures therefore calls for a careful assessment of all the energy contributions present in the system.

2.2.2.1 Short-Range Repulsion Versus Medium-Range Attraction

A *diphasic system*, that is to say one composed of two immiscible phases, is characterised by the existence of an interface between the two phases. The spatial segregation of the two chemical species present depends on the combined effect of a very short-range repulsive interaction between the molecules of each species (this is usually a hard-core repulsion, preventing the interpenetration of matter and keeping the particles spread out in space) and a longer-range attractive interaction between molecules of the same species. This attraction leads the molecules to aggregate, forming domains of one phase inside the other phase, as a drop of oil does in water (in this case the attractive interaction is the van der Waals force). The standard example of this is van der Waals gas, where the Lennard-Jones interaction energy takes into account these two contributions, allowing to write the gas-to-liquid transition.

However, the creation of boundaries between the two phases has an energy cost (*wall energy*). At the microscopic level, the wall energy derives from the fact that the isotropic attractive interaction between molecules of the same species is not offset near the wall; the molecules situated near the interface are therefore subject to a force that tends to draw them into the domain. At the macroscopic level, the wall energy is minimal when the surface area of the interface separating the two phases is as small as possible. A spherical drop (or circular if the system is two-dimensional) of one phase within the other is the form that minimises the wall energy. In the case of magnetic garnet films, the wall energy is associated with the creation of a Bloch wall; its microscopic origin lies in the ferromagnetic exchange interaction. In the case of islands of lead on a copper surface, the energy cost associated with the creation of walls is connected to interatomic attractions within the solid lead. For systems of diblock copolymers or Langmuir films, the wall energy is due to surface tension (or line tension) of the same nature as that which exists at the interface of a drop of oil in water. Lastly, in the case of type I superconductors, thermodynamic arguments can be used to demonstrate the existence of an interfacial energy between the normal phase and the superconducting phase.

2.2.2.2 Long-Range Repulsion

But why do the systems examined in this chapter form regular structures where the two phases are interlinked with each other? The reason can be found in the existence of a third energy contribution, common to all the systems that possess an internal architecture at equilibrium: an additional *repulsive energy* between the entities that make up the system. In the case of magnetic garnet films and Langmuir films, this repulsive energy is of dipolar origin and acts between two dipoles aligned in the same direction and perpendicular to the plane of the film. At the macroscopic level, this repulsive energy tends to increase the distance between the entities that repel each other, and consequently to increase the surface area of the boundaries. Thus, the formation of stripes or bubbles, or more generally of a multitude of domains of one phase within the other phase, is the result of a compromise between two competing energy contributions: a wall energy that tends to reduce the interface between the two components, and an energy of dipolar origin that tends to lengthen that interface.

However, this competition between two opposing energies does not explain the regularity of the patterns observed. To understand why the domains that emerge all have the same size and are all situated at the same distance from each other, we need one final physical ingredient. These two interactions, one attractive and the other repulsive, do not have the same range of action: the wall interaction, which is attractive, is *medium-range*, whereas the repulsive interaction is *long-range*. In other words, only close neighbours attract each other, while two molecules situated at opposite ends of the system can repel each other. This difference in the range of action naturally introduces a characteristic length scale: the distance between domains. Below this distance, the attractive energy wins out over the repulsive energy and the system tends to form a wall of minimum surface area between the domains. Beyond this characteristic distance, the repulsion is stronger than the attraction, and so the domains tend to become interspersed. We can represent this range difference between the two interactions by the diagram in Fig. 2.2. Imagine that someone is filming the system, while gradually widening the field of vision.

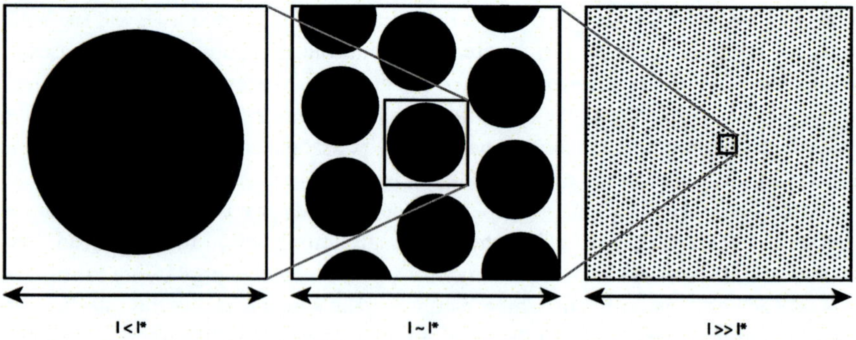

Fig. 2.2 l^* is the characteristic size of the domains, l is the size of the viewing screen through which we observe the system

When the diameter of the display screen is smaller than the characteristic distance between domains, we see one sole domain surrounded by another, which is characteristic of a diphasic system possessing wall energy. Then the field of view widens, and several domains appear, signature of the presence of a repulsive interaction counteracting the wall energy. When the field of view widens still further, we can no longer distinguish the domains: the two phases appear to be dispersed homogeneously throughout the system and the boundaries seem to have disappeared: at large length scales, the effects of the attractive interaction between molecules of the same species can no longer be perceived, only the repulsive interaction appears to be present.

The three ingredients needed to obtain self-organisation of patterns in equilibrium are therefore a very short-range repulsive interaction, a medium-range attractive interaction, and a long-range repulsive interaction. The first two interactions ensure the existence of domains of one phase inside the other, and the third is necessary to the self-organisation of those domains.

2.2.3 The Bond Number

A simple quantitative criterion can be used to determine whether the system under consideration is likely to self-organise. This criterion is provided by dimensional analysis. Based on the two energies that compete to form the patterns, we can define a dimensionless number, the *Bond number*, which is the ratio of repulsive energy to attractive energy. The system self-organises into patterns when the Bond number is greater than 1, whereas it remains divided into two spatially distinct phases when the Bond number is less than 1. The Bond number depends on the external parameter(s) of the system (external magnetic field, total volume or surface area, relative fractions of the two phases, thickness of the system in the case of a quasi two-dimensional system, etc.). We can thus estimate the order of magnitude of the control parameter for which morphologies can emerge.

2.2.4 Domain Size and Choice of Pattern

As the patterns are obtained at equilibrium, we can again make use of energy considerations to calculate the characteristic sizes of the morphologies adopted by the system. These characteristic lengths (the size of domains and the distance between them) are *internal variables*. This means that once the external parameters have been fixed, they adjust in such a way as to minimise the total energy of the system. In practice, therefore, if we write the sum of the two energy terms (repulsive and attractive) according to the internal variable under consideration, and then minimise this energy in relation to the internal variable, we end up obtaining, for example, the domain size as a function of the control parameters.

As several different architectures are possible, we must then perform this operation of minimising the energy in relation to domain size for each pattern (stripes, bubbles or other). We can then insert the equilibrium value of the internal variable obtained in the expression of the system's energy. The pattern that corresponds to the lowest level of energy is the one that will be chosen at equilibrium – provided that we allow the system the possibility to make this choice, i.e. the possibility to explore all possible morphologies. This point, which is far from evident, will be discussed later in the chapter, using a specific example.

So in theory, knowledge of the two antagonistic interactions at work in the system enables us to draw a phase diagram of possible architectures, and to know all the variables required to describe these structures.

2.2.5 *Summary*

To sum up, for a system to form regular, structured patterns in equilibrium, a very small number of very general ingredients, common to all systems, is necessary. The characteristic entities of the system must interact through:

- a very short-range repulsive interaction;
- a medium-range attractive interaction;
- a long-range repulsive interaction.

The ratio between the last two energies defines the Bond number, which provides a criterion for the emergence of these internal architectures according to the control parameters of the system. For each system, if we know the expression of these two antagonistic energies, we can establish the phase diagram of possible morphologies and calculate the characteristic size of each structure. This characteristic scale of distance is specific to each system and can vary from a nanometre to a centimetre, depending on the system.

2.3 Morphologies in Ferrofluids

2.3.1 *Ferrofluids: A Model System for Studying Structures*

Because internal morphologies in the form of stripes or bubbles are universal, and because their existence does not depend on the specific physical properties of the systems that produce them, but on the general form of the energy terms that operate in them, these architectures can be studied through a model system that presents such internal structures. Magnetic liquids, or *ferrofluids*, are one such model system. These liquids combine magnetic properties with ordinary liquid properties (see Box 1). In particular, their interface can change shape under the effect of an external magnetic field [2, 4].

Figure 2.3a shows a few examples of self-organisation where the domains of ferrofluid (in black) form equilibrium structures in a non-magnetic medium: labyrinth, parallel stripes, network of bubbles, foam structure or network of rings [3]. The

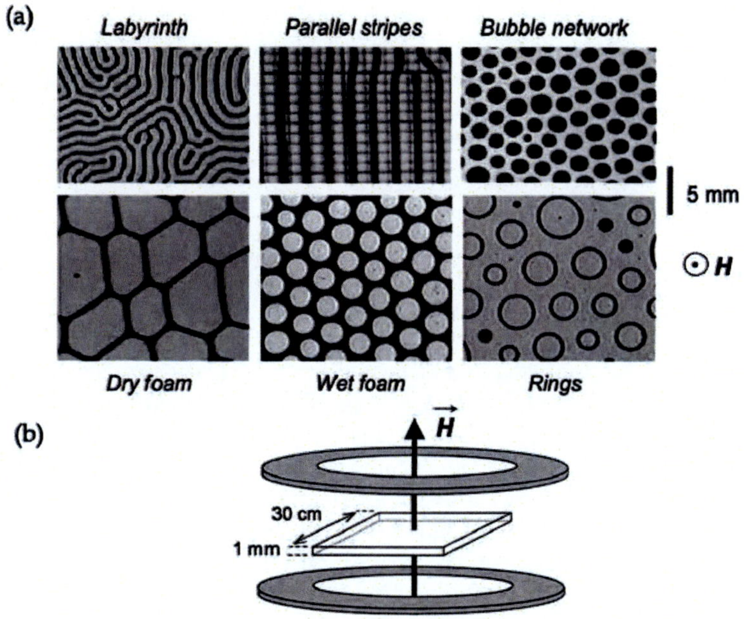

Fig. 2.3 (**a**) Patterns in 2D ferrofluids: stripes, bubbles, foams and rings. The ferrofluid is in *black*. The *grid* that can be seen under the phase in parallel stripes is a millimetre grid. (**b**) Geometry of the layer of ferrofluid in which the patterns are observed. The cell containing the ferrofluid (*in white*) is placed between two coils of wire (*in grey*) that produce a magnetic field perpendicular to the plane of the cell (a Helmholtz pair)

experimental system used here is two-dimensional: the ferrofluid and a transparent, immiscible, non-magnetic liquid are placed between two transparent plates, about thirty centimetres wide, separated by a spacer one millimetre wide that encloses the cell. The plates are placed horizontally and subjected to a uniform vertical magnetic field of a few hundredths of a Tesla (i.e. a thousand times the magnetic field of the Earth), created using two coils of copper wire in a Helmholtz configuration (Fig. 2.3b). The coils are supplied with an electric current whose intensity can be varied, thus controlling the amplitude of the magnetic field created (which is proportional to the intensity).

From an experimental point of view, there are two main reasons why this sort of layer of ferrofluid makes a suitable model system for the study of self-organised systems. Firstly, these systems are easy to manipulate. Ferrofluids, which behave like ordinary liquids in the absence of an external magnetic field, become magnetic and form structures under the effect of magnetic fields. The choice of structure depends on the amplitude of the field applied. This provides us with a control parameter that can easily be varied in the laboratory, enabling us to sample a multitude of possible morphologies. Secondly, the domain size of most self-organised systems is of the order of micrometres or nanometres, but in the case of ferrofluids it is of the order of millimetres or centimetres. Observations can therefore be made with the naked eye, avoiding the need for heavy and expensive machines.

In conclusion, the morphologies of ferrofluids are representative of the architectures observed in all self-organised systems. Easy to observe and malleable in the laboratory, they provide us with useful experimental models for understanding self-organised systems.

Box 2.1 Ferrofluids

A ferrofluid is a suspension of solid, magnetic particles in a carrier fluid. These particles are made for example of cobalt ferrite (FeCo) or maghemite (γ-Fe_2O_3). They each carry a permanent magnetic moment, and so they behave like nano-magnets (Fig. 2.4). The magnetic particles have an average size of the order of about ten nanometres. Although they are denser than the water they swim in, this small size allows them to remain in suspension, rather than sinking to the bottom of the container as sediment, because collisions against the molecules of solvent coming from all directions keep them in suspension (Brownian motion). Such homogeneous suspensions of solid particles are called *colloidal suspensions* or colloids (this is the case, for example, for India ink, certain paints, and milk). So here, we are dealing with a *magnetic colloid* [2, 4].

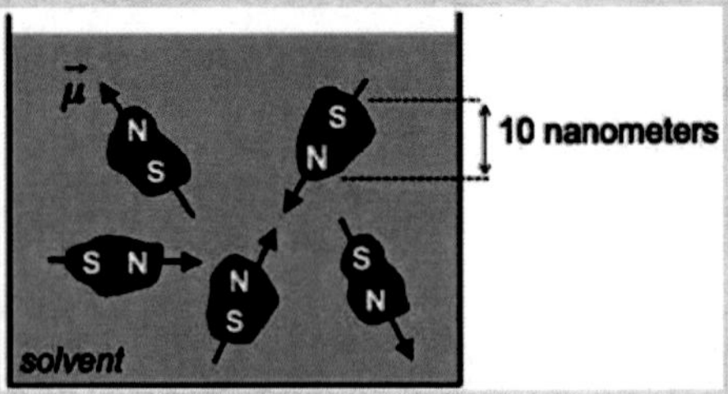

Fig. 2.4 Composition of a ferrofluid. Each particle carries a permanent magnetic moment μ, and so behaves like a miniature magnet, with a North pole and a South pole

Ferrofluids do not exist in a natural state. They are synthesised chemically in the laboratory. To make a stable suspension, it is also necessary to prevent the particles from being attracted to each other by the van der Waals force or magnetic interactions, which would result in the formation of aggregates that would sink under the influence of gravity. Two techniques can be used to achieve this, both consisting in adding a repulsive force between the particles. The first technique produces *surfacted ferrofluids*: the solid particles are covered in surface active molecules (or surfactants), the heads of which adsorb to the surface of the particles. The particles end up covered in a layer of surfactant, which acts like a sort of elastic mattress, maintaining the particles at a minimum distance from each other, a distance at which the attraction between particles is negligible. The second technique consists in introducing an electrostatic charge of the same sign on the surface of each particle. It is then the electrostatic repulsion between particles that keeps them at a certain distance from each other. To keep the solution electroneutral, counterions (ions of the opposite charge) are introduced into the solution. These counterions have the effect of screening the electrostatic repulsion between the particles, allowing to increase or decrease the distance between particles by decreasing or increasing the concentration of counterions (within the limits of stability of the suspension). This is called an *ionic ferrofluid.* The latter technique makes it possible to use a polar solvent like water, while surfacted ferrofluids are stable in organic solvents like oil.

At the macroscopic scale, in other words at our scale, a ferrofluid is a homogeneous fluid that the magnetic particles have turned black. When a magnet is brought near, the whole liquid is attracted to it. The response of the ferrofluid to a magnetic field is paramagnetic. Let us explain that in more detail. The magnetisation **M** of the ferrofluid is given by the vector sum of the individual vector moments of the particles μ_i:

$$\mathbf{M} = \frac{1}{\mu_0 V} \sum_i \mu_i$$

where μ_0 is the vacuum permeability, V is the volume of ferrofluid under consideration and the sum includes all the magnetic particles contained in the volume V. Thus, with a zero field, the magnetic moments are oriented at random because of Brownian motion, and the magnetisation of the ferrofluid is zero. When the amplitude H of the magnetic field **H** increases, the magnetic dipoles gradually align in the direction of the field, while the energy of thermal agitation tends to re-establish the disorder. M therefore increases with H. When the field is sufficiently strong, the energy of thermal agitation is defeated by the magnetic energy: all the magnetic dipoles are aligned in the direction of the applied field, and the magnetisation of the ferrofluid saturates at its maximum value. The macroscopic magnetic response of the ferrofluid to a magnetic field is therefore paramagnetic (Fig. 2.5): it is characterised by magnetic susceptibility at zero field:

$$\chi = \lim_{H \to 0} \left(\frac{\mathrm{d}M}{\mathrm{d}H} \right) = \lim_{H \to 0} \left(\frac{M}{H} \right)$$

and by saturation magnetisation M_s. When the volume concentration of magnetic particles is of the order of 10%, $M_s \approx 40$ kA.m^{-1} is reached for an applied magnetic field of the order of 80 kA.m^{-1} (equivalent to a magnetic flux intensity of the order of 0.1 Tesla), and $\chi \approx 1$, which is about one thousand times higher than the magnetic susceptibility of ordinary paramagnetic liquids. Hence the use of the term "*giant paramagnetism*" to describe the macroscopic magnetic behaviour of ferrofluids.

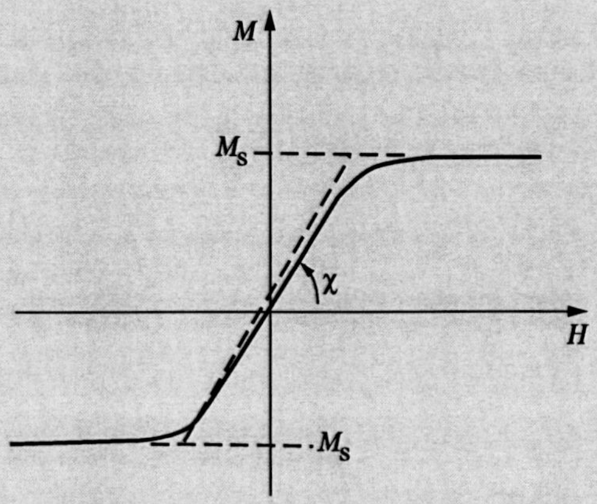

Fig. 2.5 Magnetisation curve of a ferrofluid

2.3.2 *Stripes and Bubbles, Foams and Rings in Ferrofluids*

The equilibrium patterns that emerge in a thin layer of ferrofluid also originate in a competition between short-range repulsive energy, medium-range attractive energy and long-range repulsive energy. The two phases present are a magnetic liquid and another immiscible liquid. In the absence of an external magnetic field, these two phases behave as ordinary liquids. The energies responsible for phase separation are: firstly, the Coulombian repulsion between charged colloidal particles, which plays the role of short-range repulsive energy in the case of an ionic ferrofluid (in surfacted ferrofluids, this repulsion is of steric origin), and secondly, the van der Waals interactions between the molecules of each liquid. Van der Waals forces are always attractive. Their amplitude decreases as the distance between molecules increases, but their range of action is longer than that of the short-range repulsive forces described above, and shorter than the long-range repulsive forces that we shall describe below. The van der Waals interaction is therefore a medium-range attractive interaction.

At the macroscopic level, the van der Waals interaction manifests itself in the form of interfacial tension, which tends to minimise the surface area of the domains. The surface energy is written as: $E_s = \sigma S$, where σ is the surface tension of the interface and S is the total surface area of the interface. In the case of a thin layer of ferrofluid of thickness h (here, $h = 1$ mm), we can write the surface energy as:

$$E_s = \sigma h L \, , \tag{2.1}$$

where L is the total length of the line separating the two domains. In the absence of an external magnetic field, this attractive interaction is not counterbalanced, which results in the whole system only having two domains, separated by a circular boundary.

2.3.2.1 A long-Range Repulsive Energy

The structuring of the two phases, one within the other, occurs when a magnetic field is applied perpendicular to the layer of ferrofluid. The long-range repulsive energy that opposes the interfacial energy derives from the interaction between magnetic dipoles inside the ferrofluid. This interaction can be explained as follows. The permanent magnetic moments carried by the ferrofluid particles tend to align themselves in the direction imposed by the external field, in the same way as a compass aligns itself in the direction of the magnetic field of the Earth. Now, if you take two permanent magnets, holding them firmly so that their magnetic moments both point in the same direction throughout the experiment, and move them around each other, you will find that when one is above the other they attract each other (their relative position is parallel to the direction of the magnetic moment), whereas when they are side by side they repel each other (their relative position is perpendicular to the magnetic moment). In the case of a layer of ferrofluid subjected to a perpendicular magnetic field, the number of magnetic moments placed side by side is far greater than the number placed above each other. Consequently, the dipoles repel each other, on average. The intensity of the forces of magnetic dipolar interaction decreases as the distance between particles increases, but this decrease is very slow, much slower than the decrease in the intensity of the van der Waals forces: magnetic dipolar interactions are long-range repulsive interactions. The magnetic energy of the layer of ferrofluid is expressed, for a weak magnetic field (see Box 2.2):

$$E_m = -\frac{\mu_0 \chi H_0^2}{2} \int_V \frac{1}{[1 + \chi D(\mathbf{r})]} \, \mathrm{d}^3\mathbf{r} \tag{2.2}$$

where μ_0 is the magnetic permeability of vacuum, V the volume occupied by the ferrofluid, H_0 the external magnetic field, χ the magnetic susceptibility of the ferrofluid and D the demagnetisation coefficient. Under the hypothesis that the product χD is much less than one, we obtain a simplified expression:

$$E_m = -\frac{\mu_0}{2} \chi H_0^2 V + \frac{\mu_0}{2} \chi^2 H_0^2 \int_V D(\mathbf{r}) \, \mathrm{d}^3\mathbf{r} \, . \tag{2.3}$$

The first of the terms on the right-hand side represents the interaction energy between the magnetic dipoles and the field, once the dipoles have aligned themselves in the direction of the field. This term is constant for a fixed value of the external parameters. The second term represents the interaction energy between the dipoles themselves. It depends on the morphology adopted by the ferrofluid through the factor of demagnetisation and its integral over the volume occupied by the ferrofluid. It is therefore this term that is responsible for the different architectures that the surface of a ferrofluid can adopt.

2.3.2.2 The Magnetic Bond Number and the Characteristic Size of Domains

By expressing the two energy terms whose competition generates the self-organisation, we can define the magnetic Bond number N_B that is characteristic of the system. When the χD tends to zero, it is the ratio of the repulsive energy given by the second term on the right-hand side of the (2.3) over the attractive energy given by (2.1):

$$N_B \approx \frac{\mu_0 \chi^2 H_0^2 V_{FF}}{2\sigma h L} \approx \frac{\mu_0 \chi^2 H_0^2 d}{\sigma} \tag{2.4}$$

where d is the characteristic domain size. Dimensional analysis can be used to estimate the order of magnitude of d by considering that structures emerge when $N_B > 1$: the experimental values of the external parameters ($\chi \approx 1$, $H_0 \approx 10$ kA.m^{-1}, $\sigma = 15$ mN.m^{-1}) give a value for characteristic domain size of the order of a millimetre, which corresponds well with what we actually observe. However, we can obtain an exact value for the characteristic scales of length (size of domains and distance between them) by expressing the total energy of the system, i.e. the sum of its surface energy given by (2.1) and its magnetic energy given by (2.3), as a function of the internal variables, and minimising the energy in relation to these variables. By performing this operation for all the possible patterns, and then inserting the value obtained for these internal variables into the expression of total energy, we can compare the equilibrium energies of the different morphologies with each other. The most stable equilibrium structure is the one that possesses the lowest level of energy. If several of the structures have the same level of energy, then domains of different morphologies can coexist within the same system.

2.3.3 The Influence of History: Initial Conditions and Conditions of Formation

Calculations show that the numerical value of the energy of a thin layer of ferrofluid differs little from one structure to another. But it requires the injection of a considerable amount of energy, from outside, to make it change from one morphology to another. All the possible patterns that the domains can form are called the *metastable states* of the system: each one corresponds to a local energy minimum, and they are separated by substantial energy barriers. The consequence of this metastability is a strong dependence of the morphology adopted on the initial conditions and the history of the system. To produce a particular morphology, or to change from one morphology to another, many scenarios are possible, each representing a different recipe for the preparation of the structures.

For example, if the whole ferrofluid is initially aggregated into one sole circular domain when there is zero field, then the pattern formed by increasing the amplitude of the external magnetic field is generally such that the ferrofluid remains connected: we obtain either a labyrinth of stripes of ferrofluid (see Fig. 2.3), or a foam structure. If the most stable state, for a given value of the control parameters (magnetic field, volume fraction of the ferrofluid, thickness of the layer) corresponds to a triangular network of domains in bubbles, then the system must be supplied with sufficient energy to break up the stripes and obtain domains disconnected from each other. We could, for example, wave a small permanent magnet over the layer of ferrofluid, or create cycles in the magnetic field, by rapidly varying its amplitude between zero and the desired value [3]. As the equilibrium period of the patterns is determined by the external parameters, a change in one of those parameters can entail the creation, disappearance or deformation of the domains. In the case of the phase in stripes, disappearance may occur through dismemberment followed by ejection: a stripe breaks in two, and the two halves then creep away from each other. In the case of the foam structure, the domains may coalesce. When a modification of the external parameters imposes a reduction in the period of the structure, the domains may lose their shape and produce new patterns: the bubbles of the triangular phase or the cells of the foam phase may stretch out and take the form of a bean, the stripes may grow fingers or become wavy, the phase in parallel stripes can be periodically deformed to produce a chevron structure. These examples are not exhaustive. In reality, the list of possible patterns grows from day to day: in the laboratory, experimenters regularly discover new, hitherto unobserved forms of organisation within thin layers of ferrofluid, corresponding to new forms of preparation.

Box 2.2 Magnetic Dipolar Interactions and the Magnetic Energy of an Array of Dipoles

At the Microscopic Level

One way to understand magnetic dipolar interactions is to represent the magnetic field $\mathbf{H}_m$ created by a dipole with magnetic moment μ_1 (see Fig. 2.6). The amplitude of the field $\mathbf{H}_m$ decreases spatially as the distance from the moment μ_1 increases. In the plane perpendicular to μ_1 and containing the dipole, $\mathbf{H}_m$ is parallel to μ_1 but oriented in the opposite direction. A second dipole μ_2 placed in this plane therefore feels the effect of $\mathbf{H}_m$, which tends to align μ_2 in the direction of $\mathbf{H}_m$. But if the orientation of μ_2 is fixed parallel to that of μ_1, the dipole cannot turn around. Under these conditions, μ_1 tends to push μ_2 far away, where the influence of $\mathbf{H}_m$ is negligible. On the contrary, $\mathbf{H}_m$ remains oriented in the same direction as μ_1 on a straight line containing μ_1 and parallel to μ_1. If a magnetic dipole of moment μ_3 parallel to μ_1 is placed on this straight line, then field $\mathbf{H}_m$ has the effect of reducing the interaction energy between μ_1 and μ_3. μ_3 therefore tends to move closer to μ_1 to position itself in a field $\mathbf{H}_m$ of even stronger amplitude: the interaction between μ_1 and μ_3 is therefore attractive.

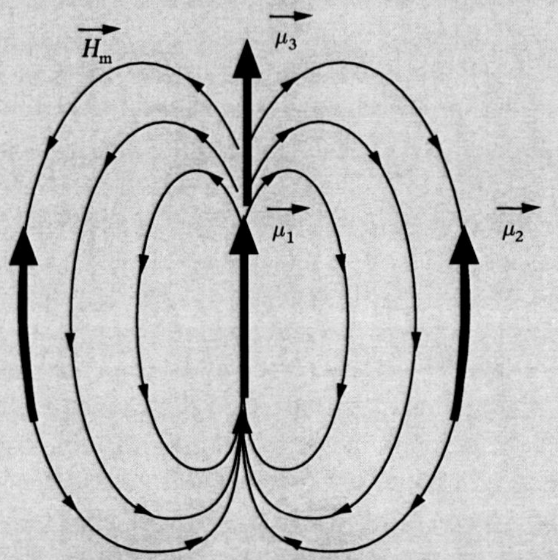

Fig. 2.6 Field lines induced by a magnetic moment

If we assume that the magnetic moments have two possible orientations, upwards or downwards, then we can distinguish four elementary figures (involving two moments) for which the magnetic dipolar interaction is either attractive or repulsive. These four configurations correspond to the different possible orientations of the magnetic dipoles relative to each other. They are represented in the diagram (Fig. 2.7). This description has the advantage of being directly applicable to other physical systems governed by dipolar interactions, magnetic or electrical, like for example magnetic garnet films and Langmuir films.

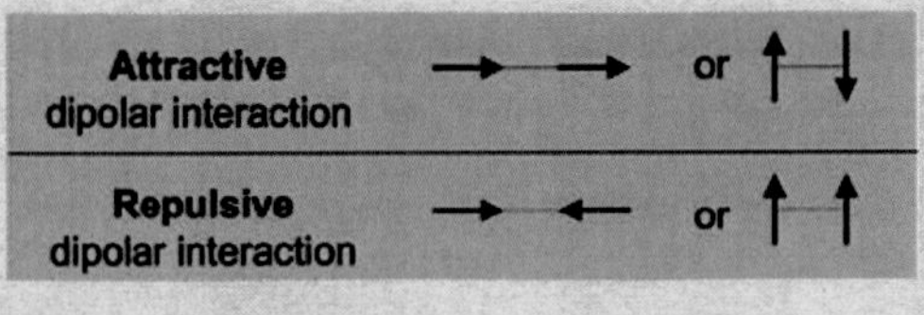

Fig. 2.7 Four possible instances of magnetic dipolar interaction

At the Macroscopic Level

The magnetic energy of a volume of ferrofluid δV is written as:

$$\delta E_m = -\mu_0 \int_0^{\mathbf{H}_0} \mathbf{M}.\mathrm{d}\mathbf{H}_0\, \delta V$$

where $\mathbf{M}$ is the magnetisation of the ferrofluid, $\mathbf{H}_0$ is the external magnetic field and μ_0 the magnetic vacuum permeability. This energy is negative, because that minimises the energy of the system when a magnetic body is placed in a magnetic field; it is all the more negative as the intensity of the field is high. The magnetisation depends on the total magnetic field $\mathbf{H}$ at the point being studied. In a weak field, $\mathbf{M}$ is proportional to $\mathbf{H}$ and the coefficient of proportionality is the magnetic susceptibility: $\mathbf{M} = \chi\mathbf{H}$. The magnetic field $\mathbf{H}$ takes into account the external field $\mathbf{H}_0$ (which aligns the magnetic dipoles) and the field induced by the dipoles: $\mathbf{H} = \mathbf{H}_0 + \mathbf{H}_d$, where $\mathbf{H}_d$ is the sum, at the point being studied, of the fields $\mathbf{H}_m$ induced by all the dipoles in the ferrofluid (see Fig. 2.6). $\mathbf{H}_d$ is known as the *demagnetisation field*. It is parallel to the magnetisation that gave rise to it, but in the opposite direction: $\mathbf{H}_d = -D\mathbf{M}$, where D is a dimensionless number between 0 and 1, known as the *coefficient of demagnetisation*. Generally, the coefficient of demagnetisation varies spatially through the layer of ferrofluid, but it is homogeneous in a number of simple cases: $D = 1/3$ for a sphere, $D = 0$ for an infinite cylinder subjected to a field parallel to its axis, and $D = 1$ for an infinitely thin plate subjected to a perpendicular field. Given the relations written above, the magnetisation can be expressed in terms of the external field: $\mathbf{M} = \chi\mathbf{H}_0/(1 + \chi D)$. The magnetic energy of a system containing a volume V of ferrofluid can therefore be written:

$$E_m = -\frac{\mu_0 \chi H_0^2}{2} \int_V \frac{\mathrm{d}^3\mathbf{r}}{1 + \chi D(\mathbf{r})} .$$

The magnetic energy E_m therefore depends on the morphology adopted by the ferrofluid, through the coefficient of demagnetisation D and its integral over the volume occupied by the ferrofluid, which takes into account the effect of magnetic dipolar interactions on the macroscopic level.

2.3.4 The Source of Patterns: Instabilities

An example of the formation of structure can be seen in the first images in Fig. 2.8. The interface of the ferrofluid, which is smooth in zero field (image a), starts to undulate with a well-defined wavelength (image b). Then the crests of the waves grow, forming fingers and then stripes. Between image (a) and image (b), an initially microscopic deformation of the interface has intensified to become visible to the naked eye, in other words macroscopic. This mechanism is due to an *instability* of

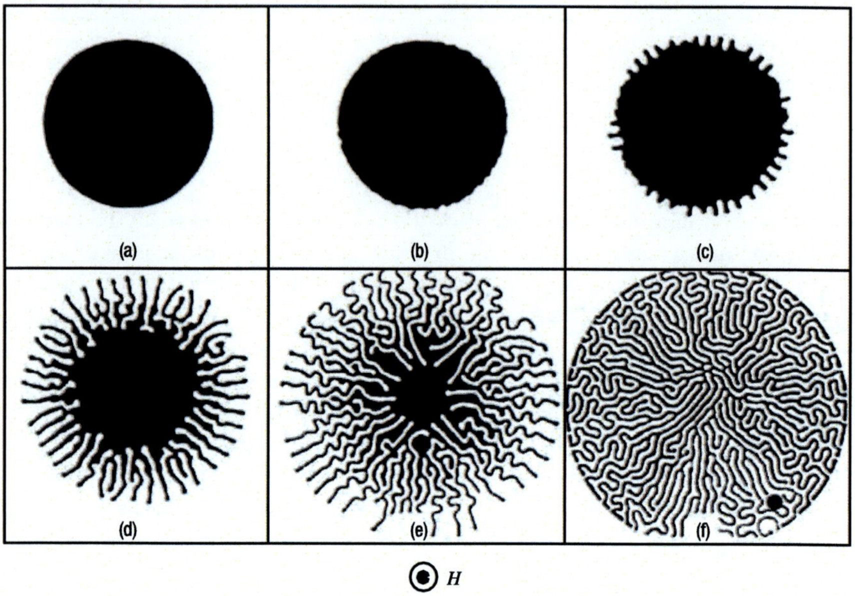

Fig. 2.8 Destabilisation of the interface of a drop of ferrofluid under the effect of a perpendicular magnetic field. (**a**) Initial situation in zero field. The field is applied from image (**b**) onwards. The time lapse from image (**b**) to image (**e**) is about 3 s. (**f**) Equilibrium pattern

the interface: the magnetic field exerts pressure on the surface of the ferrofluid, with the effect of amplifying its deformations. Below, we present a number of instabilities of the ferrofluid interface. Normal-field instability (forming spikes) is certainly the most studied and best-known of these instabilities; mathematical processing can be used to discover the value of the amplitude of the magnetic field above which instability occurs, along with the wavelength of the pattern. Lastly, by studying the deformations in a drop of ferrofluid we can bring to light the effects of confinement, which lead to the formation of stripe patterns and bubble patterns.

2.3.4.1 The Showcase of Ferrofluid Instability: Normal-Field Instability

When a ferrofluid is subjected to a homogeneous magnetic field in a direction perpendicular to the liquid-air interface, the surface of the ferrofluid forms a whole series of spikes (see Fig. 2.9a). This spectacular instability occurs when the amplitude of the magnetic field is above a certain threshold value, and the spikes emerge on the surface in a regular network, usually with triangular but sometimes with square symmetry.

Qualitatively, we can explain why this instability appears by considering the different forces present in the case where the surface experiences a weak sinusoidal disturbance. In zero field, the surface of the magnetic liquid, like that of any other liquid, is the site of small, thermally-excited surface disturbances. These are neu-

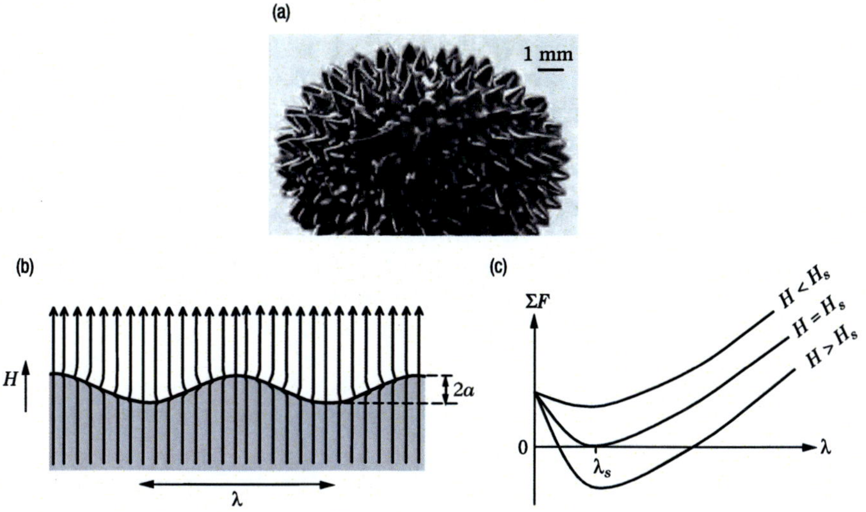

Fig. 2.9 Normal-field instability. (**a**) The ferrofluid aggregates on the end of a magnet. Its interface forms liquid spikes pointing in the direction of the magnetic field lines. (**b**) Because of the conditions under which the magnetic field passes through the interface, the magnetic field lines draw together on the crests and spread out in the troughs. (**c**) Diagram: sum of the forces exerted on a spike as a function of the wavelength of the disturbance, for different values of the amplitude of the magnetic field. The instability can develop on the condition that there exists at least one solution to the equation $\sum F = 0$. There exists, therefore, a threshold amplitude of the field H_s for the development of this instability: the wavelength at the threshold of instability, λ_s, is the capillary length

tralised by the effects of gravitational and capillary forces (linked to the surface tension). When a magnetic field is applied perpendicular to the surface, the field lines draw closer together in the neighbourhood of the wave crests and move further apart from each other in the troughs (see Fig. 2.9b). The field gradients, which are positive in the neighbourhood of the crests, induce a magnetic force (a destabilising force) that tends to amplify the crests and make them more and more pointed. The conformation of the interface results from competition between the forces of gravity and surface tension (stabilising forces) on the one hand, and the magnetic forces, increasing with the amplitude of the field applied, on the other. As long as the amplitude of the field is lower than a certain threshold value, the gravitational and capillary forces win the day, and the interface remains flat. Above the threshold, the magnetic forces are stronger and spikes develop to form a regular network (as in the case of stripe and bubble patterns, this network is linked to the repulsive magnetic interactions between the spikes).

Let us consider the case where the surface of the ferrofluid is the site of a sinusoidal deformation with amplitude a and wavelength λ. We can calculate the threshold amplitude of the field and the wavelength of the instability by means of dimensional analysis, by writing the energy balance of the vertical forces acting on a spike. The spike, comparable to a cone with height $2a$ and diameter λ, has a volume of $V \approx a\lambda^2$. It is subjected to three vertical forces: the force of gravity F_g, the capillary force F_c and the magnetic force F_m. The amplitude of the force of gravity can be written:

$$F_g \approx \rho g a \lambda^2, \tag{2.5}$$

where ρ is the density of the ferrofluid and g is the acceleration of gravity. The amplitude of the surface force is the derivative of the interfacial energy with respect to a coordinate in the plane of the surface. It is therefore of the order of: $F_s \approx (E_s' - E_s)/\lambda$ where $(E_s' - E_s)$ is the difference between the surface energy with and without spikes. After calculating the surface area of the spike, this gives us:

$$F_s \approx \sigma\lambda(a/\lambda) \approx \sigma a \tag{2.6}$$

where σ is the surface tension of the ferrofluid-air interface. Finally, the magnetic force, resulting from the gradient of the total magnetic field due to the contraction of the field lines crossing the interface, can be written: $F_m \approx -\mu_0 V\, (\mathbf{M}.\nabla)\mathbf{H}$, that is:

$$F_m \approx -\mu_0 M^2 a\lambda\,. \tag{2.7}$$

The sum of the forces exerted on the spike can therefore be written:

$$\sum F = F_g + F_m + F_s \approx \rho g a \lambda^2 + \sigma a - \mu_0 M^2 a \lambda \tag{2.8}$$

Fig. 2.9c shows the shape of $\sum F$ as a function of λ for different values of magnetisation. At equilibrium, the sum of these three forces must be zero (fundamental principle of dynamics), and this equilibrium must be stable with regard to a variation in the wavelength of the disturbance:

$$\sum F = 0 \qquad \frac{\mathrm{d}\sum F}{\mathrm{d}\lambda} = 0\,. \tag{2.9}$$

The equations (2.9) give a system of two equations with two unknowns, the solution of which gives us both the value of the threshold field amplitude H_s and the wavelength λ_s at the threshold of instability:

$$H_s = \frac{M_s}{\chi} = \frac{1}{\chi}\left(\frac{2}{\mu_0}\right)^{1/2} (\sigma\rho g)^{1/4} \qquad \lambda_s = \sqrt{\frac{\sigma}{\rho g}} \tag{2.10}$$

where λ_s is the capillary length (the length for which F_g and F_c are of the same order of magnitude).

Another interesting case is that of a thin film of ferrofluid of thickness h, with a magnetic field perpendicular to the interfaces. Experimentally, this is done by depositing a thin film of ferrofluid on the surface of another liquid, but a thin film suspended by a frame can also be used. The film is said to be thin when its thickness is much less than the capillary length $h \ll (\sigma/\rho g)^{1/2}$. Once again, a weak sinusoidal disturbance is amplified above a threshold field, and then brought back down. Two modes are then possible: the undulation mode (constant thickness) when the two interfaces of the film undulate in phase, and the bulging mode (modulated thickness) when the two disturbances undulate in opposite phase. A similar treatment can be performed by using the equilibrium of forces acting on an elementary volume. In this way, one obtains a diagram of the forces as a function of λ, with the same shape as in the previous example (Fig. 2.9c).

2.3.4.2 The Effect of Confinement in Morphogenesis

All the self-organised systems that we have described up until now have been either two-dimensional (2D) and pseudo-2D systems or patterns created on interfaces. The ferrofluid is also a model system for studying the effects of confinement.

Let us take the example of a drop of ferrofluid with radius R subjected to a magnetic field. When R is much smaller than the capillary length, the forces of gravity are negligible compared to those of the interfacial tension, and the drop is spherical in zero field. When we apply a magnetic field, the drop elongates in the direction of the field (see Fig. 2.10a): this is the effect of the magnetic spike. If we wanted to describe this effect quantitatively, our description would have to be based on the minimisation of the total energy of the system. The total energy of the drop is $E_t = E_s + E_m$ where E_s is the surface energy and E_m is the magnetic energy of the drop. Ellipsoids are magical objects in magnetism (as they are in electrostatics), because the field of demagnetisation is constant inside them. If we accept that a drop of ferrofluid is distorted into an ellipsoid of revolution under the action of a homogeneous magnetic field (which is indeed verified experimentally), then we can calculate exactly the magnetic energy as a function of the stretching of this ellipsoid:

$$E_m = -\frac{\mu_0 \chi V}{2} \frac{H_0^2}{1 + \chi D(a/b)} \tag{2.11}$$

where V is the volume of the drop and a/b is the ratio of the minor axis to the major axis of the ellipsoid. The surface area is also expressed in terms of the ratio a/b. The equilibrium form is given by the minimisation of total energy with regard to the ratio a/b:

$$\frac{\mathrm{d}E_t}{\mathrm{d}(a/b)} = 0\,. \tag{2.12}$$

Equation 2.12 produces the curves of Fig. 2.10b, which represents the variation of the ratio a/b as a function of the reduced magnetic Bond number $B_m = \chi H_0^2 R/\sigma$. Note that there is a hysteresis between the rise and fall of the magnetic field.

If we now confine the same drop lengthwise between two planes, a distance h apart from each other, it can no longer stretch in the direction of the field and tries

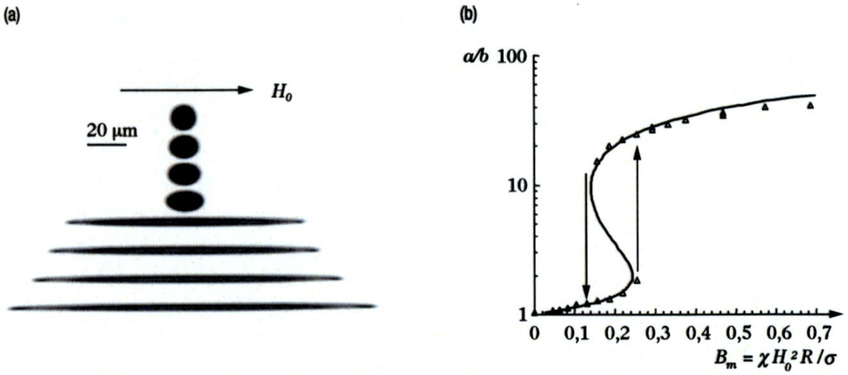

Fig. 2.10 Instability of a drop of ferrofluid subjected to a magnetic field. (**a**) The drop, spherical in zero field, stretches in the direction of the magnetic field applied and takes the form of an ellipsoid of revolution. Beyond a threshold value of the magnetic field, the drop deforms considerably into a needle, pointing in the direction of the field. (**b**) Shape of the drop (ratio of the minor axis to the major axis) as a function of the magnetic Bond number. The *triangles* represent experimental measurements and the *solid line* shows the theoretical curve. For the lower values of the ratio a/b the drop is an ellipsoid; for the higher values it takes the form of a needle. We can see that the ellipsoid-needle instability does not occur at the same value of the magnetic field when the amplitude is increasing and when it is decreasing: there is a hysteresis

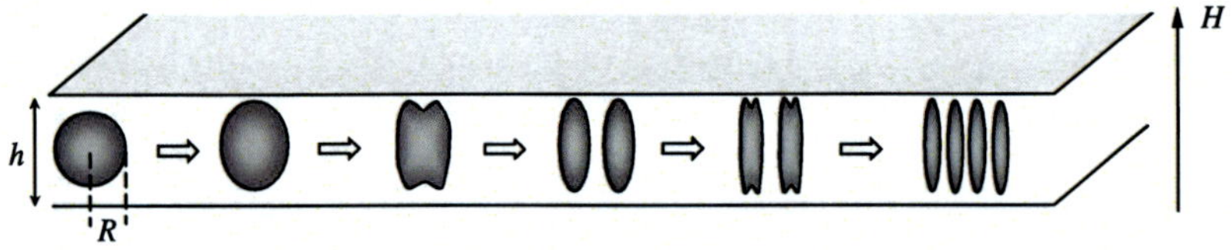

Fig. 2.11 Cascade of splitting in a ferrofluid drop confined between two planes and subjected to a magnetic field perpendicular to the planes, in the case where $R \ll h$. From *left to right*, the amplitude of the field (zero for the spherical drop) increases

to escape in other directions. There are then two possible scenarios, depending on the value of the ratio R/h: if $R/h \ll 1$, the drop separates, in a cascade of splitting (Fig. 2.11). If, on the other hand, $R/h \gg 1$, then fingers emerge on the interface of the drop (see the first images of Fig. 2.8).

In both cases, the appearance of these instabilities can be understood as the loss of stability of an ellipsoid of revolution in favour of an ordinary ellipsoid. But the mathematical treatment is not simple. The ellipsoid model may be appropriate when $R/h \ll 1$, but this is no longer the case when $R/h \gg 1$, because the drop is flattened by the planes confining it. We are then dealing with something more like a cylinder, in which the field of demagnetisation is no longer constant. Calculating the magnetic energy is then a much more complicated task than in the case of an ellipsoid drop.

When $R/h \ll 1$, we obtain a cascade of splitting that results in a network of needles with hexagonal symmetry. The distance between needles decreases as the magnetic field increases: this is the bubble pattern. As Fig. 2.11 illustrates, each division is initiated by a tip-splitting event, with the top of the ellipsoid dividing into two points. Because of the magnetic dipolar repulsion, these two points repel each other, and the splitting process begins. When $R/h \gg 1$, the formation of stripes and the repulsive interactions between stripes leads to the emergence of a labyrinth that occupies the whole space, for a sufficiently high magnetic field (Fig. 2.8). This is because the ribbon form is favourable to the minimisation of magnetic energy: the more intense the magnetic field, the thinner the ribbon.

2.4 Conclusion

We have seen that physics gives us access to a wide diversity of patterns in equilibrium: systems of bubbles, rings, foam, stripes, labyrinths and spikes, and that several ingredients are necessary and sufficient to produce them. Very short-range repulsive interactions and medium-range attractive interactions are needed to obtain a diphasic system with the two phases coexisting in interlinked domains. The equilibrium between medium-range attractive interactions and long-range repulsive interactions leads to the structuring of the domains, with a well-defined characteristic length scale. We have also seen that the effect of confinement can be a necessary condition for self-organisation, because all the systems described are either 2D and pseudo-2D or involve surface effects.

These conditions are satisfied by ferrofluids,where the source of the very short-range repulsion is either Coulombian (charged particles) or steric (surfacted particles), the medium-range attraction is induced by the van der Waals interaction and the long-range repulsion by magnetic dipolar interactions.

All these patterns also appear in the living world, as the photos in Fig. 2.12 show. These are all 2D examples (confined to the surface of the animals. 3D structuring of the interface can also be observed in the animal kingdom (for example the spikes of sea urchins).

We can rediscover some of these patterns by using a model of non-equilibrium instability (Turing instability) in reaction-diffusion systems. Turing instability has therefore become one of the explanations for these patterns (see Chap. 5 by J. Tabony in the present volume). Our conjectures are even simpler. Suppose that there is a mixture of white and black cells in a pseudo-2D system. The very short-range repulsion is caused by steric hindrance. The medium-range attraction leads to the formation of groups of white cells and groups of black cells, entailing that the white-white and black-black attractive interactions are stronger than the white-black or black-white interactions (this effect could be caused by specific adhesive proteins). This interaction will produce islands of black cells in a sea of white cells, but without any particular pattern. Now, if there exist long-range repulsive interactions (for example, a molecule with a long-range repulsive effect on the black cells), the

Fig. 2.12 Organised patterns in the animal kingdom. These morphologies display strong similarities to the equilibrium patterns observed at the interface of ferrofluids. *Stripe patterns*: zebra, tiger. *Bubble patterns*: leopard and hyena. *Foam pattern*: giraffe. ©Digital Stock

islands of black cells will then form similar patterns to those we have just described in self-organised physical systems at equilibrium.

References

1. Andelman D. and Rosensweig R.E. (2009) Modulated phases: review and recent results, *J. Phys. Chem. B* **113**, 3785–3798. Andelman D. (1996) Des zébrures aux motifs à pois, *La Recherche*, February 1996.
2. Bacri J.-C., Perzynski R., and Salin D. (1988) Magnetic Liquids, *Endeavour* **12**, 76. Bacri J.-C., Perzynski R., and Salin D. (1987) Les liquides magnétiques, *La Recherche*, October 1987.

3. Elias F., Flament C., Bacri J.-C., and Neveu S. (1997) Macro-organized patterns in a ferrofluid layer, *J. Phys. I France* **7**, 711–728.
4. Rosensweig R.E. (1985) *Ferrohydrodynamics*, Cambridge University Press (New York).
5. Seul M. and Andelman D. (1995) Domain shapes and patterns: the phenomenology of modulated phases, *Science* **267**, 476–483.

Chapter 3
Hierarchical Fracture Networks

Steffen Bohn

3.1 Introduction

During a not-very-scientific search on the Internet for the purposes of the introduction to this chapter, I came up with a few headlines from the French news: "Urban fracture" (Le Monde), "CNE: first scene of rupture in the industrial tribunal" (Libération) and "Katie Holmes and Tom Cruise, the rupture" (Madame Figaro). These are fine examples of the figurative use of the words *rupture* and *fracture*. The words are associated with abrupt and important events resulting from tensions that have often been steadily growing over a certain time. A fracture or rupture separates what was once whole, and it is usually irreversible.

When we turn to fractures in the proper sense of the word, fractures that propagate in a solid subjected to mechanical stress, we realise that these first, qualitative remarks are in fact essential to understanding their formation. Fractures are irreversible; there is no phenomenon of reorganisation. This makes the history of the system of crucial importance. Unlike systems in equilibrium (see the Chap. 2 by F. Elias and J.-C. Bacri), where the structures result from continuous and usually symmetrical interactions, the path followed by a fracture is determined during the period of its formation. Although existing fractures can influence the following ones, subsequent events can no longer change them. As a result, the interaction is hierarchical, and following the evolution over time becomes the essential element in an analysis of the system.

The situations (geometries and regimes) in which fractures form are varied, and we are going to focus on one particular situation: the hierarchical fracture network in a thin layer. Here, as we shall see, the fractures are formed successively. That is to say, the propagation time of one individual fracture is small in relation to the characteristic time period separating two nucleations. This is where the abrupt character of the rupture becomes important, and facilitates the analysis for us: the separation

S. Bohn (✉)
CNRS researcher at the Complex Matter and Systems Laboratory
(UMR 7057 CNRS-Paris-Diderot), University of Paris-Diderot, Paris, France
e-mail: steffen.bohn@univ-paris-diderot.fr

P. Bourgine, A. Lesne (eds.), *Morphogenesis*, DOI 10.1007/978-3-642-13174-5_3,

of the two time scales combined with the irreversibility of the events will enable us to consider the formation of one fracture after another.

As for the separative aspect of fractures, that will enable us to simplify the problem, by dividing it up. Two parts of a solid that are separated by a fracture are mechanically disconnected from each other, and no longer interact. But let us now turn our attention to a practical experiment.

3.2 The Formation of Hierarchical Fracture Networks

A homogeneous layer of an aqueous solution of latex particles is spread over a glass plate and left to dry. As the water evaporates, the distance between the particles decreases. When the concentration rises above a critical value, the solution changes from a liquid state to a gel state. The evaporation continues and the volume of gel diminishes. The substrate (the glass plate) to which the gel adheres opposes the contraction of the gel. This results in mechanical stresses that increase continually during the drying process. When the stresses become too strong, the gel can no longer withstand them, and it fractures.

The fractures form successively as the gel dries, and the existing fractures influence the following ones. This interaction occurs through the field of mechanical stresses. It is important to take into account the tensorial nature of the field of stresses, corresponding to the fact that the mechanical stresses vary depending on the direction. Unlike fields of temperature, concentration or pressure, where a scalar quantity is attributed to each point in the system, stresses are described by matrices, and in Fig. 3.1 we indicate their components schematically using crossed arrows.

In the absence of fractures, the stresses are homogeneous and isotropic. The fracture in the first image has locally released the stress components that are perpendicular to it. The parallel stress component is hardly affected at all. When the new fracture arrives in the vicinity of the old one, it starts to be influenced by the latter. The new fracture advances in the direction enabling it to release the most possible stress, and so it changes direction to connect up to the old fracture at right angles.

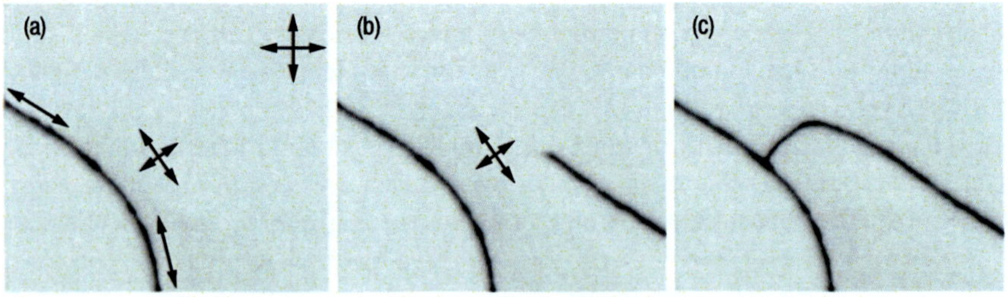

Fig. 3.1 A new fracture connects up to an existing one. (**a**) The existing fracture releases, locally, the stresses that are perpendicular to it. (**b**) The new fracture follows another direction, so as to release the stresses that remain. (**c**) It joins up with the old fracture at right angles, without disturbing it

It is because of the parallel stress component, which has not been released, that the two ends of each new fracture connect up to existing fractures, gradually dividing the surface up into cells or domains. And as the new fractures join the old ones at right angles, without disturbing them, the continuity of the old fractures is preserved. This means that in the final network, we can identify the first fractures, which cross the whole surface, and the last ones, which are very short. In this experiment, the fracturing is limited by delamination: the gel eventually loses it adherence to the substrate, and consequently the mechanical stresses disappear. This stop mechanism determines the final distance between fractures. Note that the distance between fractures gives a characteristic scale to the final geometry and the very widespread concept of fractals does not apply. One essential element of fractal geometry is the absence of any such characteristic scale. It has been shown that the characteristic distance is linearly dependent on the thickness of the layer of solution.

Another example of hierarchical fracture network is provided by enamel on stone or earthenware. Here, the fractures are caused by cooling rather than drying. The most common enamelling technique is to mix the enamel powder with water to make a paste, which is then applied to the fired earthenware. The water evaporates or is absorbed by the earthenware. The object is then put in a cold kiln, which is gradually heated up to avoid destroying the object through thermal shock. The enamel melts, and adheres to the substrate as it cools. As the kiln gradually cools down, the enamel and the earthenware both shrink. However, because they have different thermal expansion coefficients, the enamel shrinks more than its substrate, the earthenware. The first fractures form during the slow cooling in the kiln, but new fractures can continue to appear for many years.

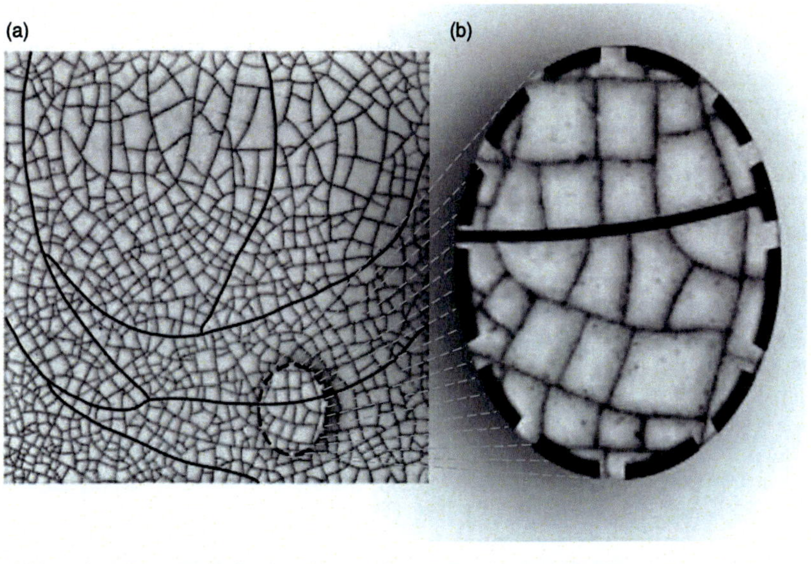

Fig. 3.2 A fracture network in the enamel of an earthenware plate. The characteristic distance between fractures is linearly dependent on the thickness of the layer of enamel

Unlike the drying gel, we cannot see any phenomenon of delamination, but there is, once again, a characteristic distance between the fractures that is proportional to the thickness of the layer of enamel. The gradient of the distance between fractures in Fig. 3.2 is the result of variations in thickness. Something to study whilst you drink your tea (...in a cup made of enamelled earthenware, of course).

3.3 The Fracture Network as a Hierarchical Division of Space

To move beyond a simple measurement of characteristic distance and grasp the physics of this system, it is worth starting by choosing a pertinent description. It happens that a description based on domains is much more pertinent than one based on fractures, because it directly reflects the physical causalities.

Let's place ourselves at a certain moment in the formation of the network. A new fracture has just divided a domain in two. By analogy with biology, where scientists talk of mother cells and daughter cells, we shall refer to the domain that has been divided as the father domain and the two resulting domains as the son domains. The two son domains will then be divided by other fractures, and so on. Knowing the history of the system, we can draw up its family tree, starting with the initial domain given by the shape of the sample.

This family tree represents the relations of causality. In the generic case, the substrate is much more rigid than the layer that fractures, and the fractures themselves extend right through the whole thickness of the layer, so that there is no longer any interaction between the different domains. As one domain is disconnected from the others, the fracture that divides it belongs exclusively to that domain. The growth of branches of the family tree corresponding to the successive divisions of domains also occurs independently.

The advantage of this description is that a large part of the physics (domain independence and succession of fractures) is already contained in the description itself, enabling us to attribute conservation laws to it. A trivially conserved quantity is the surface area. Obviously, the sum of the areas of the two son domains is equal to the area of the father domain. More interesting is another conservation law associated with the successive division of domains, concerning their form. Figure 3.3 shows the possibilities for dividing domains with three, four and five sides. We can see that the possibilities are fairly limited; a four-sided domain, for example, can be divided into either two quadrilaterals or a triangle and a pentagon. In general, the sum of the sides of the two son domains is equal to the number of sides (denoted s) of the father domain plus four:

$$s_{\text{son a}} + s_{\text{son b}} = s_{\text{father}} + 4. \tag{3.1}$$

We can introduce the concept of *geometrical charge*[1] of a domain, which we define as the difference between the number of sides and 4: $q = 4 - s$. The constraint on

[1] This is inspired by the concept of topological charge, often used in foam physics. The topological charge is equal to six minus the number of neighbours. It is conserved in all rearrangements, such as the exchange of neighbours, for example.

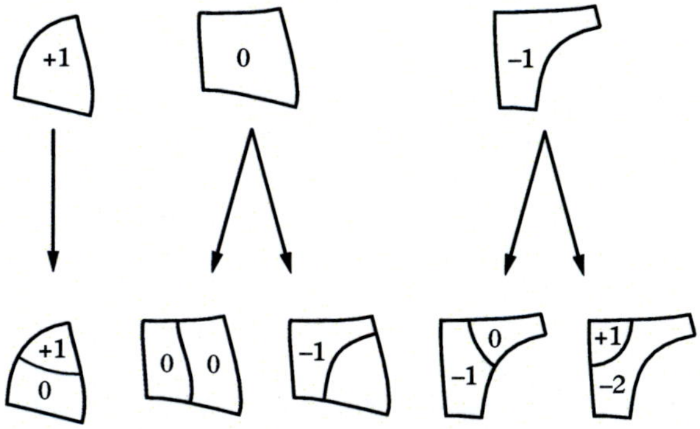

Fig. 3.3 Possibilities for dividing a *triangle*, a *quadrilateral* and a *pentagon*. We have excluded cases where the fracture ends in a corner, because they are not observed experimentally. In corners, the mechanical stresses have all been released already. The *numbers* indicate the geometrical charge of each domain

the number of sides results in the conservation of this geometrical charge: the sum of the geometrical charges of the two son domains is equal to the geometrical charge of the father domain:

$$q_{\text{son a}} + q_{\text{son b}} = q_{\text{father}}. \tag{3.2}$$

As the formation of a fracture network is nothing more than a succession of domain divisions, the total charge of any fracture network is equal to the charge of the initial domain. We can now link up the average number of sides $\langle s \rangle$ and the total geometrical charge Q_{total}:

$$\langle s \rangle = \frac{1}{N} \sum_{i=1}^{N} s_i = 4 - \frac{Q_{total}}{N} . \tag{3.3}$$

where N is the number of domains. In an extended network, this number is large, with the initial geometrical charge being of the order of 1. The average number of sides is therefore equal to four. This rather mathematical line of argument determines the average number of sides, but it does not tell us anything about the distribution around the average. On an earthenware plate like that shown in Fig. 3.2, we counted about 12% of three-sided domains, 76% with four sides and 12% with five. Out of 1,000 domains, we only found one with six sides. Other authors have found slightly different distributions, but always with an average of four, of course.

3.4 A Characteristic Scale

The question that naturally follows is this: what influence does the shape of a domain have on the fracture that divides it? This is therefore a question of shape inheritance. It is not trivial to study this question in an extended network, because the shapes of the father domains result from the history of the system, and are therefore not directly controlled by the experimenter. To get around this problem, we have found a method for forcing particular shapes. As it was impossible to explore all the possible shapes, we have limited ourselves to a rectangle.

We leave a layer of corn starch paste to dry in a rectangular box with silicone-oil coated sides and a very clean bottom. Like the latex gel, the corn starch sticks to the bottom of the box and shrinks while drying. After a certain time, the first "fractures" form: thanks to the silicone oil, the material first breaks away from the sides of the box. We now have a domain with free edges, like those demarcating a domain in an extended fracture network, but with a controlled geometry in this case. So how does this rectangular domain divide up?

As Fig. 3.4 shows, the answer depends on another parameter: the thickness of the layer of corn starch. If the thickness is about one sixth of the lateral dimensions of the domain, the fracture is perfectly deterministic: it passes right through the

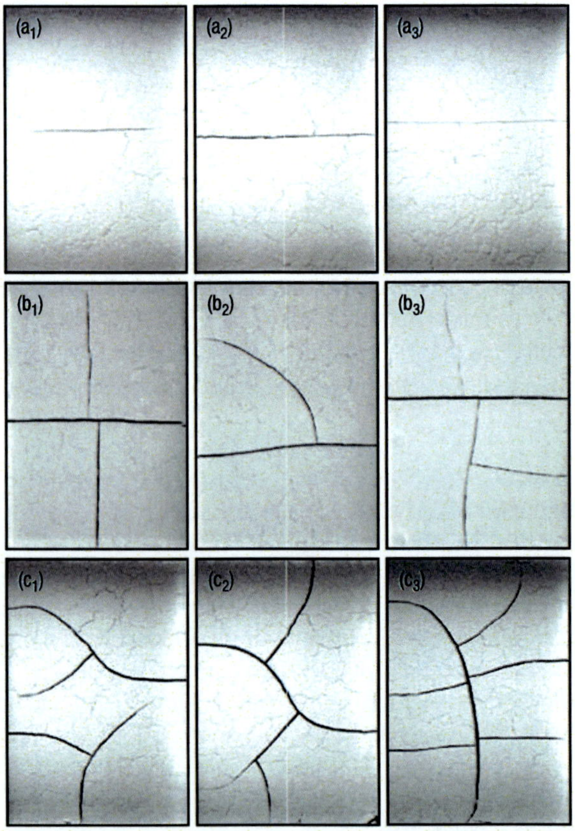

Fig. 3.4 Fractures in a controlled domain of different thicknesses e: (**a**) $e = 9.1$, (**b**) $e = 6.3$ and (**c**) $e = 4.8$ at three successive times (1, 2 and 3). For a large thickness (compared to the domain size), the fracture is entirely determined by the edges of the domain: it is deterministic. For smaller thicknesses, its form is governed by the impurities and irregularities of the material: it is non-deterministic

middle and divides the domain into two symmetrical son domains. When the layer of starch is even thicker, no fracture forms. This is in keeping with the fact that the average distance between fractures in a fracture network is linearly dependent on the thickness. When we reduce the thickness of the layer of starch, the first fracture to divide the rectangular domain begins to lose its deterministic nature. Firstly, it no longer always cuts through the middle. Then curved fractures appear. For even lower thicknesses, the fracture becomes almost completely disordered and random.

We can interpret these results in terms of characteristic lengths. Apparently, the relevant length in this system is the thickness of the layer that fractures. Not only does it determines the distance between fractures in the final network, but it also plays a role in the interaction between fractures.

We must compare this length with the second length in the problem, given by the domain size. The ratio between these two lengths is a dimensionless number. When the domain size is much greater than the thickness, which is the case for the first fractures that appear then the fractures occur randomly. But as the domains are repeatedly divided, their size decreases over time. For the last fractures, the domain size is comparable to the thickness of the layer and the fractures become completely controlled by the shape of the domains. We therefore observe a transition in the intrinsic regime during the formation of the network, from disordered behaviour to deterministic behaviour.

3.5 Conclusion

The formation of fracture networks is characterised by the dominant role of the system's history and by the tensorial nature of the field of mechanical stresses that governs the nucleation and propagation of fractures. The dominant role of history stems from the irreversibility of the rupture and from the successive nature of the fractures. It enables us to introduce the concept of a family tree for the divisions of domains, on which our whole analysis is based. In most physical and biological systems, the role of history is less evident and certainly more difficult to integrate, but it is equally important. One good example is provided by soap foams. Their dynamics is dominated above all by continual reorganisation, but it has been shown that the arrangement of bubbles – and therefore also the macroscopic behaviour of the foam – are highly dependent on the system's past. In biology, where every multicellular organ or organism stems from a succession of cell divisions, history is essential to a thorough understanding. The complexity of the task is almost certainly partly due to the impossibility of clearly separating the effects of history from the effects of instantaneous mechanisms.

The other essential aspect of fracture networks, their underlying tensorial nature, is more specific. Note that there are no known examples, either in physics or in other domains, of scalar growth (temperature, concentrations, pressure) leading to network structures. Thus, the fact that the fractures divide the surface into distinct domains appears to be directly due to the mathematical structure of the mechanical stresses.

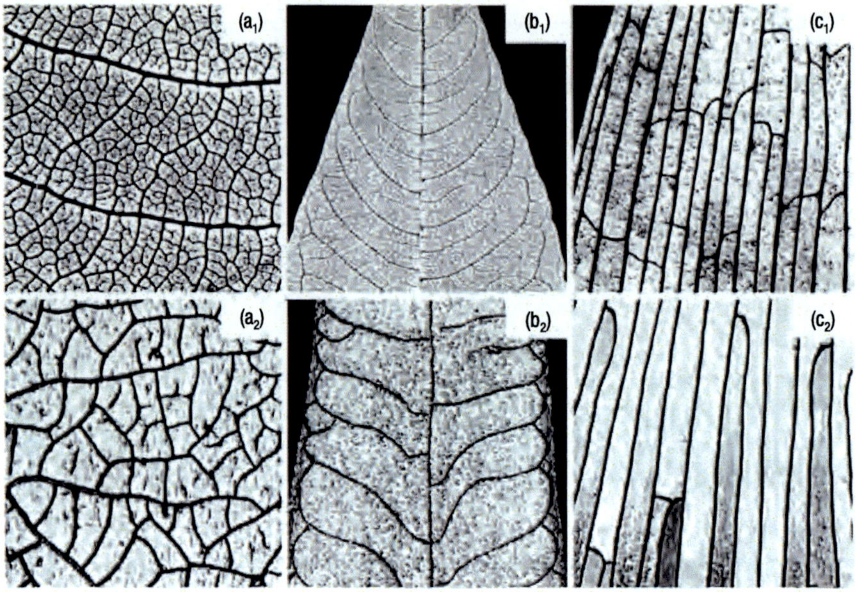

Fig. 3.5 Visual (at least) analogy between fractures (*top*) and leaf veins (*bottom*). There similarities not only in their overall structure, but also in more detailed features, like the junction of fractures/veins in a highly anisotropic regime. ©Y. Couder and L. Pauchard

However, we can also find perfectly analogous structures in biology: leaf veins (Fig. 3.5), which divide the leaf into little domains, called areolas. And like the fractures we have studied, this division of the surface occurs in successive stages during the growth of the leaf. First comes the main vein, or midrib, which divides the leaf into two halves. The secondary veins branch out from the midrib and connect up with each other. The domains are then repeatedly divided by the second-order veins, third-order veins and so on. As this sort of behaviour seems to be incompatible with growth driven by concentration gradients of chemical morphogens (such as the plant growth hormone, auxin), it is reasonable to imagine that the formation of veins is governed by a tensorial field. There are strong indications that mechanical stresses are once again involved, but we do not have the experimental proof to uphold this hypothesis.

However, even if this hypothesis is confirmed, we cannot extrapolate all the results described above. The strict hierarchical ordering of the fracture network geometry is not imposed on the formation of leaf veins. New veins can modify the existing ones, a phenomenon that manifests itself in the angles at the points of connection. Consequently, it is not easy to understand how the characteristic scale – the thickness – finds expression in the morphogenesis of the leaf. Moreover, the veins form in the leaf whilst it is growing, not in a static geometry like that of a layer of latex gel.

Chapter 4
Liquid Crystals and Morphogenesis

Yves Bouligand

At first sight, the idea of a single state of matter that is crystalline and liquid at the same time appears hard to accept, as many physicists, including Nernst himself, pointed out when the first liquid crystals were discovered more than a century ago [38]. In quite another style, Salvador Dali's melting watches, although meticulously represented, had even more incredible deformations. Dali made a work of art out of a provocation, and without really fooling the viewers, he encourages them to wander through the outer fringes of their unconscious. Dali probably did not know much about liquid crystals, but he knew that life has its own clock mechanisms, with other subtleties coming into play within a matter that is rigid here, flexible there, soft elsewhere, and liquid on other levels, all of which does not prevent the system from functioning. Quite the contrary.

Several works have shown that liquid crystals play a crucial role in biological structures (see [9, 25] and references therein), but little more is known than their applications in display screens or calculators. And yet, when left at liberty and examined under the microscope, liquid crystals, whether biological or not, generate elaborate forms, of which we shall now give a few illustrations and the beginning of an explanation.

4.1 Shells and Series of Arches

I shall approach the subject by describing the unexpected path by which I learnt what liquid crystals were, with gratitude to Dali for having prepared me. Numerous works had been published on the shells of crabs, a robust material if ever there was one. The calcium carbonate can be removed from these shells by means of acid or other compounds, leaving a fibrous and supple material made of chitin, a nitrogenous polysaccharide, associated with various proteins and other substances (in smaller quantities). Such shells are also called carapaces, exoskeletons, body walls, cuticles, and many other terms very common in scientific literature.

Y. Bouligand (✉)
École Pratique des Hautes Études, Paris, France
e-mail: yves.bouligand@univ-angers.fr

P. Bourgine, A. Lesne (eds.), *Morphogenesis*, DOI 10.1007/978-3-642-13174-5_4,

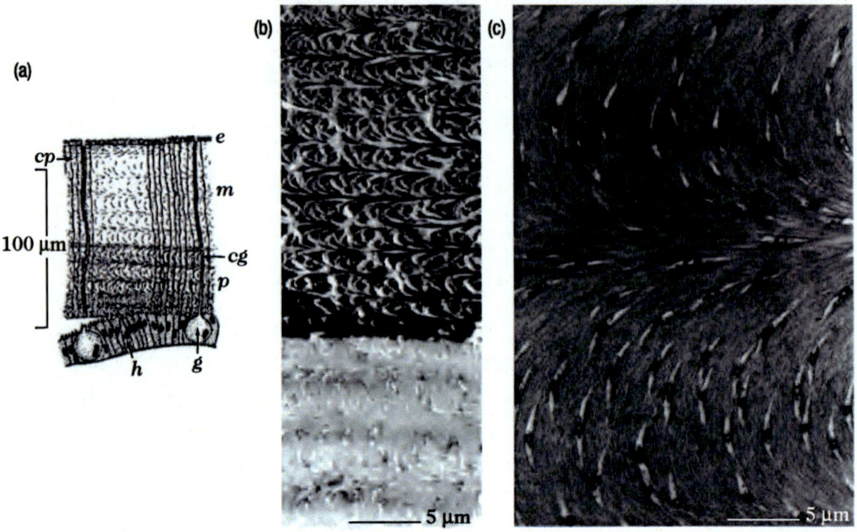

Fig. 4.1 Series of fibrillar arches, observed in an oblique cut in arthropod shells. **(a)** Example of *Julus*, a particular group of myriapods, whose calcified cuticle has been drawn here after Silvestri [54] (e: epicuticle; cg: gland duct; g and h: epidermal cells; m and p: organic fibrillar matrix apparently made of arches). **(b)** Example of crabs, with highly contrasted layers formed before the moult and lighter layers formed after it. **(c)** Magnified detail of an oblique cut even closer to the plane of the crab shell. The fibrillar order is less disturbed than in the more contrasted layers formed before the moult. Delicate cytoplasmic extensions of rectangular cross-section have worked their way between the fibrils of the shell

Using suitable microtomes, this organic matrix of the shell can be cut into thin slices, which are then examined under the microscope. The fibrous material of the shell appears to be made of superposed layers, each one formed of a series of regularly nested arches (Fig. 4.1). This is a widespread structure, observed in the shells of crustaceans [29, 30] and other arthropods, such as myriapods [54], insects [46], spiders, and all the jointed species which grow by means of successive moults [50]. Three-dimensional models were proposed involving curved filaments.

So the bodies of these animals are protected by a "cuticle", which is secreted by the epidermis. This latter is a simple epithelium, meaning that it is made of a single layer of cells. The external cuticle that it produces is supple to begin with; it subsequently hardens over almost the whole body, except at the joints, which separate the body into several parts and the legs into a series of segments. The cuticle hardening through a sort of quinonic tanning process, especially among the insects, is reminiscent of the way skins are tanned to make leather. This process also occurs among the crustaceans, transforming the cuticle into a carapace or shell, where the rigidity derives mainly from calcium mineralisation, produced and controlled by the epidermis [30, 50].

The models with curved fibrils, all pointing in the same direction, suggested that we should determine the orientations of these arches at each point on the shell, and

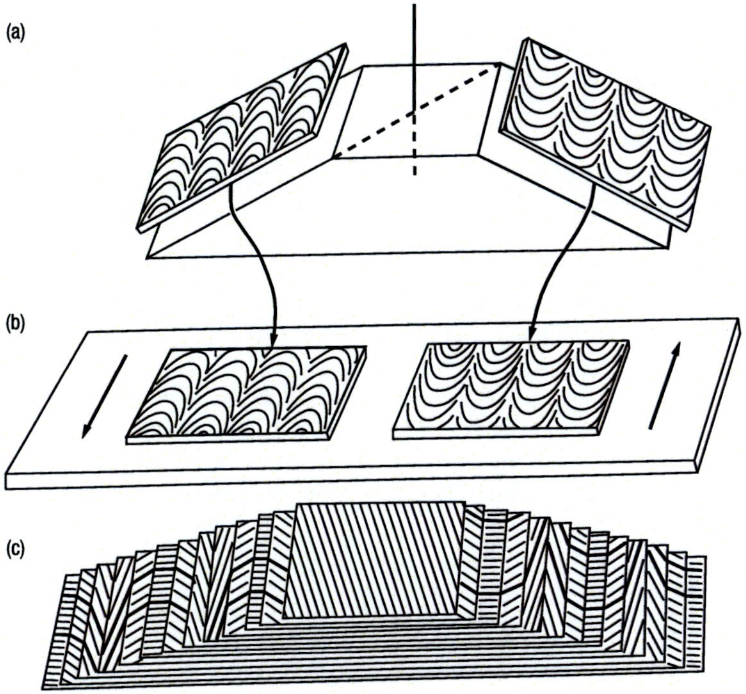

Fig. 4.2 **(a)** Diagram of the orientation of arches obtained in an oblique cut, along two planes symmetrical about a vertical axis (normal to the local plane of the shell). **(b)** Indication of the directions of the arches when the sections have been spread on a slide according to their original positions in the cuticle. **(c)** Reconstruction of the fibrillar structure in a model with equidistant layers carrying equidistant fibrils, rotating at a constant angle from one layer to the next. The arches observed in the oblique cuts are indicated in bold on the model

associate a vector with them. This should have produced a field of integral lines, the pattern of which on the surface of the shell would have been useful in biology of the development. However, our attempts to determine the orientations of the arches were unsuccessful, because we often found arches facing in opposite directions in the same area. We even showed that this result was general for oblique cuts that were symmetrical about the same normal to the shell, as Fig. 4.2a, b illustrate. Using a little bit of geometry, we were able to deduce a structure that was very different from the models with arches, but similar to that of man-made "plywood", with nothing but straight or nearly straight fibres [4].

4.2 Helicoidal Plywood

The structure of the fibrillar matrix of crustacean shells is represented schematically (Fig. 4.2c) by thin layers of equal thickness superimposed on each other; each layer contains an array of parallel, equidistant filaments, the direction of which rotates by a small, constant angle from one layer to the next, always in the same sense. When the alignments of fibrils come to the surface on the plane of the oblique cut, they form patterns of arches, underlined in Fig. 4.2c. This explains why the width of the series of arches depends on the angle of the cut, and why the arches are oriented in opposite directions in oblique cuts that are symmetrical about a vertical axis [4, 9].

The structure of these shells is therefore similar to that of a plywood sheet, except that instead of the fibrils running in just two directions, with each layer being at right angles to the one below, all the different orientations succeed each other here, covering all the different directions of the plane. In crab shells, this angular distribution is continuous, while in other cases we have observed a narrower selection of directions, close to each other, notably in the cuticles of very small crustaceans [4].

In crabs, as the structure is continuous, the successive planes of our models are simply for ease of drawing. We find the same fibrillar direction after a rotation of 180°, corresponding to a change in height of the half helicoidal pitch $p/2$, in other words a half helical turn of the structure (a few μm). The structure can be treated in a trirectangle trihedron, where Oz is normal to the plane of the shell and Oxy is chosen so that the components of the unit vector $\mathbf{n}$ giving the fibrillar direction at the point $M(x, y, z)$ are:

$$n_x = \cos(2\pi z/p), \qquad n_y = \sin(2\pi z/p) \quad \text{and} \quad n_z = 0\,. \tag{4.1}$$

As an exercise, we can look for the equations of the arched curves expected in a thin, planar, oblique slice, making an angle α with the plane Oxy. By taking the intersection between the plane Oxy and this slice as our axis Ox, we define an axis Os with a steeper angle, normal to Ox. By choosing the appropriate units on Ox and Os, we obtain: $x = \log|\sin s| + c$, where c is an arbitrary constant [5].

In fact, the fibrils diverge more or less from the directions attributed by this model, and an order parameter could be defined to take that into account. This order is already accessible by microscopic examination (Fig. 4.1b, c). For each point M, we could imagine a cone with axis $\mathbf{n}$, corresponding to an average local direction and with a half-cone angle of θ, which would contain 95% of the fibrillar directions within a spherical neighbourhood centred on M and with a radius r quite small with respect to p. The existence of a rotation around the axis z widens the cone, but this effect diminishes with r/p. There are statistical studies that I have not really tackled, which need to be carried further for these materials. Crab shells display precise variations in p and in the order parameter according to the level z in the shell. Note that the parameter of order is weaker for the layers formed before the moult, and better for those formed after it (Fig. 4.1b). In addition, this helicoidal plywood is disturbed by vertical (or almost vertical) extensions of the epidermic cytoplasm, which push aside the horizontal fibrils as they emerge (Fig. 4.1c).

Many animal and plant tissue structures have been compared to plywood, notably the arrangements of smooth muscle fibres around the stomach and the intestine, or similar arrangements of collagen fibrils in many connective tissues or in the skin of certain worms. When examined in oblique cross-section, these biological plywoods display characteristic chevron patterns, which we can obtain by reworking the model in Fig. 4.2c so that the fibres in alternate layers are perpendicular to each other.

When we published our first results on the origin of the series of arches observed in crustaceans [4, 5], we stated that they probably also existed in the cuticles of many insects or other arthropods, as the micrographs of several studies suggested. These same arch patterns could also be found in the egg envelopes of certain fish, and in compact bone tissue, where the arches also coexist with chevron patterns (see [5, 7] and references therein). This arch-chevron association is also frequent in certain worm cuticles.

The surprise was to find these same series of arches in the chromosomes of certain micro-organisms. The authors proposed models where they curved the filaments of DNA [7] as others had done with the chitin or collagen of skeletal tissues. This led us to look anew at many of these materials, especially chromosomes, to test our helicoidal plywood model on them [7, 18].

4.3 Cholesteric Liquid Crystals and Stabilised Analogues

Some time after these first publications on helicoidal plywoods of biological origin, I received a letter and several articles from C. Robinson, an English colleague who was working on "cholesteric liquid crystals", obtained with a synthetic polypeptide – polybenzyl-L-glutamate or PBLG – in a concentrated solution in certain organic solvents [51]. And in one of his works, the author indicated that he had reproduced the same type of liquid crystal with a concentrated solution of DNA in water [52]. These liquid crystals displayed a finely stratified structure, and the model used was geometrically similar to the one I had arrived at in my study of the fibrillar matrices of shells. However, the liquid crystals studied by Robinson were fluid solutions, unlike the fibrillar matrices, where the polymers were aggregated into fibrils forming quite a dense, spongy gel. Robinson informed me that very interesting works on liquid crystals had been published in France at the beginning of the twentieth century by G. Friedel, F. Grandjean and C. Mauguin, and he completed this bibliography with references to more recent papers [51]. The liquid crystals of PBLG prepared by Robinson reflected interference colours, and more precisely, circularly polarised light, a phenomenon that can also be observed with the shells of certain scarab beetles, the cetoniines, a property of these insects that was discovered by Michelson (better known for his collaboration with Morley on an essential experiment to measure the speed of light).

This all suggested that there was something in common between the helical polymer solutions and these complex gels, more or less toughened, that could be found in various tissues and shells. The liquid crystal of DNA was particularly interesting, as this same structure could be found in certain chromosomes [18]. These liquid crystals were called "cholesteric" because they were first discovered in esters of cholesterol – compounds of obvious biological interest.

I then contacted the physicists working on liquid crystals in France, notably in the Orsay group. They helped me considerably, by discussing the problems and

Fig. 4.3 Arch-like decorations on the surface of a drop of nematic liquid to which has been added some Canada balsam, a natural resin containing chiral compounds, producing the cholesteric structure. The horizontal layers come to the surface of the air-liquid interface, which is oblique at the periphery of the drop

providing me with samples, enabling me to identify more precisely the similarities and differences between liquid crystals in the strict sense of the term and the biological structures that were found to resemble them [6, 10]. In particular, I found arch patterns in certain slides of cholesteric liquid crystals (Fig. 4.3).

New molecules producing liquid crystals were synthesised at Orsay, using chemical groups that allowed cross-linking, in other words the creation of covalent links between neighbouring molecules. The liquid crystal gradually solidified, but the initial distribution of molecules remained little changed. The material obtained, similar to a resin, allowed to prepare sections that were useful for comparing the cholesteric liquid crystals stabilised in this way with the fibrillar matrix of shells [19]. We use the term *stabilised analogues of liquid crystals* to denote both these non-fluid resins, derived from liquid crystals and retaining their essential structures, and the fibrillar matrices described earlier. But let us return to the liquid crystals themselves, and recall their key characteristics.

4.4 Specificity and Diversity of Liquid Crystals

Liquids crystals are also called *mesomorphic* states, in other words intermediate states of matter. They have been the subject of numerous overviews [22, 28, 36, 38, 39]. They are *ordered liquids*. They really are liquids, because they flow and the molecules circulate between each other. They are also solvents for various compounds, notably dyes. Many liquid crystals are very fluid, and others are more viscous. In the static state, these phases are ordered, because the molecules are aligned at each point in a predominant local direction, unlike ordinary liquids, which are isotropic at rest. This molecular order in liquid crystals usually makes them birefringent, which can easily be revealed with a polarising microscope. It is often difficult to determine whether highly viscous solutions of certain polymers really are liquid crystals. If you take a drop of an isotropic solution of polymers, where the polymers are distributed in every direction, you only have to stretch the drop for the polymers to align with each other, causing a birefringence that can take a long time to relax.

4.4.1 Mesogenic Molecules

Mesogenic is the name given to molecules capable of producing liquid crystals [28, 39]. They have specific shapes, some of which are illustrated in Fig. 4.4. The rod-like shape is the most common, and the specialists agree that the central part of the molecule is generally rigid, while the extremities are flexible, as paraffinic chains can be [36, 38]. Other mesogenic molecules have disk-like shapes [22], rigid in the middle and more flexible around the edges, but there are many variations, including oblate and banana-like forms, hollow pyramids that fit into each other, etc. [28].

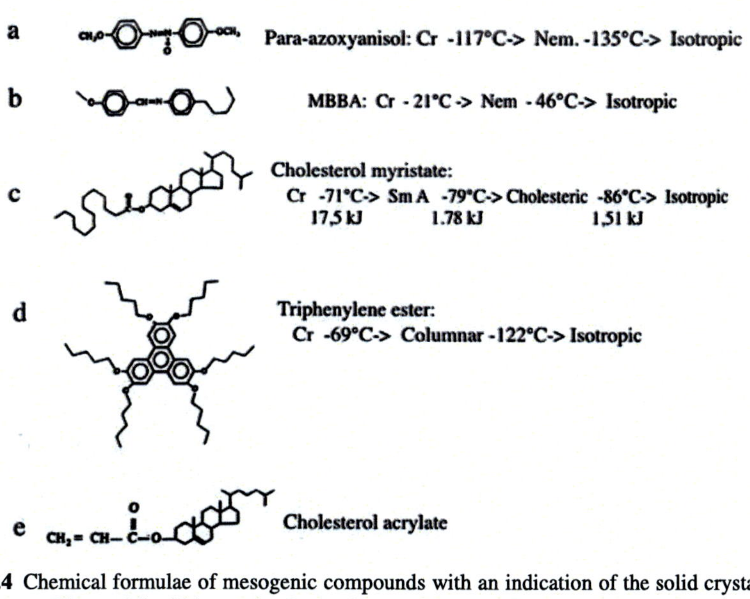

Fig. 4.4 Chemical formulae of mesogenic compounds with an indication of the solid crystalline, liquid crystalline and liquid isotropic phases they adopt according to temperature. The transition temperatures are given, as well as the enthalpies for one of the compounds. Cr: solid crystal; Nem: nematic liquid; SmA: smectic A liquid

Amphiphilic molecules, which possess two parts with different characteristics of solubility, are often mesogenic. This is the case for many surfactants and detergents. There is no lack of amphiphilic substances in biology, notably phospholipids, which are essential components of biological membranes [28, 39].

Many of these compounds have asymmetrical carbons, with a determinate orientation [36, 38]. This is true of many esters of cholesterol, which are indeed mesogenic. This asymmetry imparts a rotatory power to the solutions containing them, and this is a characteristic that plays an important role in the structure of liquid crystalline phases constructed with such molecules. Remember also that many mesogenic polymers are helicoidal, with a determinate orientation [51, 52]. We have given two examples in PBLG and DNA.

4.4.2 Structure of Liquid Crystals

We can distinguish several types of liquid crystals, according to the geometry by which the mesogenic molecules are assembled [36, 38]. When they are aligned parallel to each other, without forming layers, they are said to be *nematic* and there is no periodicity in the distances between molecules (Fig. 4.5a). Similar alignments also exist with disk-shaped molecules (Fig. 4.5b) [28]. The presence of asymmetrical molecules often leads to helicoidal modulations in the structure. The nematic phases then become *cholesteric* (Fig. 4.5c), with rod- or disk-shaped molecules. In this case, an element of large enough volume in relation to the molecular dimensions, but small enough in relation to the helicoidal pitch, is comparable to the nematic structures of Fig. 4.5a, b. The planar helicoidal structure only becomes appreciable at a greater distance.

In *smectic* structures, the rod-like molecules are aligned in different layers [36, 38]. The molecules are normal to the average direction of these layers in the *smectics A* (Fig. 4.5d), or sloping with a defined angle in the *smectics C* (Fig. 4.5e), but there are also a good number of other types of smectic structures [28, 39]. In the smectics C, the presence of asymmetrical molecules means that the direction in which the molecules are inclined on each layer rotates by a small, constant angle from one layer to the next. In this way, the alignment curves of the molecules are helices, with translation symmetry along the plane of the layers (Fig. 4.5f). A helicoidal pitch therefore comes into play, as in the cholesteric phases, and it is generally very large with respect to molecular dimensions, often of the order of a micron or more, making it observable in an optical microscope. In many phases with disk-like molecules, the molecules stack up to form columns, that group together to give a network, often hexagonal in the plane transverse to the columns [28]. These liquid crystals are said to be *columnar* (Fig. 4.5g). Other liquid crystals are known as *blue phases* [28]. They correspond to complex arrangements of twisted cylindrical domains, which we shall return to later, for many examples of their stabilised analogues exist in biology. The twisted layering that we described in the superimposed planes then occurs by means of nested cylinders. Methods were devised to produce

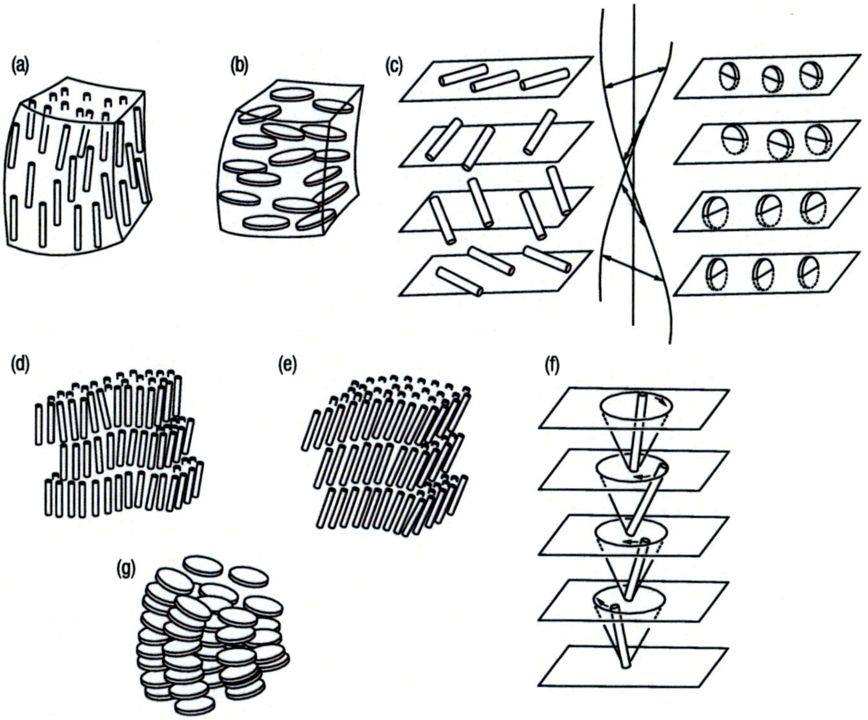

Fig. 4.5 Main structures of liquid crystalline phases: **(a)** Nematic with rod-like molecules; **(b)** Nematic with disk-shaped molecules; **(c)** Cholesteric with rod-like (*left*) or disk-shaped (*right*) molecules; **(d)** Smectic A; **(e)** Smectic C; **(f)** Chiral smectic C structure; **(g)** Hexagonal columnar with disk-shaped molecules

stabilised, non-fluid analogues of cholesteric liquid crystals [19]. They were also successfully applied to nematics and smectics [36].

4.4.3 Phase Transitions

Liquid crystals appear and disappear like ordinary crystals do, when we change the temperature of preparations or the concentration of a solvent [38]. Liquid crystals in which transitions occur with temperature changes are said to be *thermotropic* (Fig. 4.4). They are said to be *lyotropic* when these transformations are controlled by the proportion of a solvent in a preparation, which is usually the case for phases obtained with amphiphilic molecules in the presence of water, but also for PBLG in the presence of organic solvents.

Enthalpies are measured for *first-order transitions*, with clear-cut interfaces between the phases present at equilibrium. This occurs at the transition between ordinary crystal and liquid crystal, in both directions, or between liquid crystal and isotropic liquid [38]. *Weakly first-order or second-order* transitions generally occur between the different liquid crystalline phases produced by the same compound, which are also called *mesophases* [28, 39].

The alignment of molecules in a liquid takes place when the temperature is reduced or a concentration level is exceeded; the transitions are accompanied by a noticeable decrease in the volume of liquid. It seems as if the alignment of elongated molecules reduces *steric hindrance*, and this principle of *excluded volume* formed the basis for the principal theories of liquid crystalline assembly [22, 28, 36, 39]. However, steric factors are not the only ones involved, for the molecules can be ionised, carry permanent or induced dipoles, or be joined together by hydrogen bonds. Various ions, and water molecules, also contribute to these attractive or repulsive interactions, which are highly dependent on the pH, the ionic strength and many other physico-chemical factors [28].

Certain crystalline liquid phases align not just molecules but groups of molecules, in other words *micelles*. Zocher called them *liquid crystalline superphases* [59] in a work devoted primarily to mineral liquid crystals, but this concept also applies to certain lyotropic organic phases. Lastly, in a nematic medium, at the approach of a smectic transition, the molecules already form layered clusters, known as *cybotactic groups* [36, 28].

4.5 Liquid Crystals and Stabilised Analogues in Biology: A Widespread Phenomenon

After these works on shells and chromosomes, followed by an examination of various liquid crystals for the purposes of comparison, I set out to see whether other bridges had been glimpsed between biology and these intermediate states of matter. Lehmann himself, one of the first scientists to discover liquid crystals at the end of the nineteenth century, wrote a well-documented book on the subject, in which he explained that the notion of life was hardly conceivable with a matter that only manifested itself in the three states: solid, liquid and gas [42].

4.5.1 Muscles

Lehmann believed that liquid crystals were a necessary ingredient for life, and he reflected on the possibility of liquid crystal motors, evoking parallels with the functioning of muscles. He even suggested that they might be essential to the future of aviation, which was still in its infancy, so that it might one day compete with the birds.

While studying the ultrastructure of the muscles of various invertebrates, I noticed that the filaments of actin and myosin gather to form contractile fibrils, adopting geometries similar to those of the nematic and smectic A and C phases, with their helicoidal variants, and the same groups of symmetries [9, 10]. This was in line with Lehmann's ideas, but without going as far as applying them to flying machines, although the idea of artificial muscle had made a comeback in many laboratories. Of course, muscles are not liquids, and they are stabilised by transverse bridges. The filaments that align in them are made of numerous aggregated molecules, so that these systems would be *stabilised liquid crystalline superphases*.

4.5.2 Myelinic Figures and Fluid Cell Membranes

Lehmann knew a good deal about lyotropic phases, produced by fatty amphiphiles, like *lecithins*, when they are brought into contact with water [42]. It was already known how to extract them from animal and vegetable tissues or egg yolks, and today they are produced in great quantities from soya beans. They are mixtures of membranous phospholipids. The best-known of these lyotropic mesophases are *myelinic figures*, obtained in vitro with lecithin and many other amphiphilic compounds in the presence of water, or observed in vivo in diverse cells and tissues, such as the thyroid gland or the spleen, and in various plants. These myelinic figures, whether artificial or natural, are fluid and birefringent, even at rest, that is to say as crystalline liquids, in the full sense of the term, without being stabilised. Myelinic figures are produced from alcohol or ether extracts of the *white substance* of the brain, made of myelin sheaths around axons. Nageotte devoted a superbly illustrated book to them, considering myelin to be the most remarkable of "natural myelinic figures", although certain details of their organisation have led to the acceptance that there are stabilised zones [47].

When considered on their own, cell membranes are liquid crystallines, but we only know this indirectly, because it is difficult to observe their birefringence, given that they are less than 10 nm thick. It is easy to isolate the membranes of red blood cells, and in them we find the same types of compounds as in the white substance and in natural myelinic figures. Electronic microscopy has shown that myelinic figures and the *white myelin sheaths* around the axons of nerves correspond to spiralling accumulations of cell membranes [9, 10].

In the end, the liquid crystalline characteristics of cell membranes were clearly established [25]. Later works on various membranous proteins have confirmed their fluidity, at least implicitly [55]. Remember that the fluidity of cell membranes had already been suggested in the nineteenth century by Pfeffer, through the micro-manipulation of red blood cells. We shall return to that later. It was established by Chambers for the main cell types, by perfecting the techniques of micro-manipulation (see [14] and references therein). Membranes are liquid crystals, because in addition to their fluidity, the molecules are oriented transversely, due to their amphiphilic nature, and this has been confirmed by X-ray diffraction diagrams.

Other research has shown that the most visible cellular organelles in the cytoplasm are made of fluid membranes, of closely-related composition (mitochondria, endoplasmic reticulum, Golgi bodies, nuclear envelope, see Chap. 7) and also of a liquid crystalline nature (see [10, 14] and references therein).

4.5.3 *Stabilised Membranes*

Less fluid or even stabilised regions are also present in membranes, at the level where the cells are tightly apposed, for example desmosomes or other junctions between epithelial cells, or the synapses between nerve cells (see [9, 14] and references therein). Several of these stabilised zones form links between the membrane and certain filaments of the *cytoskeleton* within the cell, or with extracellular matrices, in other words the collagen fibrils of connective or bone tissues, or with the fibrils of another nature that can be found in the outer cuticles or shells of invertebrates.

4.5.4 *Nematic and Cholesteric Analogues*

Most extracellular fibrillar systems are stabilised cholesteric or nematic analogues, some examples of which we have described above, in the skeletal tissues [4, 5]. This is a general result, with very diverse features according to species [6, 7] all the more so since the cells can impose certain orientations on the fibrils during secretion. Other walls are also cholesteric analogues, like those made of cellulose associated with other compounds that enclose plant cells, especially in their early development [7, 53].

Certain chromosomes are also liquid crystalline [7, 18, 43] but this result has yet to be generalised to other types of chromosomes and chromatin, as the works of Livolant and Leforestier suggest [44, 45]. However, constraints on the architecture of the chromatin sometimes prevent the cholesteric twist from expressing itself, leading to variations of a nematic or lamellar type.

Various alignments also occur in the microtubules or filaments of the cytoskeleton. Their birefringence is often detected, notably in mitotic spindles, with nematic-type symmetries. But we are in a situation where it is still difficult to determine whether these alignments are spontaneous or of hydrodynamic origin. Cholesteric arrangements can also be found in the cytoskeleton of certain invertebrates [31].

4.5.5 *The Limits of a Widespread Phenomenon*

Liquid crystalline states, in the proper sense of the term, are therefore present in all cells, with their fluid membranes (which are largely in that state) and their chromosomes (which can be in that state, with the prospect of generalising this point with further research). The presence of membranous liquid crystals is truly massive at the level of the white matter in the brain, with the myelin of the axons, and a minimum of local stabilisation. The stabilised derivatives of liquid crystals represent an even greater mass in the skeleton and muscles of animals and in the cell walls of plants, especially in tree wood [53].

It should be noted that many biological structures are neither liquid crystals nor possible analogues, and this is the case for the best-known self-assemblies [1, 24, 32, 40]. Likewise, bacterial cells are often protected by tough walls, but no liquid crystal analogues have been detected in them. We should also recall that the three classic states of matter play their role in the construction of living beings. Solids, often crystallised, are essential in the construction of the skeleton, with minerals like calcite, aragonite, hydroxyapatite, amorphous silica, etc. The liquid part is water, mainly with blood plasma, lymph and cytoplasms, while gases are present in the lungs or in the tracheae of insects.

On the subject of the stabilised derivatives of liquid crystals, we are led to the question of the possible passage through a true liquid crystalline state during the synthesis or the assembly of the molecules involved; before, during or after the secretion itself.

This domain of self-assembly is interesting, because the stabilised analogues of liquid crystals described earlier could result from similar assemblies to those observed in the formation of liquid crystals.

4.6 Liquid Crystalline Self-Assemblies

We have seen that mesogenic molecules display specific shapes, essential to their liquid crystalline assembly, of which rods and disks are the most common. Generally speaking, biological molecules have precise shapes, which can sometimes be modulated to suit the conditions of the medium, but the appearance of these shapes is often identified with stages in the synthesis of these molecules. These are therefore chemical processes requiring the help of enzymes to catalyse them.

Whether it is thermotropic or lyotropic, the transition from an isotropic liquid to a liquid crystal is usually an exothermic change of state, as is the case for ordinary crystallisations. In living systems, these processes could be endothermic, as they are for some of the more classic self-assemblies. However, further studies need to be conducted into biological polymers and other macromolecules, because of the existence of superphases, the complex chemical context of their stabilisation, and the chemical and mechanical re-organisations introduced by the cells.

Since the 1980s, several different approaches have been made to the study of the cholesteric self-assembly of biological polymers. Livolant has developed precise methods for the preparation of cholesteric or hexagonal lyotropic mesophases of DNA [44]. She has obtained cholesteric rods of DNA from a more diluted, isotropic solution, corresponding to a self-assembly closely reproducing the main elements of the chromosome structures of prokaryotes. Likewise, the chitin in shells can be separated, purified and prepared, producing a cholesteric phase in the presence of water, under the appropriate physico-chemical conditions [48, 49]. The same can be done with the cellulose of plant cell walls [3], which are often stabilised analogues of cholesterics [53]. Liquid crystals of collagen have also been prepared in vitro, while the organic matrix of compact bones is made of collagen in the form of a stabilised cholesteric [37]. Taken together, these results constitute a strong argument in favour of a principle of liquid crystalline self-assembly of the lyotropic type, for all these biological fibrillar systems.

However, we must be careful not to make sweeping generalisations, the key example being that of alignments of microtubules, the symmetries of which recall those of nematic systems. Their alignment in the form of birefringent phase is not spontaneous, requiring the intervention of gravity. They are not, therefore, liquid crystalline, even if the principle of reduced hindrance by excluded volume may facilitate their alignment (see Chap. 5 by J. Tabony). In cells, the alignment of microtubules occurs under the effect of currents and motor phenomena, and some of these alignments are even polarised, which is not frequent among liquid crystals (except for amphiphiles, and then over very short distances.)

4.7 Curvature and Structure

The concept of *perfect crystal* is very useful for teaching, but it corresponds to an inaccessible ideal; the same is true of liquid crystals. An ordinary crystal cannot avoid various *constraints* that provoke *deformations* or *defects*. For most crystals, development requires the presence of defects known as *screw dislocations*, which determine *spiral growth*. Any idea of producing truly perfect crystals is therefore illusory [27]. But that does not prevent us from trying to approach this ideal, by monitoring certain parameters, especially the speed of crystallisation, which must be as slow as possible.

A deformation is said to be *elastic* if it completely disappears when the constraint that provoked it is removed. This does not always happen, and we then say that there is a *plastic* deformation, defects that have appeared. These concepts can be transposed to liquid crystals, but because of their fluidity, the possible deformations are much greater, and they occur within a geometrical context much closer to differentiability than is the case for real crystals.

4.7.1 Diversity of Curvatures in Liquid Crystals and Their Analogues

The main deformations observed in liquid crystals, in the absence of defects, are illustrated in Fig. 4.6. They are defined on the basis of the field of unit vectors **n**, parallel at any given point to the average local direction of the molecules. In most cases, there is no polarity in the molecular orientations, and the field **n** applies just as well as the field −**n**. These deformations in general combine several curvatures of very different natures, and to distinguish them, it is helpful to use the concept of line of force, as in other fields of vectors. These are curves along which the molecules are aligned, making allowance for thermal fluctuations. In other words, each line is

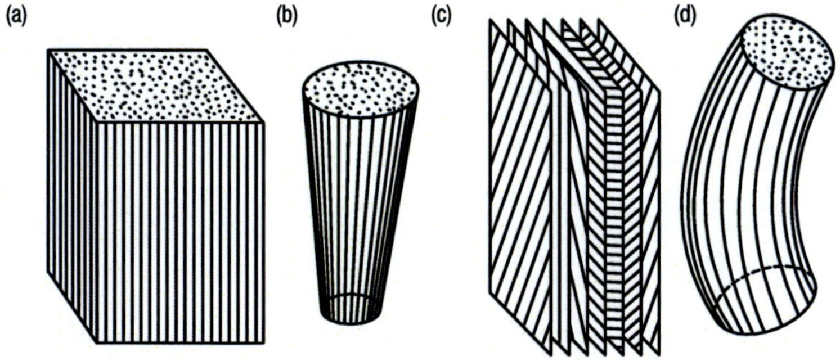

Fig. 4.6 **(a)** Representation of a liquid monocrystal and the principal deformations of liquid crystals: **(b)** splay; **(c)** twist; **(d)** bend

tangent to the average local directions of the molecules according to the vector **n** (or −**n**). The concept of *tube of force* is equally intuitive.

The lines of force of a perfect, non-deformed liquid crystal, for example nematic, would be parallel straight lines (Fig. 4.6a). This is a situation that one approaches with a microscope, between the object-slide and cover-glass, by creating the conditions that position the molecules practically perpendicular to the interfaces of the glass, in direct contact with it and in its close neighbourhood. We also know how to create the conditions to make the molecules align parallel to the glass, in a chosen direction in the plane [22, 28, 36, 39]. Outside this particular case, we can distinguish three types of curvature.

The *splay* corresponds to the tendency of the lines of force to diverge around **n** (Fig. 4.6b), which they can do in various ways. A second type of curvature (Fig. 4.6c) corresponds to the twisted arrangement already encountered in Fig. 4.2c. This is called the *twist*, to avoid confusion with the torsion specific to a line of force (that is to say, the rotation of its osculating plane, observed while moving along this curve). The twist measures the deviation in the lines of force along the normals to the field **n**. The *bend* of a liquid crystal corresponds to the curvature of the line of force at one point (Fig. 4.6d). These deformations display modulations and give rise to numerous combinations, of which we shall present a few examples in the next section.

The deformations of liquid crystals are all the more interesting as they occur in biological fibrillar structures, which are stabilised analogues: muscles, tendons, bone tissue, etc. These deformations also concern membranes, with the field of normal vectors, and intracellular structures. The precise study of these curvatures, as proposed by d'Arcy Thompson [56], is already well under way for liquid crystals and their hydrodynamic evolutions [22, 28, 36, 39], but much still remains to be explored. The work in prospect is immense in biology, because the geometries are even more complex in the cells and their organelles, in the tissues and in the organs where several different tissues are always combined and each one has its particularities. Thus, for instance, the muscles of the vocal cords and the muscles of the eyes have different histological, cytological and mechanical characteristics.

These complex morphologies have been photographed and drawn, as testified by the treatises of anatomy, histology or cytology, which are worth consulting [2, 33, 58], but all these forms change as they function, and the analysis is arduous. It is a subject that is emerging once again, thanks to advances in imaging at all scales and computer-aided microscopy.

4.7.2 Geometry of the Different Curvatures

To analyse these deformations and measure their energies, a little bit of geometry is needed. At each point M, we consider the plane normal to **n** and a small circle centred on M, on this plane and of area ΔS. The lines of force attached to this circle form a tube of force around the line L, as shown in Fig. 4.7 (but the tube of force is represented without the circle centred on M, to avoid overloading the diagram).

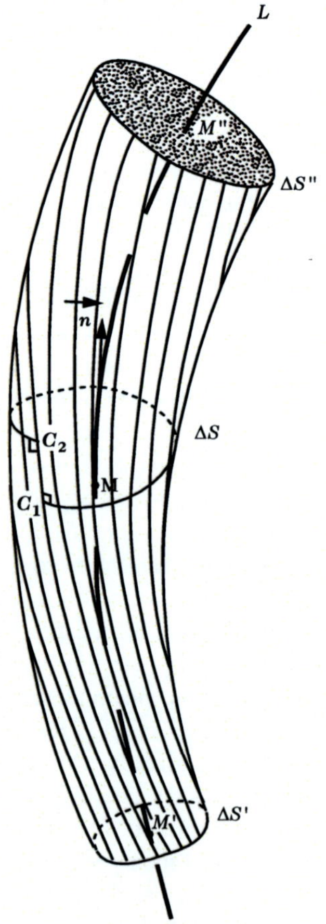

Fig. 4.7 Tube of lines of force in a liquid crystal (see text)

We choose two points M' and M'' on L, either side of M, and close neighbours, at an equal distance from M, with $\mathbf{M'M''}$ oriented the same as $\mathbf{n}$, which can be written: $\mathbf{n}.\mathbf{M'M''} > 0$. The planes normal to L in M' and M'' cut the tube along two plane contours that delimit the surfaces $\Delta S'$ and $\Delta S''$. The *splay* s is the limit of $(\Delta S'' - \Delta S')/|M'M''|.\Delta S$, when the length $|M'M''|$ and the area ΔS tend to zero, and this limit is indeed div $\mathbf{n}$, when $|\mathbf{n}| = 1$.

Now let us look at the *twist*. Let us trace, on the surface of the tube of force, a line normal to the integral curves of $\mathbf{n}$. This line generally describes a sort of more or less deformed helix (unless it closes on itself). Starting from a point C_1 on the little circle at the level of M, this line rejoins the integral curve from C_1 at a point C_2, after completing one turn. Twist t in M is the limit of $|C_1C_2|/\Delta S$, when ΔS tends to zero. This limit generally exists and does not depend on the choice of C_1 on the circle with centre M. We verify that $t = -\mathbf{n}.\text{rot}\ \mathbf{n}$, provided that the length $|C_1C_2|$ is positive, when this segment is oriented is the same direction as $\mathbf{n}$, in other words when $\mathbf{n}.\mathbf{C_1C_2} > 0$. If the field $\mathbf{n}$ presents surfaces that are normal to it, C_1 and C_2 are at the same point and we have $\mathbf{n}.\text{rot}\ \mathbf{n} = 0$.

The *bend* of the liquid crystal in M corresponds to the curvature of L at point M and we also verify that its absolute value is $b = |\mathbf{n} \wedge \text{rot}\ \mathbf{n}|$.

Each type of deformation can exist in a pure or nearly pure state. Thus, a *nearly pure splay* is produced by smectics A, since these liquids are made up of layers of equal thickness, separated by equidistant surfaces. Their normal vectors therefore align along straight lines, which are the integral lines of $\mathbf{n}$. The bend b is therefore zero, and so is the twist t, because the lines normal to $\mathbf{n}$ close up at the surface of the tubes of force. This pure splay can take two different forms (Fig. 4.8). The divergence between the straight lines of force changes along the plane containing the vector $\mathbf{n}$ at point M and varies between two extrema situated in two perpendicular planes Σ_1 and Σ_2. The divergence of the straight lines in each of these planes can

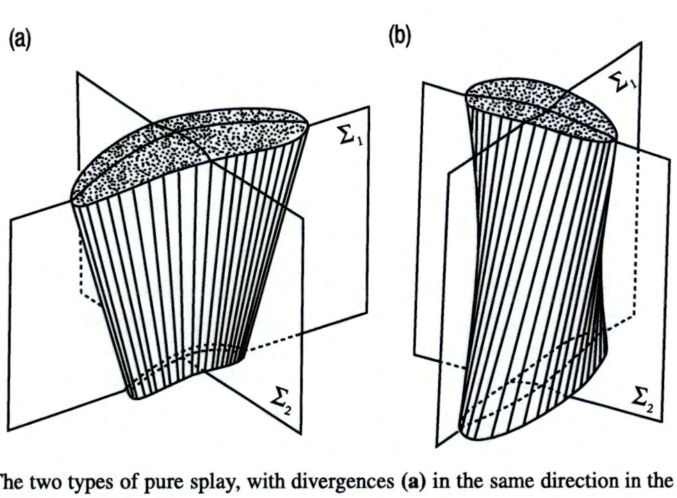

Fig. 4.8 The two types of pure splay, with divergences **(a)** in the same direction in the planes Σ_1 and Σ_2 and **(b)** in opposite directions

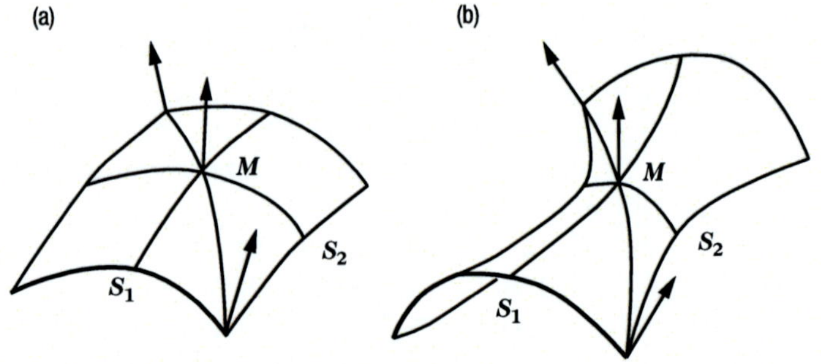

Fig. 4.9 Two surfaces with their lines of principal curvature crossing at right angles along two planes of local symmetry. The curvatures have **(a)** the same sign or **(b)** opposite signs. A twist in the normal vectors is observed along the diagonal lines

either follow the same direction (Fig. 4.8a) or opposite directions (Fig. 4.8b). As the twist is zero, there exist surfaces normal to the field **n**, and the two situations considered for the splay correspond to *elliptic surfaces*, with principal curvatures of the same sign (Fig. 4.9a) or *hyperbolic surfaces*, with curvatures of opposite signs, producing a form shaped like a saddle (Fig. 4.9b). This splay with opposite divergences is therefore called a *saddle-splay* [34]. It is the most common deformation in thermotropic smectics A, while the two types of splay are more or less equally present among lyotropic smectics, although the reason for this is not yet known.

We have just seen that smectic A systems present a pure splay, with neither twist nor bend. A situation of *pure twist* can be observed in planar cholesteric monocrystals, a situation also approached in crab shells (Figs. 4.2c and 4.6c). Likewise, the *pure bend* illustrated in Fig. 4.6d is produced in columnar liquid crystals, because the columns are normal to planes over large domains. These observations and their geometry suggest that the *structure of liquid crystalline phases largely determines the nature of the deformations observed.* This is an obvious fact, which is nonetheless a fact of morphogenesis.

In the most general case, the field **n** around a point M can be related to a trirectangle trihedron $MXYZ$, with Z oriented along **n**, and we have:

$$s = \operatorname{div} \mathbf{n} = \frac{\partial n_X}{\partial X} + \frac{\partial n_Y}{\partial Y}, \qquad \text{since} \quad \frac{\partial n_Z}{\partial Z} = 0 \quad (\text{and } |\mathbf{n}| = 1)\,. \tag{4.2}$$

The two terms $s_X = \partial n_X/\partial X$ and $s_Y = \partial n_Y/\partial Y$ each depend on the choice of axes MX and MY in the plane normal to **n**, but their sum is constant and independent of this choice. As an exercise, we can consider a fixed trihedron $MXYZ$ and take an axis Mx in the plane XY, of which we vary the angle θ with MX, to see how what we call the splay along this axis evolves, that is to say $s_x = \partial n_x/\partial x$. The result can be written $s_x = a + b\sin(2\theta + \omega)$, showing two extrema $a \pm b$ located on two rectangular axes of the plane MXY, corresponding to the directions of the lines of principal curvature of the surface normal to **n**, when it exists. The extrema may or may not have the same sign, as in the Fig. 4.8a, b. The different scenarios for the curve $s_x = f(\theta)$ can be represented using polar coordinates.

In a similar fashion, we can study the twist in each plane around **n**, by changing the direction Mx around **n** and calculating $\partial n_x/\partial y$, My being the moving rectangular axis associated with Mx [8]. We obtain two extrema for two perpendicular directions, bisecting those we found for the splay. Whatever the choice of Mx normal to **n**, we have:

$$t = \frac{\partial n_x}{\partial y} - \frac{\partial n_y}{\partial x} = t_x + t_y = -\mathbf{n}.\mathrm{rot}\,\mathbf{n}\ . \qquad (4.3)$$

We have an illustration of this when the overall twist t is zero, which is generally not the case for a twist studied along one sole axis normal to **n**. If t is zero, this means that there exist surfaces normal to the field **n**, as in the Fig. 4.9a, b. The surface is locally symmetrical along its two planes of principal curvature. The normal vector does not twist along the lines of principal curvature. On the other hand, this twist is very visible along the bisecting lines, turning to the left along the bisecting line illustrated on Figs. 4.9a, b, and to the right along the other bisecting line.

The different types of curvature can be superimposed, as shown in Fig. 4.10, but there are also cases where one type is excluded. This is the case for the bend in Fig. 4.10a, but we do not know of any examples among liquid crystals. Only the splay is excluded in Fig. 4.10b, which occurs in a chiral smectic C monocrystal, as in most liquid crystals with long polymers, and also in certain columnar phases. Only the twist is excluded in Fig. 4.10c, while all three curvatures exist in Fig. 4.10d, which is often the case for nematics.

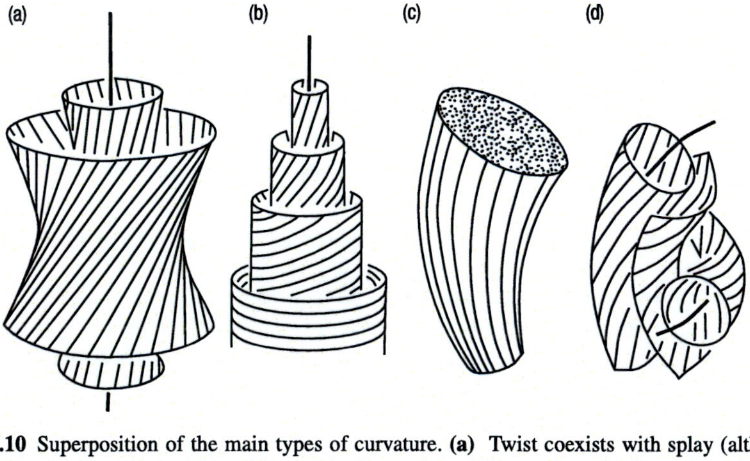

Fig. 4.10 Superposition of the main types of curvature. **(a)** Twist coexists with splay (although this is zero in the centre) and bend is zero in these tubes of force, where the lines of forces generate hyperboloids of one sheet. **(b)** Cylindrical twist, producing bend, but the splay is zero. **(c)** Splay and bend are superimposed in the absence of twist. **(d)** Superposition of the three types of curvature: splay, twist and bend

4.7.3 Elastic Coefficients and Spontaneous Curvatures

For many liquid crystals, it can be shown, by means of a quadratic approximation and considerations of symmetry [34], that the density of elastic energy F per unit of volume v has three terms, corresponding to the three types of deformation: splay s, twist t and bend b:

$$\frac{\mathrm{d}F}{\mathrm{d}v} = [k_s(s - s_0)^2 + k_t(t - t_0)^2 + k_b(b - b_0)^2]/2 . \tag{4.4}$$

The terms k_s, k_t and k_b are elastic constants specific to the different curvatures, while s_0, t_0 and b_0 are spontaneous curvatures, which can arise owing to the shape of the molecules. The energy of a specific type of deformation is proportional to the square of the difference between the value of its curvature and that of the corresponding spontaneous curvature of the perfect monocrystal. The spontaneous twist t_0 is frequent in liquid crystals containing an asymmetrical compound, in the absence of its enantiomer, or in a non-racemic situation. This is because many molecules display helicity or a simple asymmetry in a given direction, and some of them, instead of aligning with the others, follow the oblique direction of a helicoidal groove, during closer contacts with neighbouring molecules. Even if these oblique contacts are short-lived, they nevertheless generate a twist that is more or less developed according to their frequency.

A term s_0 operates in the same way in cell membranes, owing to the physico-chemical differences between the two constituent monolayers. Examples of the term b_0 are rare.

Equation 4.4 has an essential significance in morphogenesis, because it gives the density of elastic energy. If a liquid crystal satisfies this equation in one domain of space, and if limit conditions are imposed, notably molecular orientations at each point of the border surface, then we can seek to determine the molecule distributions that minimise the elastic energy over the whole domain. This is a variational problem, for which the Lagrange equations are often difficult to establish. We have looked for solutions in some of the most useful and yet relatively simple cases, but approximations are generally necessary. We have often used simulation to approach the solutions. The important thing is to understand that elastic coefficients and spontaneous curvatures are ingredients of the form adopted at equilibrium and therefore of morphogenesis itself.

The concept of spontaneous curvature is at the origin of many morphologies. We can show, for example, that if div $\mathbf{n} = 0$ and $\mathbf{n} \wedge \text{rot } \mathbf{n} = 0$, with $|\mathbf{n}| = 1$, that is to say if the lines of force are rectilinear and if the flow is conservative, then the vectors $\mathbf{n}$ are coplanar [8, 13]. This is a useful result for understanding the planar twist in cholesterics and in shells and other analogues. The result becomes intuitive, if we take a bundle of uncooked spaghetti in one hand, holding them straight and parallel, and then twist them with the other hand. Splay appears at the two extremities of the bundle. The only way to conserve zero splay in the presence of a twist with rectilinear spaghetti is to layer them, as illustrated in Figs. 4.2c and 4.6c. In a cholesteric, the term t_0 is very developed and gives a twist of 180° over very short distances, if we compare them to the long, quasi-rectilinear lines of force, with a quasi-conservative flow of **n**, simply because of sufficiently high rigidities k_s and k_b. On the other hand, if the bend was easier, because of a weaker elastic coefficient k_b, we would obtain a cylindrical twist like the one illustrated in Fig. 4.10b, where the twist can occur in the same way in every direction of the plane normal to a molecule. This happens in blue phases, often confined to a narrow temperature range, close to the isotropic transition.

4.8 Lyotropic Systems and Cell Fluidity

The essential biological example is that of membranes made of amphiphilic molecules – mainly phospholipids – with cholesterol and diverse proteins, where a lipophilic region is enclosed by two polar regions. The minimisation of interface energies leads these molecules to assemble not only in the form of bilayers, but also by forming diverse spherical, cylindrical or ribbon-like micelles, which can themselves assemble into liquid crystalline superphases. We can also obtain them in vitro from other, non-biological amphiphilic compounds. The main types are presented briefly in Fig. 4.11. This is a new illustration of the morphogenetic richness of these lyotropic self-assemblies, constantly at work in living cells. From the single, isolated micelle to superphases, all these structures are ordered and fluid, as evidenced by the nuclear magnetic resonance (NMR) studies that allow to estimate the diffusion speeds of molecules among each other [26]. The geometry of the rectangular and hexagonal arrangements of micelles is reminiscent of the columnar phases of Fig. 4.5g, just as the stacks of bilayers resemble smectic liquids. Following the nature of the molecules, the bilayers can form closed, spherical or extended vesicles, in the shape of flattened, superimposed sacks, or more or less ramified tubes, as can be found in the cytoplasm, at the level of the main cellular organelles. The cell membranes never present free edges, thus forming closed surfaces. However, exceptions might exist, notably for the pulmonary surfactant.

Bilayers are prone to more complex morphologies, like those illustrated in Fig. 4.12, when they spontaneously form saddles, already presented in Fig. 4.9b. This generally produces phases with cubic symmetry, where the bilayer separates two inter-penetrating aqueous compartments (Fig. 4.12a), and the median surface of this bilayer is close to a periodic minimal surface (Fig. 4.12b), especially for phases produced in vitro, with two identical monolayers. On the contrary, biological membranes present marked differences between the two monolayers, which slightly increases the volume of one of the two compartments, the one that communicates with the cytoplasm.

The fact that membranes and membranous organelles are both fluid and ordered at the same time is essential to many cellular processes. For example, various white blood cells and numerous protozoa are capable of *phagocytosing* bacteria. Their outer membrane adheres to the prey, envelopes it and internalises it by forming a vesicle which sinks into the cytoplasm and then closes up, leaving the outer

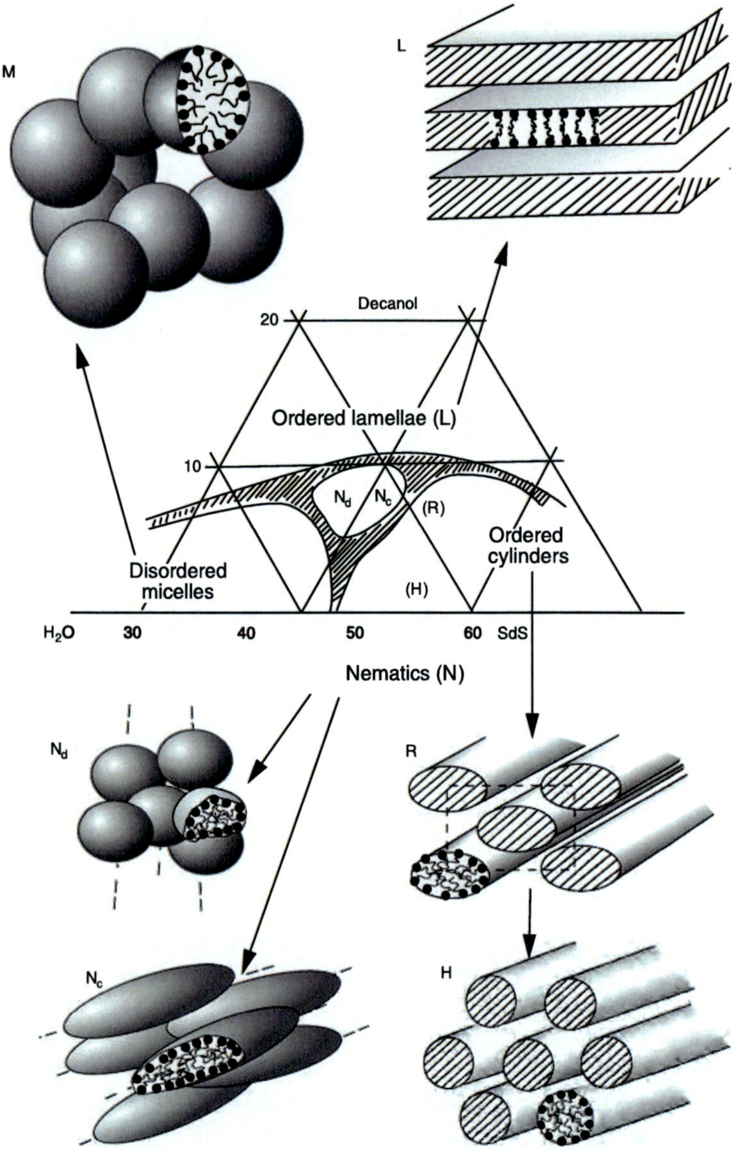

Fig. 4.11 Main types of assembly of amphiphilic molecules (reproduced after [26] with courtesy of J. Charvolin). M: Spherical micelles in water, with the lipophilic parts on the periphery. L: bilayer. In the *middle*, a partial view of a ternary diagram indicating the distribution of phases obtained in the water, decanol and sodium dodecyl sulfate (Sds) system. N_d: Nematic phase with disk-shaped micelles. N_c: Nematic phase of cigar-shaped micelles. R: Rectangular phase of ribbon-like micelles. H: Hexagonal phase of cylindrical micelles

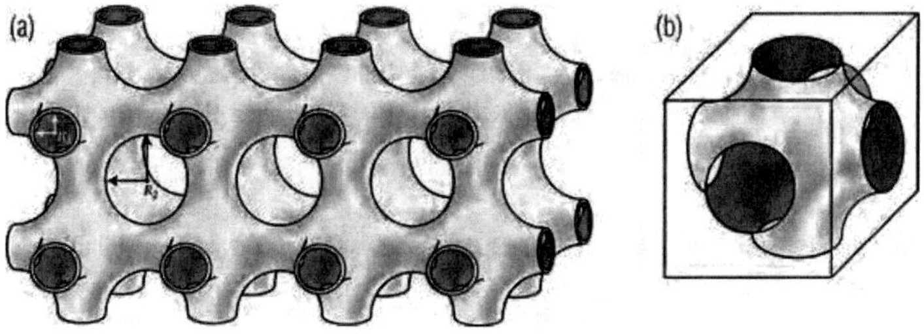

Fig. 4.12 **(a)** Cubic structure of a bilayer separating two multiply-connected aqueous phases. The radii of curvature R_1 and R_2 of the circular junctions are different in biological examples, but these systems are close to similar systems obtained in vitro where $R_1 = R_2$. **(b)** Example of minimal surface that verifies this condition

membrane intact. During the separation, there are no enzymes present to help seal the holes in the vesicle and the outer membrane. It is the fluidity of the cell membranes that allows these *recombinations*, like those observed in a film of soapy water in a ring, on which one blows to separate the bubbles. The soap molecules are amphiphilic and form two external monolayers, enclosing the water with micelles or bilayers. These sometimes align along the plane of the film, being separated by layers of water that are thicker than the bilayers themselves, a subject formerly studied by Jean Perrin.

As we have already stated, the fluidity of cell membranes was discovered by Pfeffer, better known for his work on osmosis. By manipulating red blood cells under the microscope with glass microneedles, he succeeded in separating them into several vesicles without any leakage of the red pigment (haemoglobin). So there had to be a recombination of the membranes without any holes appearing. Only liquids like soapy water in air, or the vesicles of amphiphiles in water, are capable of recombining like that.

It goes without saying that membrane recombination enables the creation of a diversified combinatory topology, but one that is rigorously controlled by the cellular machinery. Proteins are involved, but their function is not to repair the holes as one might darn a sock. These proteins sometimes align to form a collar within the membrane, delimiting a small area where the fusion with the vesicle will take place. It appears that their chief role is to reduce local rigidity, as the perforations do around a postage stamp, and so to help in the fusion of the vesicle, which would be harder to obtain with higher elastic coefficients. Cell fluidity is also essential to secretion, the products of which are firstly wrapped up in an intracytoplasmic vesicle, before being expelled from the cell when the vesicle recombines with the cell membrane. Extracellular matrices are one example. They are initially produced in a liquid crystalline or slightly stabilised state. The reorganisation imposed by the cell during this liquid crystalline self-assembly takes place at the level of the membrane, which is itself fluid or stabilised, linked to its cytoskeleton.

This constitutes a very suitable compromise between partially gelified structures: those of the matrix that has just been secreted and that of the cytoplasm containing

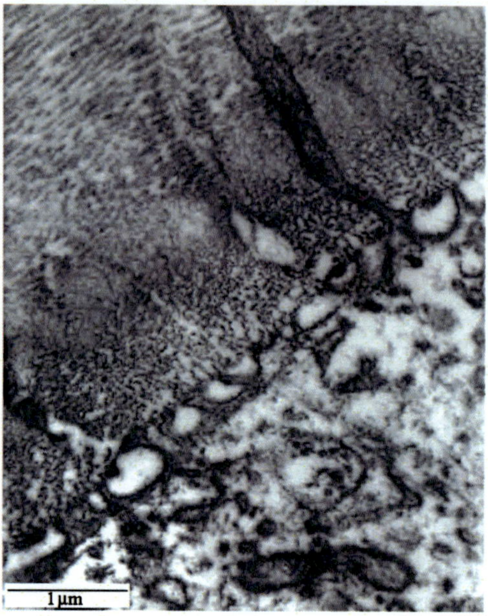

Fig. 4.13 Contact zone between the cytoplasm of the crab's epidermis and the shell in the process of being secreted (the fibrous shell lies at *top left* and the epidermal cytoplasm, below on the *right*)

the motor elements, which enable some control of the emerging morphology. Thus, a cell membrane, more or less stabilised itself, establishes contact between a growing matrix and its secreting cytoplasm. This contact is presented in Fig. 4.13 for the example of the crab shell. The ordered, fluid (or slightly gelified) state appears to be indispensable to morphogenesis in the usual form it takes among living beings.

As we have shown in Figs. 4.5–4.12, many liquid crystals contain parallel curves or surfaces, and the extension of certain layers compared to others has precise morphological consequences. Defects can also appear with regular distributions, because of the fluidity.

4.9 Liquids with Parallel Surfaces and the Geometrical Origin of Forms

The concept of parallel surfaces is used above all in the field of optics, and one of the classic figures concerns the geometry of parallel, equidistant waves in the neighbourhood of a focus. If the waves are perfectly spherical, they converge on the focus F (Fig. 4.14a). If they are cylindrical, and if their cross-sections normal to the generators are not circular, then the light rays can concentrate on what are called *caustic* surfaces, which are the envelopes of these light rays (Fig. 4.14b).

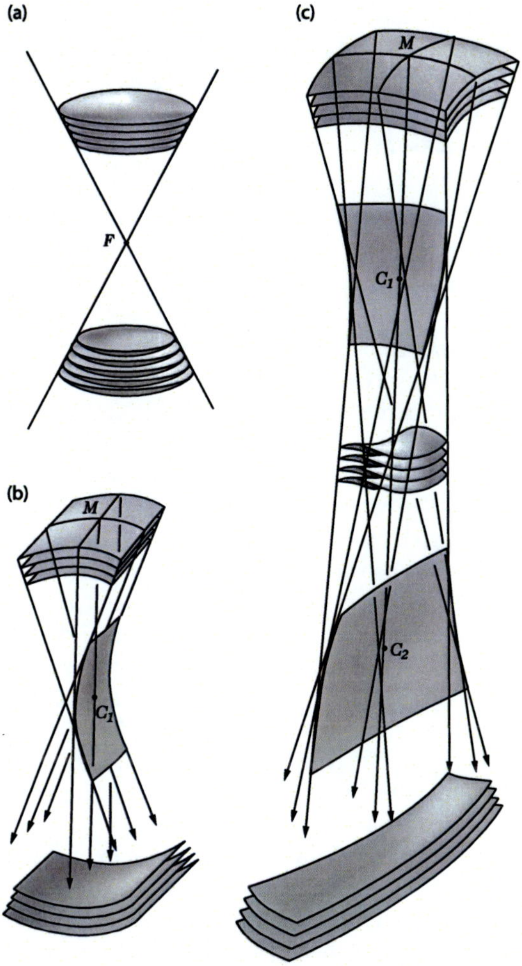

Fig. 4.14 Figures drawn from treatises on optics, representing parallel surfaces (waves) or their normals (rays). **(a)** Concentric spherical surfaces (waves and focus F). **(b)** Cylindrical surfaces with one sole caustic. **(c)** Surfaces with two caustics

More generally, if we start with a cap-shaped surface, the waves keep this form before the rays run along a first caustic; the waves then take the form of a saddle, before returning to a cap shape with reversed concavity after the rays have run along a second caustic, globally normal to the first one (Fig. 4.14c).

4.9.1 Caps and Saddles: Elliptic or Hyperbolic Surfaces

We can verify that these surfaces share, for the same normal, the two same centres of curvature C_1 and C_2. We have represented equidistant surfaces, separated from each other by a constant distance e. Let us consider three successive surfaces, limited by curvilinear rectangles, as illustrated, with their sides parallel to the lines of principal curvature, and let us denote S_{+e}, S_0 and S_{-e} the areas of three of these successive rectangles. If c_1 and c_2 are the two principal curvatures at the centre of the middle rectangle, we can verify that if we reduce the two dimensions of these rectangles towards zero, and if e remains much smaller than the two radii of curvature, we obtain at the limit:

$$c_1 + c_2 = \frac{S_{+e} - S_{-e}}{2eS_0} \quad \text{and} \quad c_1 c_2 = \frac{S_{+e} + S_{-e} - 2S_0}{2e^2 S_0} . \tag{4.5}$$

Now let us apply these results to a bilayer or a membrane, the polar heads of which are aligned on two parallel surfaces at a distance of $2e$ apart, enclosing a median surface (a paraffinic layer) in the lipophilic region. The values of e, S_{+e}, S_{-e} and S_0 for a given mass of a small disk of membrane or bilayer are determined by the physical chemistry of the system [16], and the same is true for the sum and product of c_1 and c_2, because of (4.5). The two spontaneous principal curvatures are therefore the two roots of a second degree equation. More precisely, the second equation shows that the spontaneous form is cap-shaped if $S_{+e} + S_{-e} \geq 2S_0$, and saddle-shaped if $S_{+e}+S_{-e} \leq 2S_0$. *In other words, if the area of the paraffinic layer is smaller than the average of the two areas occupied by the polar heads, then the cap-shaped form is favoured, whereas if the paraffinic layer takes up too much space, the saddle-shaped form prevails*. This second situation prevails in bicontinuous liquid crystals, close to periodic minimal surfaces, as in the example in Fig. 4.12.

Similar models are applicable to monolayers of which we consider three levels, to examine the case of micelles. Looking at the phase diagram in Fig. 4.11, we can see that the variations in curvature are dictated by the physical chemistry, for the extension of the polar layers depends on the concentrations of ions in the hydrophilic regions, and the extension of the lipophile part depends on the presence of an organic solvent. The physical chemistry governs the relative areas and the thicknesses of the layers. These geometrical data being determined, the morphology derives from them. This is a morphogenetic principle at work here at the scale of micelles or bilayers. We shall see that other mechanisms come into play at higher scales.

4.9.2 Dupin Cyclides in Liquid Crystals

Let us now consider sets of large numbers of parallel layers, as can be found in smectics A, and let us assume that these layers are represented in two dimensions, like the cylindrical layers of Fig. 4.14b. We once again find the concepts of parallel curves in the plane [56], with their common normals enveloping a curve called the *evolute*, while the parallel curves are the *involutes* of this envelope (Fig. 4.15a, b). Evolutes often present *cusps*, corresponding to the *curvature extrema* of the parallel lines, these cusps being oriented towards the concavity for the maxima, and in the opposite direction for the minima.

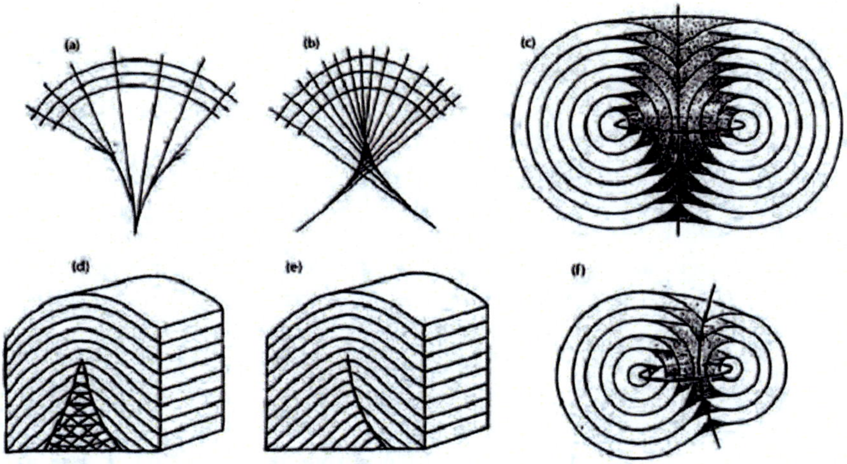

Fig. 4.15 **(a), (b)** Parallel equidistant curves in a plane. The envelope of common normals or "evolute" often presents cusps. **(c)** Concentric circular tori, with their normals passing through two curves: an axis and a circle; part of these tori has not been represented. **(d)** Cylindrical light waves intersecting between two caustic surfaces, which they have generated from a cuspidal edge. **(e)** Such intersections are impossible for equidistant parallel layers, like those existing in liquid crystals, where one sole caustic surface is retained. **(f)** The most general surfaces with normals resting on two curves are Dupin cyclides, of which (c) is an example with parallel tori. In the general case, the circle and its axis are replaced by a branch of hyperbola and its focal ellipse

In the smectics A, one situation that often occurs is that of *nested toroidal surfaces*, a simple example of which is illustrated in Fig. 4.15c. The tori intersect at conic points along an axis that represents a singularity of the smectic structure. Another singular line is the locus circle of the centres of the circular cross-sections of the tori. We can see immediately that the normals to the surfaces passing through two *singular lines*: a circle and the axis rising from its centre, perpendicular to its plane. The molecules normal to smectic A layers are aligned along the segments of straight line that connect the axis to the circle. What could be the origin of such morphologies?

Let us return to the caustic surfaces studied in optics and the systems with parallel surfaces, with Fig. 4.15d, copied from Fig. 4.15b but with cylindrical layers. In space, the light waves intersect between two caustic surfaces that meet at the level of their cuspidal edges. In a smectic A, on the contrary, the layers cannot intersect, and one sole layer direction is retained, along with one sole caustic surface, as shown in Fig. 4.15e. Although caustics exist in optics, they never materialise in liquid crystals, or only approximately. They would require prohibitive levels of energy, because the curvature increases indefinitely in the neighbourhood of one of the faces of the caustic surface, and the layers do not correspond to each other between the two faces; there is a strong change in orientation. The fluidity of the phase allows to reorganise these energy-costly walls, and all that remains is the cuspidal edge [16, 12]. As there are two caustic surfaces, we obtain two singular lines and the layers adopt the form of *surfaces with normals resting on two curves*.

Surfaces that have this property are Dupin cyclides [27], of which Fig. 4.15c, f give examples. In the most common case, the circle is replaced by an ellipse and the axis by a branch of hyperbola, the apex of which coincides with one of the two foci of the ellipse, while the corresponding apex of the ellipse is at the focus of the branch of hyperbola (Fig. 4.15f). The two conic sections are said to be focal. Any cone of which the apex is located on one of the conic sections and of which the generators rest on the other conic section is a cone of revolution (Fig. 4.16a). Historically, these focal conics were interpreted by Grandjean and Friedel in 1910 [35] as being due to the stacking of liquid layers in the form of Dupin cyclides, which led them to suggest the smectic structure, long before they could establish it definitely by X-ray diffraction in 1923 [21].

The smectic molecules align along segments connecting the two conics. The Dupin cyclides are therefore surfaces normal to the field $\mathbf{n}$ aligned along these segments of straight line, and they form cones of revolution of one sheet, in the neighbourhood of the two conics. Obviously, we have rot $\mathbf{n} = 0$, i.e. $t = b = 0$ and, as already stated, the only deformation allowed is the splay s, but it tends to infinity at the level of the conic branches. At this level, the layers form conic points, as indicated in Fig. 4.15c, f, but the tips of the cones are probably blunt (Fig. 4.16b) or spiral around, following a screw dislocation (Fig. 4.16c). This is often observed in electron microscopy, when freeze-etching techniques are used to examine lyotropic smectics. The separating surfaces of the layers therefore cease to be equidistant at the level of the conics, or *focal lines*, which are very frequent singular lines in smectics A. These variations in the thickness of layers, and the strong (but not infinite) curvature require a lot less energy than singular walls like caustic surfaces, even when the latter have modifications of thickness or attenuated curvature.

These descriptions do not exhaust the list of singular lines in smectics or in cholesterics, which are also liquid crystals with parallel surfaces [8, 12]. There are edge and screw dislocations (Fig. 4.16d) like in true crystals, and the screw dislocations are not necessarily associated with focal lines, as they are in Fig. 4.16c. Here and there, a layer "comes unstuck" from another (Fig. 4.16e) and forms an *elementary pinch*, introducing two singular lines known as *disclinations*, which can also involve much larger groups of layers.

The arrangements of smectic layers in Dupin cyclides are limited to domains, each of which contains two conics in focal position, limited to an arc for at least one of the two. The domain is constituted of molecules aligned on rectilinear segments that connect the two conics or their arcs. The domains are therefore limited by conic surfaces of revolution and they are tangent to each other along generators, allowing the layers to be connected without changing orientation (Fig. 4.16f). Between three tangent domains, the layers can stack up spherically around a centre that is a point of intersection of the conic arcs of the domains in question [41].

The variations in thickness of the layer allow the development of other defects such as disclinations (Fig. 4.16g) or very diverse singular points underlying elaborate morphologies like those sketched out in Fig. 4.16h, i. Cholesterics also give

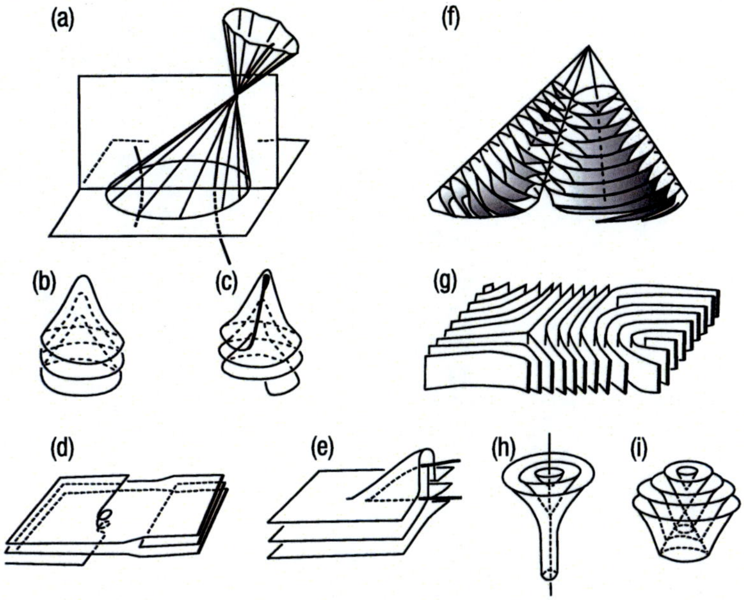

Fig. 4.16 **(a)** Focal conics: the planes of the two conics are perpendicular and intersect along a common axis, where the focal points of one conic coincide with the apexes of the other. The straight lines emanating from a point on one conic and resting on the other conic form a cone of revolution. **(b), (c)** Conic layers with blunt apexes or forming a screw dislocation along the axis of the conic stacking. **(d)** Presence of a screw dislocation or edge dislocation in practically horizontal layers. **(e)** Elementary pinch. **(f)** The smectic layers are grouped into domains limited by cones of revolution tangent to each other along certain generators. **(g)** Two disclinations $+\pi$ and $-\pi$ seen in a system of vertical layers. **(h), (i)** Two plausible pointwise singularities in smectics

rise to equidistant parallel surfaces, with tolerance in the helicoidal pitch [8], and a large variety of defects, but the topology is much more complex at the level of the singularities [8, 12]. Let us start by looking at a few examples of morphologies imposed by the parallel surfaces.

4.10 Germs and Textures of Liquid Crystals: Their Biological Analogues

By way of a riddle, one could ask scientists what they think the drawings in Fig. 4.17a, drawn from [35], represent. They might well suggest that they are biological structures, flower pistils, for instance. In fact, they are *thermotropic smectic rods*, in equilibrium with their isotropic phase. The second of these rods displays a focal domain with its two associated conics: a circle seen in perspective, associated with a segment perpendicular in the centre and aligned along the axis of the rod. A large proportion of these defects are situated on the surface of the rod, in contact with the isotropic phase, because this divides roughly by two the energy of their elastic

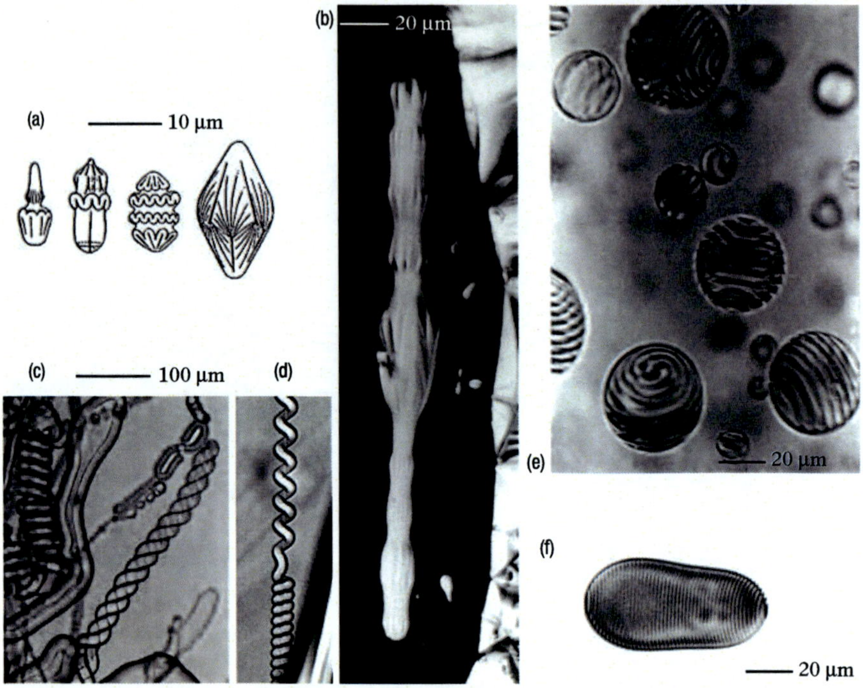

Fig. 4.17 **(a)** Figures of smectic rods of ethyl para-azoxybenzoate drawn by F. Grandjean. **(b)** Smectic rod in the presence of the isotropic phase (cyano-octyl-biphenyl, with Canada balsam, crossed polarisers). **(c), (d)** Myelinic figures observed in lecithin in the presence of water. **(e)** Cholesteric spherulite immersed in its isotropic phase (MBBA and cholesterol benzoate). **(f)** A more developed cholesteric germ, in a rod shape extended perpendicular to the layers (MBBA and Canada balsam)

deformations. By sticking on the interface, these defects give it a raised texture. The fluidity allows access to positions that minimise the global energy and these minima, even relative, usually give rise to symmetrical and highly decorative figures.

In Fig. 4.17b, a much more developed smectic A rod is in equilibrium with its isotropic phase; it is observed between crossed polarisers, with a fairly uniform colour, corresponding to a good alignment of layers. We can verify optically that they are practically normal to the axis of the rod. The perpendicular orientation of the thermotropic smectic layers to the interface separating them from their isotropic phase corresponds to an energy minimum. Layers that were oblique or parallel to the interface would lead to higher interfacial energies. This *surface tension anisotropy* is at the origin of the rod-like shapes of smectic germs.

Lecithin is a mixture of membranous phospholipids. On contact with water, it gives what are called *myelinic figures* (Fig. 4.17c, d). These are cylindrical ropes, which twist around themselves in single, double or more complex helices. They have a lyotropic lamellar structure with smectic symmetry, but here the bilayers are arranged concentrically to the interfaces between the cylinders and the water, which is normal, since the two outer faces of the bilayers are hydrophilic. Here again, the morphology is determined at least partly by the interfacial tension. Moreover, the helical spirals reduces the area of this interface. But there are many other examples, and for every sort of liquid crystal.

Cholesteric liquids are also systems with parallel surfaces, forming droplets in equilibrium with their isotropic phase (Fig. 4.17e, f). They often present disclinations (Fig. 4.17e), especially after the coalescence of several droplets, but these defects sometimes reach the isotropic interface, so reducing their energy, and once there they may even disappear. In the absence of defects, the layers are practically unidirectional, give or take a few minor deformations. In many cases, the layers are normal to the interfaces, while obliqueness or parallelism are only tolerated. The droplets are round to begin with, which reduces the surface energy per unit of volume, but when the layers become more numerous, the germ shape stretches out, which facilitates the perpendicularity of layers to the interface.

Chromosomes generally take the form of rods, and there are some, made up almost entirely of DNA, with a morphology where the local directions of double helices are the same as those of the molecules of cholesteric rods. The anisotropy of interfacial tension is therefore a plausible factor in the elongated shape of chromosomes, but it is certainly not the only one, because other axial structures exist. The morphogenesis involved in living processes is too essential to be entrusted to one sole mechanism. When one fails to function correctly, others must be there to take over. Simply put, life do not put all its eggs in one basket. Every programmed developmental or physiological event is accomplished through highly diverse mechanisms, controlled to function in perfect synergy.

But let us return to the cholesteric germs suspended in their isotropic phase. Layers that are parallel to the interface are often favoured, producing spherulite forms [17, 52], but this occurs for other chemical compositions than the ones shown in Fig. 4.17e, f. We can even, in certain cases, observe that the favoured orientation of the cholesteric layers at the isotropic interface changes with temperature. Thus, these initially planar layers become concentric, through a process resembling certain embryonic gastrulations [17], but it is quite clear that the mechanisms involved are completely different.

We define a *texture* in terms of the defects encountered in an ordered phase, by their nature as singular points, lines or walls, and by their distribution, which often causes domains to appear where the structure is more regular, with a reduced density of defects. Because of their fluidity, the textures of liquid crystals sometimes display repetitive arrangements, suggesting the idea of an ordered system at a higher level of organisation. Elastic deformations and defect structures depend on the structure of the phase, as we have explained above. A wide variety of textures coexist in each phase, connected to the combinations of the diverse types of defects. The singular points may be isolated, or stem from the convergence of several lines. These are edge or screw dislocations, focal conics or disclinations, all capable of being superimposed. The combinatory possibilities are enormous, and this is an as-yet uncharted domain.

The focal conic texture of smectics A contains numerous elliptic and hyperbolic arcs. That of Fig. 4.18 shows a predominance of ellipses that are often entire, each

Fig. 4.18 Focal domains in a smectic A (cyano-octyl-biphenyl, with Canada balsam, crossed polarisers)

with a focus from which the branch of hyperbola starts, but the focusing would have to be adjusted to display that. It is possible to take three-dimensional shots [15].

Other textures with focal lines display stacks of conic points. They can be found in cholesteric analogues such as the shells of crabs or other arthropods, with similar or even more refined regularities among certain insects. However, one essential difference is that these lines are not paired as they are in focal conic textures. The reason is that the secretory epidermis imposes constraints, during the assembly at the level of its outer membrane.

The first constraint is that the secretion will be stabilised only a few hours after its production, thus excluding any future reorganisation; the bulk of the organic matrix of the shell is therefore never fluid. The second constraint imposed on secretion is that the filaments must be deposited in parallel to the outer membrane of the epidermis, whereas the differentiation of a focal ellipse would oblige the epidermis to become locally perpendicular to the cholesteric layers.

However, these constraints do not prevent the layers from forming Dupin cyclides in vast areas of the shell, and even to produce this geometry almost everywhere, if we consider the spheres and planes as particular case of Dupin cyclides. The focal domains that we know in smectics A and cholesterics are therefore incomplete in cholesteric analogues like shells, where the layers do, however, form Dupin cyclides and focal lines, but which are not associated in pairs. The influence of the cells and tissues on the texture appears to be essential, and precise control of the membranous morphology is therefore necessary during secretion of the shell. The cytoskeleton obviously plays a role in this. These questions need to be raised in their geometrical context, with all due precision, if they are to be tackled usefully on the biophysical, biochemical and genetic levels.

4.11 Topological Nature of Liquid Crystalline Textures

Up until now, we have mainly considered the geometrical aspects of liquid crystals and their biological counterparts, but these states of matter also lend themselves to an extremely rich combinatory topology, through their large-scale deformations and their wide variety of singular lines, illustrated in a number of works [8, 12, 15, 17, 23, 41]. We have already mentioned the recombinations of membranes or bilayers, but the topological evolutions of cholesterics are much more complex. A brief glimpse is given in Fig. 4.19, where the same slightly twisted cholesteric liquid produces Möbius strips in (a), and a pair of interlocking rings (c). The Möbius strip illustrates the non-polarity of the liquid and, in this case, even if the molecules have chemically different extremities A and B, it appears that they can align just as well in the same direction (AB with AB) as in opposing directions (AB with BA). In other words, even if we use vectors to describe the alignments, the integral lines are not physically arrowed.

4.11.1 Möbius Strips

Let us consider a simple closed circuit, a small circle in the liquid at rest, and the ribbon generated all along this circuit by all the short segments, of the same length and centred on the circuit, each of which representing the local direction of the molecules. There are then two possibilities for the ribbon, because of the non-polarity of the liquid: either its two faces do not communicate, or they communicate as in a Möbius strip. In this second case, we can verify that there exists at least one singular line of the liquid that passes by the circuit, i.e. which cuts across a minimal surface on the circuit.

This result is intuitive, if we look at what happens when we reduce the length of the circuit carrying a Möbius strip in a nematic liquid. We end up concentrating in one point the multiplicity of molecular directions represented by this ribbon, a singular situation that extends along a line which crosses the surfaces resting on the circuit, at least among those closest to the minimal surface.

What is the case in Fig. 4.19a? It shows an image taken using phase contrast microscopy, which darkens the molecules that are horizontal, i.e. perpendicular to the optical axis of the microscope. In a field of vectors $\mathbf{n}$, the locus of points where the field is horizontal corresponds to the surface $n_z(x, y, z) = 0$. This generally cuts across the planes of the slide and cover glass along two lines that are

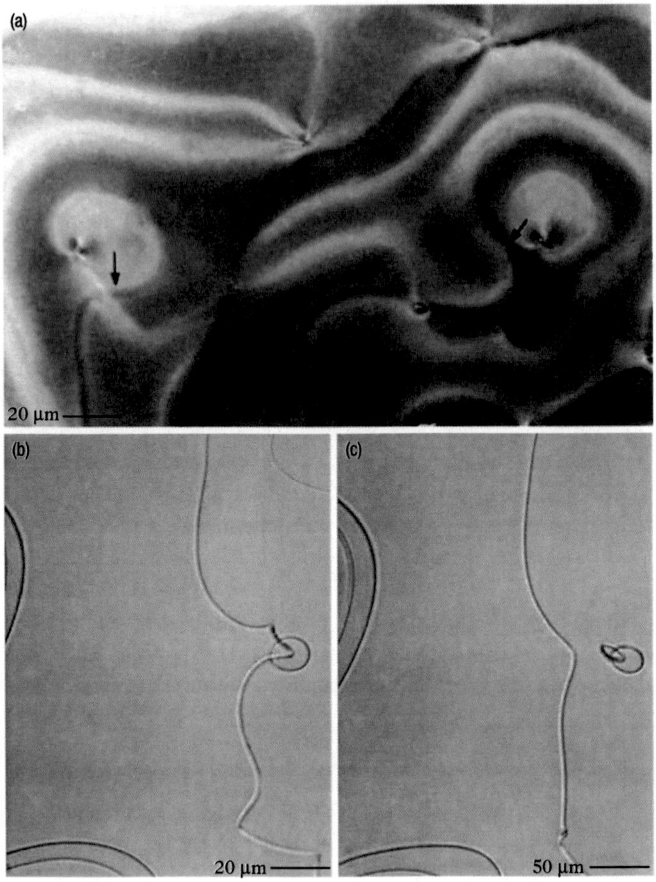

Fig. 4.19 Combinatory topology in liquid crystals of MBBA slightly twisted by cholesterol benzoate. **(a)** Möbius strip. **(b), (c)** Thread encircled by a ring transforming into a pair of rings, as a result of hydrodynamic movements imposed on the preparation

sharply contrasted, because they are singular, due to the orientations imposed on the molecules in the neighbourhood of these glass plates, and which are not horizontal. Here, the glass plates favour obliqueness, or even homeotropy. The ribbon, itself contrasting, is therefore reinforced by its two even more contrasting edges, despite the optical complexity of the system. We shall not go into this feature in any greater detail here, but a more complete study is presented in [8].

4.11.2 Pairs of Interlocking Rings

Another original topological situation arises in these slightly twisted nematic liquids when the molecules are aligned parallel to the slide and cover-glass, along the same horizontal direction. By pressing lightly on the cover-glass with a needle, we can produce hydrodynamic movements, allowing recombinations of the threads contained in these liquids, as shown in Fig. 4.19b, c. This results in the creation of a pair of interlocking rings. Here again, we shall not describe the system in its entirety.

These threads represent the locus of points where the molecules are vertical, in other words parallel to the optical axis of the microscope. They correspond to the intersection of the surfaces $n_x(x, y, z) = n_y(x, y, z) = 0$. These are thick threads, in the nomenclature of nematics, along which the field **n** is continuous and differentiable. There are also thinner threads, along which there is a discontinuity in molecular orientations. These are the singular lines described above that give rise to Möbius strips. At the level of our two interlocking rings of thick thread, on the contrary, the distribution of molecular directions is continuous and differentiable; the vector **n** is vertical, but oriented upwards for one of the rings and downwards for the other. On a path allowing to go from one thread to the other, **n** changes orientation, passing at least once through the horizontal.

To better understand this situation, let us consider a continuous and differentiable field of unit vectors **n**, in a simply connected domain D of $\mathbf{R}^3$, subjected to precise conditions of orientation at the level of its boundary surface S, for example $\mathbf{n} = \mathbf{n}_s$ a constant field over the whole of S. Let us assume, on the contrary, that the field **n** is quite complex within D. We can then seek to determine whether it is possible to modify this field to make it uniform through a continuous sequence of differentiable transformations, with the end result that $\mathbf{n} = \mathbf{n}_s$ throughout D and on S. This is not always possible, and one of the cases corresponds to the situation that we have just described, where **n** is constant, equal to $\mathbf{n}_1$ over a first closed circuit and equal to $\mathbf{n}_2$ over the other closed circuit. To simplify, $\mathbf{n}_1$ and $\mathbf{n}_2$ are both assumed to be different from $\mathbf{n}_s$. If these two circuits are separate, it is possible, during the gradual alignment of **n** with $\mathbf{n}_s$, for both of them to shrink and disappear without any problem.

On the other hand, this alignment cannot be achieved through continuous deformation if the two rings are interlocking. The two circuits cannot separate by crossing each other, because that would give a singular point with two different vectors $\mathbf{n}_1$ and $\mathbf{n}_2$. One of the circuits could reduce itself to a point on the second, or the two could shrink simultaneously, but the problem would remain the same. These pairs of interlocking rings are frequent in slightly twisted nematics [8]. They can often be observed to shrink homothetically towards zero, in an ever tighter embrace, until at the last moment they disappear, producing an extremely short-lived singularity that is pointwise both in space and time [11]. But we can also observe certain pairs of rings that grow bigger, instead of disappearing [8].

These works have allowed to introduce the concept of homotopy into the study of the textures of liquid crystals [57], classifying their defects, pointwise or linear, on the basis of their topological properties, and to discover the illustration of one of the classes of homotopy, occurring without any molecular discontinuity, except at the moment of transition to another class [20].

4.12 Liquid Crystals and Mechanical Clock Movements

At the end of this little journey, we have seen that most of the materials of living things are formed out of liquid crystalline assemblies, many of which are then stabilised. This is true not only for skeletal tissues, both bones and shells, but also for many other tissues, which present strong analogies with liquid crystals. The outer membrane of cells and of the main organelles are largely fluid, but the molecules in them present an orientational order, thus meeting the complete definition of liquid crystals, and displaying all their geometric properties with great refinement. We have briefly visited the geometry of the curvatures observed in these liquids, some of which produce perfect conics, as ellipses and hyperbolas, which evolve without ever seriously contradicting the rigour of the theorems. We hope that our illustrations have provided sufficient evidence of these geometric properties.

Clockmakers use movements prepared with the meticulousness of a goldsmith, but the mechanisms at work in liquid crystals are even more refined. The study of these intermediate states of matter combines fluid mechanics with crystallography and other disciplines. Through their infinite capacity to be deformed and reorganised, they lend themselves to the richest topological recombinations. They also allow the diffusion of chemical substances that they can dissolve, with the prospect of elaborate chemical differentiation. Liquid crystals display an extraordinary capacity to produce forms that can be stabilised, a capacity far beyond that of the other mechanisms usually considered in the realm of morphogenesis, and life has obviously taken full advantage of it.

References

1. Agard D.A (1993) To fold or not to fold, *Science* **260**, 1903–1904.
2. Alberts B., Bray D., Lewis J., Raff M., Roberts K., and Watson J.D. (1998) *Molecular biology of the cell*, 2nd edition, Garland Publishing (New York).
3. Belamie E., Davidson P. and Giraud-Guille M.-M. (2004) Structure and chirality of the nematic phase in a-chitin suspensions, *J. Phys. Chem.* **108**, 14991–15000.
4. Bouligand Y. (1965) Sur une architecture torsadée répandue dans de nombreuses cuticules d'Arthropodes, *C. R. Hebd. Acad. Sci. Paris* **261**, 3665–3668, in French.
5. Bouligand Y. (1965) Sur une disposition fibrillaire torsadée commune à plusieurs structures biologiques, *C. R. Hebd. Acad. Sci. Paris* **261**, 4864–4867, in French.
6. Bouligand Y. (1969) Sur l'existence de pseudomorphoses cholestériques chez divers organismes vivants, *J. Physique* **30**, Colloque C4, 90–103, in French.
7. Bouligand Y. (1972) Twisted fibrous arrangements in biological materials and twisted mesophases, *Tissue & Cell* **4**, 189–217.
8. Bouligand Y. (1972–1974) Recherches sur les textures des états mésomorphes, *J. Physique, Paris* **33**, 525–547; **33**, 715–736; **34**, 603–614; **34**, 1011– 1020; **35**, 215–235; **35**, 959–981, in French.
9. Bouligand Y. (1978) Liquid crystalline order in biological materials, in *Liquid Crystalline Order in Polymers*, edited by A. Blumstein, Academic Press (New York), pp. 261–297.
10. Bouligand Y. (1978) Liquid crystals and their analogs in biological systems, in *Solid State Physics*, Suppl. 14, *Liquid Crystals*, edited by L. Liébert, pp. 259–294.
11. Bouligand Y. (1978) Aspects expérimentaux des défauts dans les structures mésomorphes, *Journal de Microscopie et de Spectroscopie Electronique* **3**, 373–386.

12. Bouligand Y. (1980) Defects and textures in liquid crystals, in *Dislocations in Solids*, edited by F.R.N. Nabarro, North Holland (Amsterdam), vol. 5, pp. 298–347.
13. Bouligand Y. (1983) Some geometrical problems in liquid crystals, in *Bifurcation Theory, Mechanics and Physics*, edited by C.P. Bruter, A. Aragnol and A. Lichnerowicz, Reidel Pub. Co. (Dordrecht), pp. 357–381.
14. Bouligand Y. (1990) Geometry and topology of cell membranes, in *Geometry in Condensed Matter Physics*, edited by J.-F. Sadoc, World Scientific (Singapore), pp. 193–231.
15. Bouligand Y. (1998) Defects and Textures, in *Handbook of Liquid Crystals*, edited by D. Demus et al., Wiley-VCH (Weinheim), vol. 1, pp. 406–453.
16. Bouligand Y. (1999) Remarks on the geometry of micelles, bilayers and cell membranes, *Liq. Cryst.* **26**, 501–515.
17. Bouligand Y. and Livolant F. (1984) The organization of cholesteric spherulites, *J. Physique Paris* **45**, 1899–1923.
18. Bouligand Y., Soyer M.-O., and Puiseux-Dao S. (1968) La structure fibrillaire et l'orientation des chromosomes des Dinoflagellés, *Chromosoma* **24**, 251–287, in French.
19. Bouligand Y., Cladis P.E., Liébert L., and Strzlecki L. (1974) Study of sections of polymerized liquid crystals, *Mol. Cryst. Liq. Cryst.* **25**, 233–252.
20. Bouligand Y., Derrida B., Poenaru V., Pomeau Y., and Toulouse G. (1978) Distorsion with double topological character, *J. Physique* **39**, 863–867.
21. Broglie M. de (1923) La diffraction des rayons X par les corps smectiques, *C. R. Acad. Sci. Paris* **176**, 738.
22. Chandrasekhar S. (1977) *Liquid Crystals*, Cambridge University Press (Cambridge).
23. Chandrasekhar S. (1986) The structure and energetics of defects in liquid crystals, *Adv. Phys.* **35**, 507–596.
24. Chapeville F. and Clauser H. (eds.) (1974) *Biochimie*, Hermann (Paris), in French.
25. Chapman D. (1966) Liquid crystals and cell membranes, *Ann. N. Y. Acad. Sci.* **137**, 745–754.
26. Charvolin J. and Tardieu A. (1978) Lyotropic liquid crystals: structures and molecular motions, in *Solid State Physics*, Suppl. 14, *Liquid Crystals*, edited by L. Liébert, pp. 209–257.
27. Dekeyser W. and Amelinckx S. (1955) *Les Dislocations et la Croissance des Cristaux*, Masson (Paris), in French.
28. Demus D., Goodby J., Gray G.W., Spiess H.-W., and Vill V. (eds.) (1998) *Handbook of Liquid Crystals*, Wiley-VCH (Weinheim).
29. Drach P. (1939) Mue et cycle d'intermue chez les Crustacés Décapodes, *Ann. Inst. Océan.* **19**, 103–392, in French.
30. Drach P. (1953) Structure des lamelles cuticulaires chez les Crustacés, *C. R. Hebd. Acad. Sci.* **237**, 1772–1774, in French.
31. Duvert M., Bouligand Y., and Salat C. (1984) The liquid crystalline nature of the cytoskeleton in epidermal cells of the chaetognath Sagitta, *Tissue & Cell* **16**, 469–481.
32. Favard P. and Bouligand Y. (1981) La phénoménologie des auto-assemblages biologiques, in *La Morphogenèse, de la Biologie aux Mathématiques*, edited by Y. Bouligand, Maloine (Paris), pp. 101–113, in French.
33. Fawcett D.W. (1981) *The Cell*, 2nd edition, Saunders (Philadelphia).
34. Frank C.F. (1958) On the theory of liquid crystals, *Disc. Faraday Soc.* **25**, 19–28.
35. Friedel G. and Grandjean F. (1910) Observations géométriques sur les liquides à coniques focales, *Bull. Soc. Fr. Minéralogie* **33**, 409–465.
36. Gennes P.-G. de (1974) *The Physics of Liquid Crystals*, Clarendon (Oxford).
37. Giraud-Guille M.-M. (1996) Twisted liquid crystalline supramolecular arrangements in morphogenesis, *Int. Rev. Cytol.* **166**, 59–101.
38. Gray G.W. (1962) *Molecular Structure and the Properties of Liquid Crystals*, Academic Press (New York).
39. Kelker H. and Hatz W. (1980) *Handbook of Liquid Crystals*, Verlag Chemie (Weinheim).
40. Kellenberger E. (1980) Control mechanisms governing protein-protein interactions in assemblies, *Endeavour, New Series* **4**, 2–13.

41. Kléman M. (1983) *Points, Lines and Walls in Liquid Crystals, Magnetic Systems and Various Ordered Media*, Wiley (Chichester).
42. Lehmann O. (1908) *Flüssige Kristalle und die Theorien des Lebens*, J. A. Barth (Leipzig), in German.
43. Livolant F. and Bouligand Y. (1978) New observations on the twisted arrangement of Dinoflagellate chromosomes, *Chromosoma* **68**, 21–44.
44. Livolant F. and Leforestier A. (1996) Condensed phases of DNA: structure and phase transitions, *Prog. Polym. Sci.* **21**, 1115–1164.
45. Livolant F. and Leforestier A. (2000) Chiral discotic columnar germs of nucleosome core particles, *Biophys. J.* **78**, 2716–2729.
46. Locke M. (1964) The structure and formation of the integument of insects, in *Physiology of Insecta*, edited by M. Rockstein, Academic Press (New York), vol. 3, pp. 379–470.
47. Nageotte J. (1937) *Morphologie des Gels Lipoïdes – Myéline – Cristaux Liquides -Vacuoles*, Hermann (Paris), in French.
48. Revol J.-F., Bradford H., Giasson J., Marchessault R.H., and Gray D.G. (1992) Helicoidal self-ordering of cellulose microfibrils in aqueous suspensions, *Int. J. Biol. Macromol.* **14**, 170–172.
49. Revol J.-F. and Marchessault R.H. (1993), *In vitro* chiral nematic ordering of chitin crystallites, *Int. J. Biol. Macromol.* **15**, 339–335.
50. Richards G.A. (1951) *The Integument of Arthropods*, Minnesota Press (Minneapolis).
51. Robinson C. (1961) Liquid crystalline structures in polypeptide solutions, *Tetrahedron* **13**, 219–234.
52. Robinson C. (1966) The cholesteric phase in polypeptide solutions and biological structures, *Mol. Crystals* **1**, 467–494.
53. Roland J.C. and Vian B. (1979) The wall of the growing plant cell: its three-dimensional organization, *Int. Rev. Cytol.* **61**, 129–166.
54. Silvestri F. (1901) *Acari, Myriapoda et Scorpiones hucusque in Italia reperta, 1. Segmenta tegumentorum, musculi*, Vesuviano (Portici).
55. Singer S.J. and Nicolson G.L. (1972) The fluid mosaic model of the structure of cell membrane, *Science* **175**, 720–731.
56. Thompson, D'Arcy W. (1917 and 1942) *On Growth and Form*, two editions from Cambridge University Press (Cambridge).
57. Toulouse G. and Kléman M. (1976) Principles of a classification of defects in ordered media, *J. Physique Lettres* **37**, 149–151.
58. Weiss L. (1983) *Cell and Tissue Biology, A textbook of Histology*, 6th edition, Urban & Schwarzenberg (Baltimore).
59. Zocher H. and Törok C. (1967) Crystals of higher order and their relations to other superphases, *Acta Cryst.* **22**, 751–755.

Chapter 5
Biological Self-Organisation by Way of the Dynamics of Reactive Processes

James Tabony

In the Beginning how the Heav'ns and Earth
Rose out of Chaos

a dark
Illimitable Ocean without bound,
Without dimension, where length breadth, and height,
And time and place are lost; where eldest Night
And Chaos, Ancestors of Nature, hold
Eternal anarchie

darkness fled,
Light shone, and order from disorder sprung

Anon out of the earth a fabrick huge
Rose like an exhalation

Paradise Lost (1667)
John Milton (1608–1674)

The physical-chemical processes by which order and form spontaneously develop in an initially largely unstructured biological object, such as an egg or a seed, remain uncertain. At present, there are two quite different approaches that may account for biological self-organisation and pattern formation. One approach: that based upon static interactions between non-reacting species is outlined in another chapter of this book (see the introduction, Chap. 1, and Chap. 4 on liquid crystals). In the other approach, which is the subject of this chapter, self-organisation develops by way of the non-linear dynamics of reactive processes. Because in most cases, solutions of reacting chemicals or biochemicals in a test-tube do not self-organise, it was for a long time believed that the dynamics of reactive processes could not result in self-organisation.

However, very slowly over the last hundred years, both theorists and experimenters have progressively shown, in the face of much opposition, that this is not necessarily the case. In particular, since the late 1930s, some theorists (Kolmogorov, Rashevsky, Turing, and Prigogine and co-workers) have proposed that some

J. Tabony (✉)
Direction des Sciences du Vivant, Institut en Recherches et Technologies des Sciences du Vivant, Laboratoire Biopuce, CEA Grenoble, 17 Avenue des Martyrs, 38054 Grenoble, France
e-mail: james.tabony@cea.fr

P. Bourgine, A. Lesne (eds.), *Morphogenesis*, DOI 10.1007/978-3-642-13174-5_5,

particular types of chemical reaction might, due to being sufficiently far-from-equilibrium, show strongly non-linear reaction phenomena [10, 15, 28, 36]. They predicted that such systems could show macroscopic self-ordering in which a chemical pattern spontaneously develops from an initially homogeneous solution. At a molecular level, this phenomenon results from a coupling of reaction and diffusion, and the patterns that arise are comprised of periodic variations in the concentration of some of the reactants. Such structures are often called reaction-diffusion or Turing structures; the latter after the British mathematician who was one of the first persons to propose such a mechanism in 1952. Prigogine and co-workers called them *dissipative structures* [27], because a dissipation of chemical energy is required to drive and maintain the system sufficiently far-from-equilibrium such that self-organisation occurs. It is this flux or dissipation of chemical energy that provides the thermodynamic driving force for self-organisation. Rashevsky, Turing, Prigogine et al., and others, all proposed that biochemical mechanisms of this type might provide an underlying physical chemical explanation for biological pattern formation and morphogenesis.

Since then, some chemical reactions, based upon reactions initially discovered in the 1920s [3] and early 1950s [1], have at last been recognised to self-organise in this way. Here, I will give a personal overview of the conceptual and historical background of this approach to biological self-organisation and illustrate it with some of our own work on the in vitro self-organisation of microtubules, a major element of the cellular cytoskeleton.

The extracts from John Milton's epic poem *Paradise Lost* quoted above provide a powerful and succinct description of the Creation. Not only does Milton make us fully aware of an initially homogenous state devoid of order and positional information:

> a dark illimitable ocean without bound, without dimension, where length breadth, and height, and time and place are lost,

but he also claims that self-organisation arises spontaneously from this state. He implies that the development of order from the void is a fundamental process central to the creation of the world as we know it. Since the nineteenth century, the laws of thermodynamics have taught us that the natural tendency of the Universe is towards a progressive loss of order. To a person of our times familiar with the concepts of chaos theory [11] and *complex* systems [4, 7, 24], in which self-organisation spontaneously arises as an *emergent* phenomenon, the image Milton evokes is one of "order at the edge of chaos". The parallels between his vision of creation and the approach of self-organisation in *complex* systems is particularly striking due to the fact that a very different outlook has dominated much of scientific thinking over the intervening centuries. John Milton (1608–1674), a contemporary of René Descartes (1596–1650) and William Harvey (1578–1657), claims to have visited Galileo (1564–1642) in 1638 when he was under house arrest by the Catholic Church. Milton was considered one of the most erudite persons in the England of his time and was reputed to have read all the classical works then available. He would certainly have been familiar with the writings of the Greek philosopher Anaxagoras (circa 500–428 BC). According to Anaxagoras, in the cosmos in its original state "All things were together". Then at some time, this original mixture is set into a rotation; as a result of which things are separated out and recombine with one another to produce the world we perceive through our senses. And so the idea that dynamic processes can give rise to order and form are of very long date. However, from the eighteenth century, this outlook has played a progressively decreasing role. To a large extent, this came about because until the advent of the computer, it was mostly only possible to quantify scientific problems in terms of linear or close-to linear relationships. Most strongly coupled many-bodied problems were considered un-solvable. It became customary to use a strongly reductionist approach in which as many phenomena as possible were approximated as being the sum of the properties of the individual elements. For a long time, it was widely believed that in populations of strongly interacting elements, the overall dynamics would be so complicated as to prohibit the development of any type of spatial or temporal order.

Within the last 50 years, however, workers in different disciplines have progressively discovered that this is not necessarily the case, and that under some circumstances new phenomena, in particular spatial and temporal self-organisation, can arise in populations of strongly interacting elements. These new phenomena are not the properties of the individual elements as such, but develop by way of the non-linear dynamics of the interactions by which the elements "talk and communicate with one another". In recent years, systems of this type have been called *complex* and the phenomena that arise are known as *emergent* [4, 7, 14, 24]. Complex systems show certain general types of behaviour that are largely independent of the detailed properties of the individual elements as such, be they molecules, water droplets, cells, bacteria, ant colonies, or galaxies. In most *complex* systems, self-organisation, either spatial or temporal, occurs as a major *emergent* property.

The development of this approach is causing a change of scientific paradigm; for in a *complex* system, the behaviour of the whole can no longer be considered as being simply the sum of the parts. The approach thus imposes a different level of reductionism in which frequently the notion of cause and effect is strongly attenuated. Some scientists are asking whether some of the global properties of biological systems can be accounted for in terms of *emergent* properties and even whether life itself should be considered as such. An increasing number of researchers consider the development of "*complexity*" as the most significant scientific development for many years. In any case, it is a new way of viewing many aspects of the world in which we live.

The idea that change and dynamics may result in pattern and order date back to antiquity. However, a major objection to the possibility that this premise might apply to biological processes involving biochemical reactions, is that in most practical test-tube cases, chemical or biochemical reactions do not self-organise. In 1917 D'Arcy Thompson in the introduction to his influential book *On Growth and Form* [35] wrote:

> the road of dynamical investigation in morphology has found few to follow it; but the pathway is old. The way of the old Ionian physicians, of Anaxagoras, of Empedocles and in his disciples in the days before Aristotle, lay just by that highway side. It was Galileo and Borelli's way; and Harvey's way, when he discovered the circulation of the blood. It was little trodden for long afterwards.

D'Arcy Thomson also wrote:

> M. Dunan, discussing the Problème de la Vie, in an essay which M. Bergson greatly commends, declares that "les lois physico chimiques sont aveugles et brutales; là où elles règnent seules, au lieu d'un ordre et d'un concert, il ne peut y avoir qu'incohérence et chaos."

This position summarises the opinion prevailing at that time, and which is still strongly entrenched; namely that the dynamics of reactive processes cannot lead to self-organisation.

5.1 Self-Organisation by Dynamic Processes in Physical Systems

Hydrodynamics is one area where it has been established for many years that macroscopic self-organisation occurs by way of non-linear dynamics. Familiar examples are the anti-cyclonic variations in atmospheric pressure represented on meteorological maps and the spiral shaped vortex that arises when a washing basin is emptied. Well-studied cases of hydrodynamic ordering are the Bénard-Rayleigh [2, 30, 37] and the Taylor instability experiments [34]. In the Bénard-Rayleigh experiment, liquid, several millimetres deep in a Pétri dish, is heated from below to produce convection. Under suitable conditions, the convection currents spontaneously self-organise, and a macroscopic hexagonal pattern develops. The explanation proposed by Rayleigh [30] was the following. As the sample is heated from below, the density of the liquid close to the bottom decreases. Initially the buoyancy force due to the density difference is insufficient to overcome viscous drag and there is no net movement of liquid. With increased heating, the density difference becomes large enough to overcome this; the homogeneous equilibrium state becomes unstable; less dense liquid rises from the bottom whilst more dense liquid descends from the top. The warmer liquid at the top then in its turn, cools and falls to the bottom, whilst colder liquid at the bottom warms and rises to the top. Because this cannot happen simultaneously over the entire sample, there is a spontaneous division of the layer into convection cells. The width of the individual convection cell is approximately the thickness of the liquid layer and the fluid circulates in a closed orbit. In neighbouring cells, the directions of circulation have to be in opposite directions. This constraint when combined with the dimension of the convection roll, and which is itself determined by the height of the liquid layer, results in the macroscopic pattern. Self-organisation arises from the dynamics of the process by which convection cells are strongly coupled to one another in a non-linear manner. The pattern is dependent on the flow of energy through the liquid implicit in heating it from below. When heating stops, there is no further energy dissipation, the system returns to equilibrium, the dynamic behaviour is no longer non-linear and the self-ordered structure disappears.

In the Taylor instability experiment, a viscous liquid contained between two cylinders is set into motion by rotating the inner cylinder with respect to the outer one. At rest, and low rates of rotation, the liquid appears uniform. With increasing rate of rotation, changes occur, and at certain values, equidistant stationary bands form. Although the bands are stationary, they are obviously comprised of liquid in movement. In the regime corresponding to the banded structure, the equations describing the dynamics of this movement are non-linear. The solution to these equations, give rise to periodic variations in the parameters describing the motion of the liquid, and correspond to the banded appearance. The driving force for this dynamic pattern – and which keeps the liquid out-of-equilibrium and in motion – is the energy put into the system through rotation. As in the Rayleigh-Bénard experiment, when rotation ceases, energy dissipation stops, there is no further movement of the liquid, the dynamics are no longer non-linear, and the ordering disappears.

If the rate of rotation is increased beyond that necessary for the formation of the banded structure, then this structure is in its turn destabilised and a new arrangement develops. With increasing rates of rotation, there are further instabilities associated with bifurcations and the development of other types of spatial and temporal order, until finally the movement of the liquid becomes random and is described by chaotic dynamics.

Bifurcation properties frequently occur in dynamic systems. Systems at, or close to, equilibrium, are frequently described by linear equations. The unique solution to these equations leads to the equilibrium state as the only stable state. When a system is progressively moved away from equilibrium, a point is reached where its dynamics are no longer approximated by linear relationships but become strongly non-linear. Non-linear systems can exist in multiple stationary states; for example, a non-linear equation such as the quadratic, $y = ax^2 + bx + c$, may, depending on the parameter values, have two real solutions in x for a given value in y. Hence, under given experimental conditions, a system showing non-linear dynamics can show more than one stationary state. In populations of strongly interacting elements, some types of non-linear dynamics result in self-organisation. Thus, in appropriate cases, different types of self-organising behaviour, corresponding to the different solutions to the non-linear equations, are allowed. As a system is progressively displaced from equilibrium, at the point where the linear equilibrium system becomes unstable, it adopts one of the different non-linear dynamic pathways open to it, leading to the self-organised morphology that subsequently develops. At this critical instant, or bifurcation point, there is little to choose between the different pathways open to the system. Hence, the presence of a weak external factor, otherwise too weak to affect an equilibrium state, can favour one of these pathways, and so determine the system's future development. Furthermore, this factor need only be present at the critical moment when the initial state is unstable. For, once the system has bifurcated, it subsequently follows the selected dynamic pathway to the pre-determined state. After the bifurcation has occurred, the critical determining factor may be removed without any further effect on the system, which behaves as though it retained a memory of the conditions prevailing at the bifurcation.

5.2 Self-Organisation in Colonies of Living Organisms

Colonies of living organisms provide many examples of self-organisation by collective processes [4]. Fish schools, bird clouds, wasp swarms, ant colonies, and colonies of certain types of unicellular organisms and bacteria, all self-organise in this way (see Chap. 12). In these cases, structures and organisations develop, not by action at the level of the individual, but rather by way of dynamic processes in which the individuals, strongly coupled to one another in a non-linear manner, behave as a collective ensemble. Similar types of morphology often develop in spite of large differences in the nature and size of the individual element. Striped arrangements frequently arise; when they do, they are nearly always the result of an outside external perturbation that induces a directional bias on the actions of the individual.

A well-studied example of this type of behaviour is that of ant colonies and other social insects. None of the individual ants has an architectural plan in its head. The behaviour of the population results uniquely from the actions of individuals strongly coupled to one another by a form of chemical communication. A moving ant leaves behind itself trails of chemicals known as pheromones, which attract or repel other ants. An ant encountering a trail of an attractive pheromone will change its direction to follow the trail. This ant, will in its turn, deposit more pheromone on the trail thus reinforcing it. The self-amplification of these chemical trails leads to the self-organisation of the ant population. Although the rules governing the behaviour of individual ants are relatively simple, the overall behaviour is extremely sophisticated.

One of the advantages of this type of process is that ants rapidly establish the shortest route between a food source and the nest. Consider a situation where there are two food sources close to a population of ants, but where one of the food sources is closer to the colony than the other. As ants return to the nest with food, they leave chemical trails behind themselves. These trails are then followed by other ants who in their turn, deposit chemicals which reinforce the original trails. In such a way, progressively more and more ants follow the paths to the food sources. However, because the trail from the closer of the two sources is shorter, it takes less time for an ant to return to the colony. This results in a slightly larger number of ants taking the path to this food source, thus reinforcing the strength of the chemical trail of the shorter path at the expense of the longer path. Hence, progressively more and more ants take the shorter path to the closer food supply until they nearly all follow this route. This illustrates how self-organisation results from the progressive reinforcement of chemical trails by moving objects which themselves produce these trails. If, the two food sources are at approximately equal distance from the nest, then the ants still mostly accumulate on the path to one of the food sources. This comes about because any small factor which early in the process favours the reinforcement of one of the chemical trails over the other, will progressively lead to nearly all the ants using this pathway. Once the reinforcement of one pathway has gone sufficiently far, then the determining factor may be removed without affecting the subsequent behaviour. This is a simple example of a bifurcation due to a weak external factor in a self-organising system.

Another striking example of dynamic self-organisation in a living system is the macroscopic patterns generated by certain colonies of bacteria. Over a distance scale of several centimetres, the bacteria form a stationary pattern. Observations at higher magnification show that the pattern is comprised of regions containing differing bacterial densities. At even higher magnification, where the individual bacteria are observed, the bacteria are seen to be undergoing a rapid, seemingly random movement. It is out of this dynamic process involving the collective movement of many bacteria that the stationary pattern arises. Individual bacteria interact indirectly with one another via trails in the concentration of chemical attractants and repellents that they themselves produce. The energy source is the chemicals consumed by the bacteria. When it runs out, the bacteria stop moving and the pattern disappears.

Hydrodynamic patterns, and ant and bacteria colonies, are all examples of macroscopic self-organisation by way of non-linear dynamics. Nevertheless, the systems are very different; one is a liquid in motion, the others are populations of living organisms. The question obviously arises as to how such processes might occur in systems intermediate between these extremes. In particular, how might they come about at a molecular level by way of biochemical reactions within a biological object such as a cell, or an egg.

5.3 Self-Organisation by Reaction and Diffusion: Stripes in a Test-Tube

Already in 1896, Liesegang [18] had reported the formation of a striped structure by a process now known to involve a combination of reaction, diffusion, and precipitation. In 1925, Alfred Lotka published a book entitled *Elements of Physical Biology* [20]. Lotka showed how regular oscillations in animal populations might arise by way of the dynamics of the interactions between individuals of different species. He was very much aware that similar considerations could also apply to chemical reactions involving populations of reacting species and in 1910 had already published an article entitled "Contribution to the theory of periodic reactions" [19]. However, discouraged by the reception his work received, he left academic research to work for a life insurance company. In 1921, an American chemist, William Bray, described a chemical reaction showing regular oscillations [3]. In his article, he cited the 1910 article of Lotka as a possible theoretical explanation for his observations. Like Lotka, his work was unfavourably received, and his findings were dismissed out of hand as being due to poor experimental work. This hostile intellectual climate seems to have played a major role in discouraging any further developments.

In 1937, the Soviet physicists, Kolmogorov, Petrovsky and Piskunov published a theoretical article pointing out that under appropriate conditions a coupling of molecular diffusion with certain types of reactive processes could lead to self-organisation [15]. Likewise, in 1940, Nicolas Rashevsky at the University of Chicago briefly outlined how a combination of reaction and diffusion might lead to different chemicals concentrations developing at different positions in a developing embryo [28, 29]. At the time of their publications, these articles appear to have limited impact. A more significant advance occurred in 1952 with the publication of an article by the British mathematician, Alan Turing, entitled "The chemical basis of morphogenesis" [36]. Turing proposed in a concise mathematical form a theory by which, in certain types of reaction scheme containing an autocatalytic step, a combination of reaction and diffusion could result in a pattern of chemical concentrations. He also proposed that such physical chemical processes might underlie biological morphogenesis.

About 1950, when Turing, one of the conceptual inventors of the computer, was developing his theory, a Soviet researcher, Boris Belousov, whilst seeking to mimic the Krebs's cycle in a test-tube, devised a chemical system that showed regular oscillations. Unfortunately, when he tried to publish his findings, the referees rejected his manuscript on the basis that his supposed discovery' was quite impossible (the referees do not seem to have been aware of the findings of either Lotka or Bray). He tried again, 6 years later, to publish his findings, but without success [6]. Like Lotka before him, Belousov took this refusal very much to heart and retired from academic research. A two-page report eventually appeared in 1958 in the proceedings of an obscure symposium on radiation chemistry [1]. Subsequently in the 1960s, Zhabotinsky and co-workers [38–40] demonstrated that Belousov's reaction could also develop a striped pattern corresponding to regions of different chemical composition (Fig. 5.1). The stripes arise not by way of static interactions but because different chemicals are produced at different overall rates at different positions in the test-tube. When the reaction ceases, due to consumption of the chemicals, then diffusion and convection cause the structure to disappear. In 1980, Belousov (together with Zhabotinsky and Zaikin) was posthumously awarded the Lenin prize for his discovery. Belousov and Turing never met. Turing [13] committed suicide in 1954. During the Second World War, he played a major role in breaking the Enigma code used by the Nazi military machine. In the early 1950s, when the cold war was at its height, his contribution was still classified top secret. Even had the referees allowed Belousov to publish his findings, there would have been little chance of a meeting between himself and Belousov [6].

Subsequently, during the 1960s, the concepts and theory of chemical self-organisation was the subject of considerable theoretical development by Prigogine and co workers [10, 23]. In their book published in 1971, Glansdorff and Prigogine presented the Belousov-Zhabotinsky reaction as an example of a dissipative structure [10]. Unfortunately, many scientists chose not to accept this explanation and it was not until 1990, when researchers presented evidence obtained on a similar reaction carried out in a gel [5], that scientists were convinced that such structures

Fig. 5.1 Stripes in a test tube: the Belousov-Zhabotinsky reaction. © J. Tabony

could actually exist. The contribution of Turing, Rashevsky and Kolmogorov were each limited to a single article. On the contrary, the contribution of Prigogine and co-workers encompass several books. In addition, in books such as *Order out of Chaos*, Prigogine, who in 1973 was awarded the Nobel Prize for chemistry, played a major role in propagating the far-reaching implications of self-organisation in out-of-equilibrium systems to a more general public [27].

What I personally find amazing in the works of Lotka, Kolmogorov, Rashevsky, Turing, and Prigogine, is that they were able to arrive at conclusions contrary to much conventional thinking, concerning what are essentially non-linear processes, without the aid of a computer. Although such terms were not used at the time, what these theorists predicted was that biological self-organisation could arise as an *emergent* phenomenon in a *complex* system by molecular processes of reaction and diffusion. Unfortunately, it has taken almost 70 years for the approach first proposed by them to be accepted. Their conclusions, were revolutionary because they implied a change in paradigm. Until then, it was not believed possible that solutions of chemicals or biochemicals could self-organise by reactive processes. The 2nd law of thermodynamics teaches us that order is progressively and ineluctably lost with time. In particular, at a molecular level, an existing macroscopic order is gradually attenuated by molecular diffusion. Two miscible liquids, initially separated from one another, slowly mix by way of diffusion, and the existing order is progressively lost. Kolmogorov, Rashevsky, Turing, Prigogine and co-workers, predicted that under some circumstances the contrary might occur, that is, reactive processes when combined with diffusion can lead to the partial separation of initially homogeneous chemicals.

One of the conditions for this is an autocatalytic reaction that produces other chemicals which both activate and inhibit the formation of the reaction product. If a local fluctuation gives rise to slight excess of activator, then the reaction accelerates, and at this point even more activator is produced. Consider a situation in which the inhibitor diffuses at a much faster rate than the activator. This results, with increasing distance from the source of the fluctuation, in progressively more inhibitor compared to activator. As a function of distance from the fluctuation, this in turn leads to a decreased production of both activator and inhibitor, and a progressive slowing down of the reaction. Hence at a certain distance from the first fluctuation, another small fluctuation in activator concentration will once again lead to an acceleration in the reaction and to a repetition of the behaviour [21]. Stationary periodic variations in the concentration of the reactants can hence spontaneously emerge from the initially homogenous solution. This pattern arises not from static interactions but is a consequence of the non-linear dynamics describing the way that rates of chemical reaction are coupled to one another and modified by molecular diffusion.

A different aspect of these systems is the manner by which reaction-diffusion systems may show bifurcation properties and hence be sensitive to weak external factors. For some types of reaction-diffusion system, Prigogine and Kondepudi [16] explicitly calculated that the presence at a crucial moment early in the process, of an external factor, such as gravity or electric and magnetic fields, can determine the self-organised morphology which subsequently developed.

The pioneer workers in this field were fully aware of the possible implications that their approach might have towards some problems in biology, and the concepts described above have, at various times over the last 50 years, aroused a considerable amount of interest and debate. However, the majority of chemists and biologists have not adopted them. Although there are several reasons for this, the main reason is conceptual. Many chemists and biologists are unfamiliar with the concepts of non-linear dynamics. Even now there is still a great reluctance on the part of many older scientists to accept to the possibility that self-organisation can arise by way of the dynamics of reactive processes. This scepticism has in its turn been aggravated by the scarcity of experimental systems recognised as behaving this way. However, this scarcity of experimental systems is itself largely a result of scientific opposition to this approach. Like Belousov, any researcher who claimed such a behaviour was likely to find his publications and grant proposals refused out of hand by his reviewers. For example, in chemistry, it was not until 1990 [5] that a chemical reaction, similar to those first discovered long ago by Bray (1921) [3] and Belousov (1951) [1] was finally accepted as the first example of a Turing-like structure. To my mind, this change seems to have as much to do with the beginning of a general acceptance of such an approach, concomitant with the development of the science of *complexity*, rather than with the appearance of a substantial new body of experimental evidence.

The same situation has prevailed in biology. Since the work of Turing and Prigogine and co-workers, many authors have compared the morphologies that occur in biological organisms with the mathematical predictions of reaction-diffusion theories. There is a whole body of literature in this area [12, 21, 22]. More recently, other workers [8, 17] have demonstrated that the patterns of calcium waves that observed in vivo in the cytosol arise from reaction-diffusion processes. In spite of these advances, one of the elements lacking, has been an in vitro example of a simple biochemical system in a test-tube which self-organises in the general manner outlined above.

Under suitable conditions, we have found that the formation in vitro of microtubules, a major component of the cellular cytoskeleton, behaves in this way. These preparations self-organise by reaction and diffusion and the morphology that develops is determined, at a critical early moment, by weak external factors, such as gravity and magnetic fields.

5.4 Microtubule Self-Organisation

The interior of the cell is organised by the cytoskeleton. The latter is composed of three filamentary components; microtubules, actin and intermediate filaments. Microtubules are long tubular shaped objects, with inner and outer diameters of about 16 nm and 24 nm respectively. They arise from the self-assembly of a protein, tubulin, by way of reactions involving the hydrolysis of a nucleotide, guanosine triphosphate (GTP), to guanosine diphosphate (GDP). Their length is variable; but they are often several microns long. Once microtubules are formed, they continually grow and shrink by processes in which tubulin molecules are added to one end of a microtubule whilst other tubulin molecules are lost from the other shrinking end. This process is likewise associated with the hydrolysis of GTP to GDP. The system is hence chemically irreversible and there is a continual consumption and dissipation of chemical energy.

Microtubules have two major cellular roles; they organise the cell interior, and they permit and control the directional movement of intracellular particles and organelles from one part of the cell to another. They participate in many fundamental cellular functions including, the maintenance of shape, motility, and signal transduction. Microtubules are a significant component of brain neurone cells, they make up the mitotic spindles that separate chromosomes during cell division, and they play a determining role in the organisational changes that occur during the early stages of embryogenesis. Microtubule organisation is a fundamental cellular property affecting numerous biological functions and the viability of a cell is compromised when it does not occur correctly. Biologists have established in living cells that microtubule organisation, and reorganisation, results from the chemical dynamics of the reactive processes associated with their formation and maintenance. Another of the characteristic features of in vivo cellular microtubules is that they frequently organise or reorganise in response to weak internal and external stimuli of either physical or biochemical nature.

Microtubules can be readily formed and studied in vitro. A solution of purified tubulin, in the presence of an excess of GTP, when warmed from about 7°C to 35°C, assembles within a few minutes into microtubules. After the microtubules have formed, this reaction continues by processes in which the complex, tubulin-GTP, is added to the growing (+) end of a microtubule and tubulin-GDP is lost from the opposite shrinking (−) end. Due to differences in reactivity at opposing ends, microtubules often grow from one end (+), whilst shrinking from the other end (−). When the rates of growth and shrinking are comparable, individual microtubules retain the same approximate length but change position at speeds of several microns per minute. This type of behaviour is termed *treadmilling*. Another type of behaviour termed *dynamic instability* occurs when individual microtubules either shrink or grow abruptly at much faster rates. By modifying experimental conditions, such as buffer composition, it is possible to observe, in vitro, a very large range of microtubule reaction dynamics.

Under many conditions, microtubule solutions do not show either temporal or spatial self-organisation. However, in 1987, it was reported that they could show regular damped oscillations of assembly and disassembly [26]. In 1990, we reported experiments under different buffer conditions in which macroscopic self-ordering occurred [32]. When assembled in glass containers, measuring 4 cm by 1 cm by 1 mm, the microtubule solution progressively self-organises over approximately 5 h to form a series of periodic horizontal stripes of about 0.5 mm separation. Once formed, the striped pattern remains stationary and it is stable for between 48 and 72 h. After this, the system runs progressively out of reactants. In each striped band, all the microtubules are very highly oriented with respect to one another. The direction of orientation is at about either 45° or 135° to the direction of the stripe, but adjacent stripes differ from one another in having a different orientation from their neighbours. Hence, the microtubule orientation flips from left to right periodically up to the length of the sample container. In addition to this orientational pattern, a pattern of variations of microtubule concentration is also present and which coincide with the changes in orientation. The microtubule concentration drops by about 30% and then rises again each time the microtubule orientation flips from acute to obtuse or vice versa.

The structure is complicated, for each 0.5-mm stripe also contains within it another series of stripes of about 100 μm separation. These, in their turn contain other sets of stripes of about 20 μm, 5 μm and 1 μm separation. In samples made up in a 15 mm diameter test-tube, an additional level of ordering of several mm arises. These large stripes, in turn, contain the lower levels of organisation already mentioned. Hence, similar types of pattern spontaneously arise over distances ranging from a few microns up to several centimetres. The range of dimension over which these microtubule structures occur is typical of those found in many types of higher organisms. Cells are about 10 μm in size, eggs are often about 1 mm, and a developing mammalian embryo is several centimetres long. Self-organisation also arises when samples are prepared in small containers (50–200 μm) of dimensions comparable to those of cells and embryos.

Experiments, that will not be described here, show that self-organisation contains both reactive and diffusive contributions and arises from reactive processes involving the continual growth and shrinking of individual microtubules. Striped morphologies occur when the microtubules are prepared in upright sample containers (Fig. 5.2, left), but a different pattern, consisting of concentric circles, arises when they are prepared in the same containers lying horizontal, flat down (Fig. 5.2, mid-

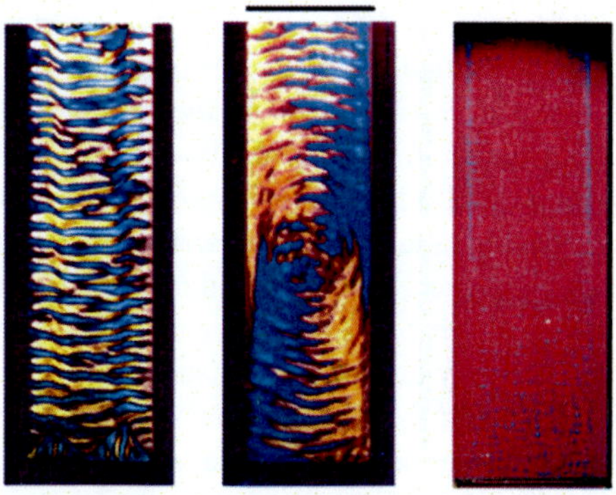

Fig. 5.2 Self-organised microtubule structures formed in spectrophotometer cells 4 cm by 1cm by 0.1 cm. In the presence of gravity, the solution spontaneously self-organises. (*left*) Stripes form when the sample container is upright but (*middle*) circles arise when the microtubules are assembled with the container flat. Self-organisation takes about 5 h. The morphology that forms depends on the orientation of the container at a critical moment (6 min), early in the self-organising process, and before any pattern has developed. (*right*) Self-organisation does not occur when microtubules are assembled under conditions of weightlessness for the first 13 min of the process. © J. Tabony

dle). This indicates that gravity in some way intervenes in the self-organising process. Once formed, the structures are stationary and independent of their orientation with respect to gravity [33]. To establish at what moment during self-organisation the sample morphology depended on the gravity direction, we carried out the following simple experiment [31]. Twenty samples of purified tubulin in the presence of GTP and at 4°C, were placed vertical, then simultaneously warmed to 35°C so as to instigate microtubule formation. Consecutive cells were then turned from vertical to horizontal at intervals of one minute, and the samples examined 12 h later after the structures had formed. Twenty minutes after instigating microtubule formation, when the last sample was rotated from vertical to horizontal, there are no obvious signs of any striped structure. Since the structures form while the cells are flat, one might expect that they would all form the horizontal pattern. This is the case for samples turned during the first few minutes. However, samples which were upright for six minutes or more all formed striped morphologies similar to preparations that remained vertical all the time. The final morphology depends upon whether the sample container was horizontal or vertical, at a critical time six minutes after instigating assembly, early in the self-organising process. This can be described as a bifurcation between pathways leading to two different morphological states, and in which the direction of the sample with respect to gravity determines the morphology that subsequently forms.

A question which arises is what would be the result of an experiment in which gravity was absent at the bifurcation time. To answer this question, we carried out an experiment under conditions of weightlessness, produced in a free-falling rocket of the European Space Agency, for the first 13 min of the process [25]. We found that, contrary to the reference samples assembled on an onboard-1g centrifuge, samples assembled under conditions of weightlessness, did not self-organise (Fig. 5.2, right).

This result shows that under the conditions used in this experiment, the presence of gravity at the bifurcation time actually triggers self-organisation. To study the effects of weightlessness, it is not necessary to go to the trouble, expense, and risk-to-life, of carrying out experiments in space. Gravity effects can also be substantially reduced in ground-based laboratories using simple inexpensive methods, such as clinorotation and magnetic levitation. We also carried out experiments using these methods and observed a behaviour very close to that observed in space-flight.

In far-from-equilibrium systems which self-organise, bifurcations are associated with an instability in the initially homogenous state. When self-organisation arises from a chemical processes, as in the present case, then this instability will involve reactive elements. For the microtubule case, we would hence expect a chemical instability to occur close to the bifurcation time, involving the relative concentrations of microtubules and free tubulin. This is the case. Frequently, the kinetics of microtubule self-assembly, after an initial increase due to the formation of microtubules from the tubulin solution, remains at a stationary level. In general, microtubule solutions showing this type of behaviour do not self-organise. However, microtubule preparations that do self-organise do not show this type of assembly kinetics. Instead, after an initial rapid increase corresponding to the formation of microtubules from tubulin, the microtubule concentration shows an overshoot and progressively decreases over the next 30 min to a value about 20% lower than the maximum. The maximum in the microtubule concentration occurs approximately six minutes after instigating microtubule assembly, and coincides with the bifurcation time when self-organisation is determined by gravity.

Microtubule self-organisation is not just dependent upon the presence of gravity at an early critical moment, other experiments show that it also depends on other weak external factors, such as magnetic fields, sheering and weak vibrations. These experiments strongly suggest that any factor, which at the bifurcation time, leads to a privileged direction of microtubule orientation, will trigger self-organisation; and this observation gives us an important clue to the molecular mechanism by which self-organisation comes about in this system. Microtubules are continually growing from one end and shrinking from the other. For appropriate values of reaction dynamics, the shrinking end of a microtubule will leave behind itself a chemical trail of high tubulin-GDP concentration. Excess GTP in the reaction mixture then converts tubulin-GTP back to tubulin-GTP. At this point, the tubulin-GTP is again available either to be incorporated in the growing end of a neighbouring microtubule, or to nucleate with other tubulin-GTP molecules to form a new microtubule. During this time, the tubulin freely diffuses into the surrounding solution. Likewise, growing microtubule ends produce regions depleted in tubulin-GTP. Because reaction rates increase with increasing concentration, neighbouring microtubules will preferentially grow into regions of high tubulin-GTP concentration whilst avoiding those of low concentration. We postulated that for appropriate reaction dynamics, the chemical trails produced by individual microtubules, can modify and determine the direction of growth of their neighbours. Thus neighbouring microtubules will "talk to each other" by depleting and accentuating the local concentration of active chemical. Under such circumstances, the coupling of reaction with diffusion will progressively lead to macroscopic variations in microtubule orientation and concentration. The chemical trails formed in this way play a self-organising role similar to those of the chemical trails laid down by ants.

When the microtubules first form from the tubulin solution, they are in a phase of growth and are distributed uniformly through the solution in an isotropic manner. At this stage, there is almost no disassembly from their shrinking ends. However, the rapid initial growth of the microtubules depletes the concentration of free tubulin in solution and this in turn provokes the partial disassembly of the microtubules. This partial disassembly manifests itself as the "overshoot" in the assembly kinetics. When partial disassembly starts to occur, just prior to the bifurcation time, it leads to the formation of the chemical trails outlined above. The isotropic arrangement of microtubules is now unstable, for at this time, orienting just a few microtubules will induce their neighbours to grow along the same orientation. Once some microtubules have take up a specific orientation, then neighbouring microtubules will also grow into the same direction. Orientational order will then spread from neighbour to neighbour, and so on. The process mutually reinforces itself with time and leads to self-organisation. Hence, in agreement with experiments, any small factor that at the instability (bifurcation time) directly orients microtubules, or leads to a privileged direction of microtubule growth, will trigger self-organisation.

To investigate whether such an explanation is realistic we carried out computer simulations, incorporating experimentally realistic microtubule reaction dynamics, of a population of growing and shrinking microtubules [9]. Simulations involving just a few microtubules, demonstrated both the formation of the tubulin trails outlined above and the promotion of the growth of neighbouring microtubules into them, along their direction. When the simulations were extended to a population of about 10^4 microtubules on a two-dimensional reaction space, 100 μm by 100 μm, then after 2–3 h of reaction time, a self-organised structure comprised of regular bands of about 5 μm separation developed (Fig. 5.3). This structure is comparable with the experimental self-organised structure that arises over a similar distance scale. In addition, the simulations also predict an "overshoot" in the microtubule assembly kinetics and which coincides with the development of strong fluctuations of concentration and density (3%). For self-organisation to occur, the algorithm also required the presence, at a critical moment early on, of a small asymmetry in the reaction-diffusion process, which either orients some of the microtubules, or makes tubulin diffusion faster along one direction than the others. The latter favours the growth of microtubules along this direction and thus triggers self-organisation. Gravity, by interacting with these density fluctuations, gives rise to increased molecular transport along the vertical, and so triggers self-organisation by this indirect orientational effect. Magnetic fields and sheering, on the other hand, act by directly orienting microtubules at the bifurcation time.

Gravity and magnetic fields break the symmetry of the initially homogenous state and lead to the emergence of form and pattern. Such processes may have played a role in the development of life on earth. Gravity and magnetic fields can thus intervene in a fundamental cellular process and will indirectly affect other cellular processes that are in their turn dependent upon microtubule self-organisation. Other

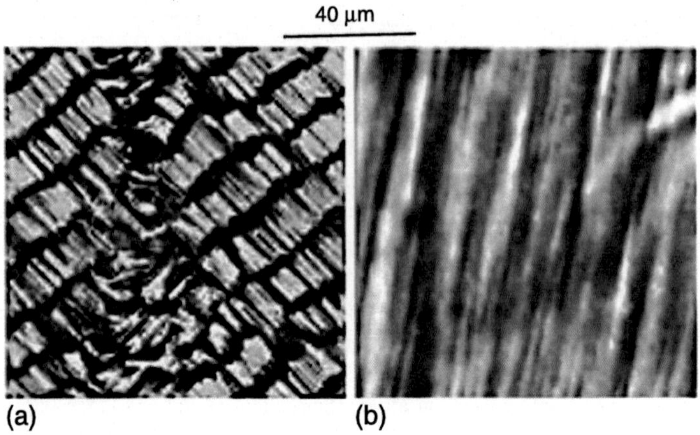

Fig. 5.3 **(a)** Numerical simulations containing only reactive and diffusive terms predict macroscopic self-organisation comparable with **(b)** experiment. © J. Tabony

external factors, such as vibrations, have the same effect. Processes of this type could form a general type of mechanism by which outside environmental factors are transduced into living systems.

Microtubule self-organisation shows strong analogies with the manner by which ants self-organise. On the contrary, there are significant differences from the type of reaction-diffusion scheme originally proposed by Turing. In the Turing system, the molecules communicate with one another by diffusion (fast diffusion of the inhibitor and slow diffusion of the activator). In the microtubule system, on the other hand, as for ants, communication occurs essentially by way of the chemical trails that the microtubules produce by their own reactivity. It is a reaction-diffusion system, since without tubulin diffusion at the appropriate rate, self-organisation would not occur. Another difference with the Turing scheme is the reactive anisotropy and heterogeneity of the microtubule system. In a normal reaction-diffusion scheme, there is no inherent anisotropy in the reactive process. This is not the case for an individual microtubule, in which reactive growing and shrinking can only lead to chemical trails along one specific direction. The system has an in-built propensity for symmetry breaking under the effect of a weak external factor. In addition, in a microtubule preparation, chemical reactions can only occur at the ends of individual microtubules, and these ends are often several microns apart. The solution, once microtubules have assembled, is hence chemically heterogeneous; and this factor also favours self-organisation. It may be that the specific type of reaction-diffusion mechanism encountered here, based on reactive growth and shortening of tubes or rods, is a mechanism particularly suited to self-organisation. At this stage, it is not clear whether these processes are widespread in biology, or if they are limited to microtubules.

The overall phenomenological behaviour of the microtubule preparations shows a qualitative resemblance to some aspects of living organisms in the following ways. Firstly, macroscopic ordering appears spontaneously from an initially homogenous starting point. Secondly, the final state depends upon small differences in conditions at a critical moment at an early stage in the process. This is reminiscent of what occurs during biological development, when after a certain stage, cells of identical genetic content take different developmental pathways to form different cell types. Just after bifurcating, a non-linear system could be described in biological vocabulary as being "determined but not yet differentiated".

Rashevsky, Turing, Prigogine and co-workers, first developed their theories as a possible underlying physical-chemical explanation for biological self-organisation during embryogenesis. They predicted a way by which macroscopic chemical patterns could spontaneously develop from an initially unstructured egg. The results obtained on the in vitro microtubule system demonstrate that reaction-diffusion processes involving biochemical reactions can result in self-organisation. The question obviously arises as to whether these processes also occur in vivo; in particular do they occur during embryogenesis and during the cell cycle.

Biologists have long known that the cellular self-organising behaviour of microtubules arises from their reaction dynamics. One of the characteristic properties of microtubule self-organisation by reaction diffusion is its dependence on various external factors such as gravity. It is established that cellular functions are modified when cells are cultured under conditions of weightlessness. Moreover, recent experiments under conditions of weightlessness show a disorganised microtubule network compared to control experiments under normal gravity conditions. This behaviour is consistent with the in vitro observations reported here and raises the possibility that the processes outlined above might occur also in the cell. Likewise, there is evidence that microtubule self-organisation by reaction and diffusion occurs during embryogenesis. At the moment, however, it is too early to affirm whether or not this process plays a role in determining the body plan of the resulting organism. What we can say is that the non-linear reaction dynamics can in principle account for biological self-organisation, and that an important cellular component, microtubules, behaves this way in a test tube.

The microtubule results demonstrate how a very simple biological system comprised initially of just a protein and GTP, and without DNA, can a show a *complex* behaviour reminiscent of certain aspects of living systems. These phenomena, which are of considerable biological importance, appear simply as a consequence of non-linear reaction dynamics.

References

1. Belousov B.P. (1958) A periodic reaction and its mechanism, in *Sbornik Referatov po Radiatsonno Meditsine 1958*, Medgiz (Moscow), pp. 145–147 (in Russian).
2. Bénard H. (1900) Les tourbillons cellulaires dans une nappe liquide, *Rev. Gen. Sci. Pure Appl.* **11**, 1261–1271 and 1309–1328.
3. Bray W.C. (1921) A periodic reaction in homogenous solution and its relation to catalysis, *J. Am. Chem. Soc.* **43**, 1262–1267.
4. Camazine S. (2001) *Self-organization in biological systems*, Princeton University Press (Princeton).

5. Castets V.V. et al. (1990) Experimental evidence of a sustained standing Turing-type nonequilibrium chemical pattern, *Phys. Rev. Lett.*, **64**, 2953–2956.
6. Coveney P. and Highfield R. (1990) *The arrow of time : a voyage through science to solve time's greatest mystery*, W.H. Allen (London).
7. Coveney P. and Highfield R. (1995) *Frontiers of complexity : the search for order in a chaotic world*, Fawcett Columbine (New York).
8. Dupont G. and Goldbeter A. (1992), Oscillations and waves of cytosolic calcium: insights from theoretical models, *Bioessays* **14**, 485–493.
9. Glade N., Demongeot J., and Tabony J. (2002) Numerical simulations of microtubule self-organisation by reaction and diffusion, *Acta Biotheoretica* **50**, 239–268.
10. Glansdorff P. and Prigogine I. (1971) *Thermodynamic theory of structure, stability and fluctuations*, Wiley-Interscience (London).
11. Gleick J. (1987) *Chaos*, Penguin Books USA (New York).
12. Harrison L.G. (1993) *Kinetic theory of living pattern*, Cambridge University Press (New York).
13. Hodges A. (1983) *Alan Turing : the enigma*, Simon & Schuster (New York).
14. Kauffman S.A. (1993) *The origins of order : self-organization and selection in evolution*, Oxford University Press (New York).
15. Kolmogorov A., Petrovsky L., and Piskunov N. (1937) An investigation of the diffusion equation combined with an increase in mass and its application to a biological problem. *Bull. Uni. Moscow. Ser. Int. A1*, **6**, 1–26, in Russian.
16. Kondepudi D.K. and Prigogine I. (1981) Sensitivity of non-equilibrium systems, *Physica A* **107**, 1–24.
17. Lechleiter J. et al. (1991) Spiral calcium wave propagation and annihilation in *Xenopus laevis* oocytes, *Science* **252**, 123–126.
18. Liesegang R. (1896) Ueber einige Eigenschaften von Gallerten, *Naturwissenschaftliche Wochenschrift* **11**, 353–362, in German.
19. Lotka A. (1910) Contribution to the theory of periodic reactions, *J. Phys. Chem.* **14**, 271–274.
20. Lotka A. (1925) *Elements of Physical Biology*, Williams & Wilkins (Baltimore).
21. Meinhardt H. (1982) *Models of biological pattern formation*, Academic Press (New York).
22. Murray J.D. (2002) *Mathematical biology*, 3rd edition, Springer-Verlag (New York).
23. Nicolis G. and Prigogine I. (1977) *Self-organization in nonequilibrium systems : from dissipative structures to order through fluctuations*, Wiley (New York).
24. Nicolis G. and Prigogine I. (1989) *Exploring complexity : an introduction*, W.H. Freeman (New York), p. 313.
25. Papaseit C., Pochon N., and Tabony, J. (2000) Microtubule self-organization is gravity-dependent, *Proc. Natl. Acad. Sci. USA* **97**, 8364–8368.
26. Pirollet F. et al. (1987) An oscillatory mode for microtubule assembly, *EMBO J.* **6**, 3247–3252.
27. Prigogine I. and Stengers I. (1894) *Order out of chaos : man's new dialogue with nature*, Bantam Books (New York).
28. Rashevsky N. (1940) An approach to the mathematical biophysics of biological self-regulation and of cell polarity, *Bull. Math. Biophys.* **2**, 15–25.
29. Rashevsky N. (1948) *Mathematical biophysics*, Univ. of Chicago Press (Chicago).
30. Rayleigh L. (1916) On convective currents in a horizontal layer of fluid when the higher temparture is on the under side, *Phil. Mag.* **32**, 529.
31. Tabony J. (1994) Morphological bifurcations involving reaction-diffusion processes during microtubule formation, *Science* **264**, 245–248.
32. Tabony J. and Job D. (1990) Spatial structures in microtubular solutions requiring a sustained energy source, *Nature* **346**, 448–451.
33. Tabony J. and Job D. (1992) Gravitational symmetry breaking in microtubular dissipative structures, *Proc. Natl. Acad. Sci. USA* **89**, 6948–6952.
34. Taylor G.I. (1923) Stability of a viscous fluid contained between two rotating cylinders, *Phil. Trans. Roy. Soc. A* **223**, 289–343.

35. Thompson, sir D'Arcy W. (1917) *On growth and form*, Cambridge University Press (Cambridge).
36. Turing, A.M. (1952) The chemical basis of morphogenesis, *Phil. Trans. Roy. Soc.* **237**, 37–72.
37. Velarde M. and Normand, C. (1980) Convection, *Sci. Am.* **243**, 78–93.
38. Zaikin A.N. and Zhabotinsky A.M. (1970) Concentration wave propagation in two-dimensional liquid phase self-oscillating system, *Nature* **225**, 535–537.
39. Zhabotinsky A.M. (1964) Periodical oxidation of malonic acid in solution (a study of the Belousov reaction kinetics), *Biofizika* **9**, 306–311 (in Russian).
40. Zhabotinsky A.M. (1985) The early period of systematic studies of oscillations and waves in chemical systems, in *Oscillations and travelling waves in chemical systems*, edited by R. Field and M. Burger, Wiley (New York), pp. 1–6.

Chapter 6
Dunes, the Collective Behaviour of Wind and Sand, or: Are Dunes Living Beings?

Stéphane Douady and Pascal Hersen

6.1 Discovery

According to the dictionary, a dune is *"a mound or ridge of wind-blown sand"*. But the shape of a dune is more than just a "mound", even if it is often difficult to interpret the contours at first glance. Expanses of dunes are often compared to the sea. Firstly because of the multitude of crests, so similar to waves. But also because, just like waves, we sense intuitively that sand dunes are in movement, and it is their movement that determines their form. For if you look closer, the shape of dunes is not so random ... But how can we describe this shape, understand it, determine its origin and dynamics? How can we bring to light the logic that links these forms to the countless grains of sand swept up by the wind, or understand what it is that gives this collective phenomenon its coherence?

When the wind blows hard enough, it can whip up the sand, carrying it some distance and then dropping it again, ultimately forming those great mounds of sand. When described like this, the shape of the dune would appear to be controlled entirely by the wind. But at the same time, these dunes are big enough to affect the wind itself. The wind carries the sand, but the pile of sand disturbs the wind. So we are dealing with a coupled system of action and reaction, for which we cannot, as a rule, predict the outcome – in this case the shape of the dune – a priori.

6.2 The Wind Drives the Sand ... Which Steals the Wind's Force as It Flies

Because of friction, the wind weakens as it nears the ground. But if it is strong enough, it will still have enough speed left either to push the grains of sand or, in avoiding them, to create a depression above them (through the Bernoulli

S. Douady (✉)
CNRS research director at the Complex Matter and Systems Laboratory
(UMR 7057 CNRS–Paris-Diderot). University of Paris-Diderot Paris, France
e-mail: douady@lps.ens.fr

P. Bourgine, A. Lesne (eds.), *Morphogenesis*, DOI 10.1007/978-3-642-13174-5_6,

effect[1]), that will suck them up. When the grains are sucked up, they rise and find themselves in a stronger wind that carries them along and increases their speed. When these grains fall back down under the effect of gravity, they hit others, pushing them forwards and throwing some of them up into the wind ... The process is therefore rapidly self-multiplying; a few grains of sand flying high take a lot of energy from the wind, and when they fall back to the ground they cause a lot of other grains to roll and fly.[2] However, there is a limit to the process, because as the sand takes energy from the wind, the wind speed falls. An equilibrium becomes established, because there is a maximum flow of sand for any given wind force [1].

But this flow of sand takes a certain time, or a certain distance, to reach equilibrium. If we assume that quite a strong wind (in the absence of sand) moves from hard ground onto a sand-covered surface, then the first grains will start to move; but it will take a certain distance for the wind and the bounce to get the maximum flow of grains into motion. Although the details of the movement of the grains of sand (their jumps and bounces) and their interaction with the wind are complicated, we can nevertheless gain a simple idea of this distance, which we call the *"saturation length"*. Roughly speaking, it corresponds to a few times the distance needed for a grain of sand released in the air to acquire the same horizontal speed as the wind. By balancing the weight of the grain, which pulls it towards the ground – proportional to $\rho_s d^3$, if d is the average diameter of a grain and ρ_s its density – and the force of the wind that is driving it – proportional to $\rho_a d^2$ where ρ_a is the air density – then we find a saturation length proportional to $(\rho_s/\rho_a)d$. This distance may seem tiny, for d is of the order of a few tenths of a millimetre, but as the density of the sand is typically 2,000 times that of the air, the saturation length is easily of the order of a metre.

6.3 The Minimal Dune

Geomorphologists – specialists in the study of land forms – have observed that there is a minimum size for sand dunes [4]. We all know the ripples formed by the wind on the surface of dunes or simply on the beach. But although they are also caused by the wind, these forms are different to dunes. They are formed solely by the impact of falling grains, and they are not large enough to affect the wind in return. These ripples are just a sub-structure due to the transport of grains by saltation (small bounds). For this reason, there is a distinct separation between the biggest ripples

---

[1] The Bernoulli effect is an expression of the conservation of energy. When the wind becomes more concentrated, for example in a tube that gets narrower or in avoiding an obstacle, the kinetic energy (wind speed) increases, and the potential energy (pressure) decreases.

[2] The fact that the grains dislodge others when they fall makes the process "sub-critical": if the wind is just below the threshold needed to suck up some grains of sand, the process can still be triggered by throwing a few grains into the wind. The wind must then fall to an even lower value before the process of collision and ejection will stop. The phenomenon is therefore said to be sub-critical, or that it presents a hysteresis.

and the smallest dunes, showing clearly that this is neither the same phenomenon nor the same mechanism.

The saturation length helps explain the existence of a minimum size for dunes [3, 11]. If we consider the windward side of a dune (the backslope), the wind accelerates up the slope to avoid the dune. While accelerating, it can pick up more sand, and so it becomes more loaded. This results in erosion of the back of the dune (the windward side). When the wind gets to the top of the dune, it has "more space", and so it slows down. Consequently, it cannot carry as much sand as it did over the backslope of the dune. If the dune is larger than the saturation length, the wind on the top of the dune will already be more or less saturated. When it slows down, it therefore finds itself carrying too much sand, and some of the grains will be deposited. So if the dune is large enough, the sand eroded from the back is deposited on the leeward side, and the dune advances without losing any sand.

When the backslope of the dune is too short, on the other hand, the wind does not have time to pick up all the sand that it could. When it slows down on the crest of the dune, it may not have reached saturation point, or only just. Consequently, it will deposit little or none of the sand it is carrying on the lee side of the dune. The dune loses sand, and starts to melt away. As it shrinks, it loses even more sand, and ends up disappearing. Small dunes are never observed to last very long.

6.4 The Wind Runs Over the Dune ... and Pushes It Along

With relatively large dunes, the wind no longer follows the form of the dune, but carries straight on, by inertia. This results in the formation of an eddy or recirculation zone, and a slipface. Once the wind has come off the dune, the grains fall in a zone with little or no wind, the eddy. The sand settles, and a drift forms on the top of the dune. When the gradient of this drift exceeds the gradient of the dune face – about 33 degrees – the drift slides down the face. For steeper gradients, the drift is unstable and forms an avalanche. This is how an active face, the slipface, is formed in larger dunes. On the contrary, for smaller dunes, close to the minimum size, the wind continues to follow the contours of the dune and no slipface is formed.

But in both cases, by eroding the sand from the backslope and depositing it on the lee side, the wind moves the whole dune forwards, in a sort of rolling movement. Each grain of sand moves from the back of the dune to the crest and then to the slipface, before being buried under the dune, reappearing much later to start a new cycle.

6.5 Does the Wind Flow Make the Dune?

It is obvious that the wind, which picks up, moves and deposits the sand, is essential to the formation of dunes. It is therefore natural that the shape of the dunes should depend on the type of wind flow. Since the great explorers like Bagnold, and more recently with the growing use of satellite observations, the basic dune forms have been recorded and linked to dominant wind flows by geomorphologists [4, 5, 12].

When there is little sand on hard ground, and the wind always blows in the same direction, a crescent-shaped form can be observed, with the horns pointing in the direction of wind movement and the slipface in the hollow between them. This is known as a *"barchan"* (Fig. 6.1) [2, 3, 14]. This is a fascinating form, because it is isolated, individualised. The edges of the dune can be clearly distinguished against the hard, sand-free ground around it.

When there is more sand, the barchans move closer together, the slipfaces meet up and the crests align in one long ridge perpendicular to the wind. This creates "transverse" dunes (Fig. 6.2, top and middle). Constant wind flows are quite rare and are found near the tropics, with the trade winds, for example.

More often, the wind turns from one season to the next (summer and winter, for example), thus presenting two dominant directions over the course of the year. In this case, another type of dune appears, stretching out in the average direction of the wind, along which the crest oscillates. In each season, small slipfaces form on one side of the dune, to disappear and reappear on the other side when the wind changes during the next season. These "longitudinal" dunes are quite high and often very regularly spaced out. This is the most common type of dune (Fig. 6.2, bottom).

Lastly, when the wind blows in several different directions during different periods (seasons, years), we can observe another system of dunes, "star" dunes, with a central peak and multiple arms (Fig. 6.3). They are globally immobile, as there is no dominant wind direction. Each directional component of the wind generates an oscillating crest that forms one arm of the star. These are the biggest dunes, easily several hundred metres high.

If there is a well-established link between wind flow and dune form, how does it operate? Does the wind directly control the shape of the dune, or is the shape of the dune not rather an adaptive response of the mound of sand to the wind?

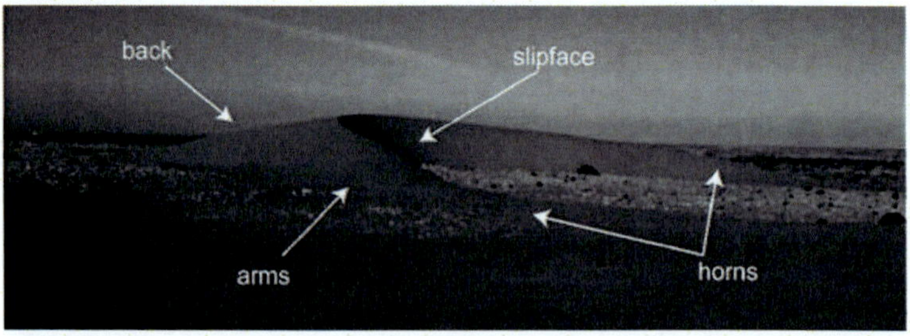

Fig. 6.1 Barchan: an isolated, crescent-shaped dune on hard ground, which forms under constant wind conditions. Note the use of terms borrowed from animal or even human anatomy to describe the different parts of the dune. © S. Douady and P. Hersen

Fig. 6.2 Transverse dunes (*top and middle*) and longitudinal dunes (*bottom*). The very movement of the dune invites an animalistic description, and the way it oscillates in the direction of the wind evokes the adjective "snaking". *Top*: © S. Douady and P. Hersen. *Middle*: © Goodshot. *Bottom*: © NASA/USGS/E Tad Nichols

6.6 Understanding the Barchan Shape

Before embarking on the analysis of complex forms, let us start with the easiest case, that of the barchan. As we have said, it has the advantage of being a simple shape, and what is even better, an isolated shape, which we should be able to consider individually, in itself, independent from the rest of the environment. As it is the wind that drives the sand, and the wind has a limited transport capacity, the speed of

Fig. 6.3 Star dune: a multitude of "arms", which sometimes subdivide, come together to form a star shape. *Top*: © S. Douady and P. Hersen. *Bottom*: © NASA/USGS/US Department of Defense

a dune depends very much on its size. What is more, as the dunes all have roughly the same shape, whatever their size, the wind does not accelerate any faster on a large dune than it does on a small one. So we can see that it takes a lot more time to move a large dune than it does to move a small one. More precisely, the speed of a dune is inversely proportional to its height [4, 5, 12].

This result, established for an individual dune, can be used a priori for the different parts of the dune, if we assume that the wind and sand continue to move forwards without being much affected by the form. In a pile of sand, of whatever shape, the edges are thinner than the central part, and so they will advance faster. An ordinary mound of sand will therefore immediately start stretching out to form a crescent, with the horns pointing downwind.

In itself, therefore, this crescent shape is not surprising. What is surprising is that the dune actually continues to move forwards without changing shape, as an individual. If we follow the reasoning outlined above, the horns should continue to move faster, and the crescent should stretch out more and more, becoming incredibly elongated. And yet barchans keep the same overall shape, with their length remaining roughly equal to their height. How can we explain this?

The preservation of the global form expresses a coherence of behaviour, and therefore a linkage between the different parts of the dune. This linkage derives from the fact that the flow of sand does not follow perfectly the direction of the wind: like the wind, it is sensitive to the shape of the dune. If one part of the dune is further forwards, then the backslope does not remain level; it slopes towards the part that is more advanced. The flow of sand is then quite simply diverted (by gravity) towards the side. This extra flow feeds more sand into the wind blowing over the sides of the dune (Fig. 6.4). Consequently, the wind here is closer to its saturation point, so it erodes less sand from this part of the dune. The erosive power of the wind on this part of the dune is diminished, and the movement of this part slows down.

So that is how the "logic" of the shape of the barchan operates: without lateral flow, the sides, which are thinner, advance more quickly and form the crescent shape. But when they move too far forwards, an ever larger flow of sand is diverted from the middle to the sides, limited their erosion and slowing down their progress [6]. The form is self-regulating, with the sides remaining far enough forwards to receive enough lateral flow to advance at the same speed as the centre. This logic is quite particular, deriving from the influence of the dune's shape on the dune itself, and not from the direct influence of the wind, unlike ripples of sand. It also explains how barchans aggregate into transverse dunes, because this linkage prevents one part from advancing too fast in relation to the rest.

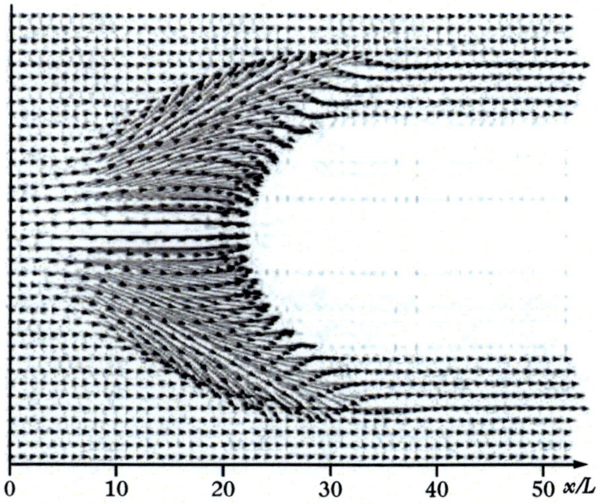

Fig. 6.4 Flow of sand on a barchan. The sand is trapped in the slipface, but diverted from the back towards the arms, where some of it escapes through the horns

6.7 The Paradox of Corridors ... or the Problem of Dunes Among Themselves

This may explain the form of the barchan and its overall behaviour, but in fact it leads us to a paradox. Much of the sand of the dune, pushed by the wind, finds itself trapped on the slipface. But the sides of the dune are necessarily shorter than the saturation length, and so they lose sand. All the more so since some of the sand is diverted onto the sides, towards the horns.

An isolated barchan should therefore lose sand, and slowly but surely melt away. To explain the paradox, we must now consider the environment of the dune. Like any "individual", the dune can only survive with the help of external inputs. In the present case, the incoming flow of sand comes from beaches and other sources of coarse sand, or losses from other dunes (collective recycling). Fortunately for the dunes, lateral sand losses do not increase very much with the size of the dune. If we assume that the flow of sand from outside sources is relatively homogeneous, which happens quite fast with the disordered rebounding of grains of sand on hard ground, a large enough dune can capture enough sand to offset its losses, and thus survive. As in any living system, equilibrium is only established through the constant interplay of gains and losses.

But this situation is unstable. If the dune is bigger than its equilibrium size, it captures more sand than it loses and grows even bigger. As a result, it captures even more sand, and continues to grow indefinitely. Conversely, a dune that is too small loses more sand than it can recover, and therefore becomes even smaller. As a result, it captures even less sand, and quickly disappears.

According to this line of reasoning, we should only observe bigger and bigger barchans in the desert, ending up fairly quickly with just one giant barchan capturing all the sand and hardly moving at all [10]. But this is not the case. On the contrary, the barchans we observe nearly all have the same size, over hundreds of kilometres ...

6.8 The Wind is Never Constant

To explain this, we must once again return to the environment and its interactions with the dunes. Winds are never really constant. Even the trade winds weaken in winter. And in winter there are storms. Storm winds are very powerful and do not blow in the usual direction of the wind. While barchans have a very aerodynamic form, well-adapted to "sliding" through the normal wind, the storm winds literally rub them up the wrong way. The flow of sand carried away by the wind is enormous, and the dune can lose a lot in a short time. It will re-adapt to the new wind, but it sometimes loses much sand in the process, for example by losing an arm that immediately moves away as a new little barchan [7]. The same phenomenon is repeated when the wind returns to its normal direction. The details of the dune's reaction depend on its inertia (its size) and the duration of the disturbance in relation to its internal response time. But one is certain: if the dunes are large enough to fatten peacefully during the summer, the winter storms will subject them to a slimming diet.

6.9 Dunes are Not Isolated

By considering barchans as isolated individuals, we have also neglected the interactions between dunes, other than those consisting simply in transfers of sand. Now, we have seen that small dunes move faster than big ones. If there are different-sized dunes in a corridor, then the faster, smaller ones will necessarily catch up and collide with the larger, slower ones. This is what we call dune collisions [8, 13]. Up until now, given the difficulty in observing this kind of event in the field (which can take several years), it has been generally accepted that the small dune merges into the large one, which absorbs it. But this fails to take into account the eddy that forms on the lee side of the dune. If the impacting dune is large enough, the eddy may be strong enough to destabilise the "target" dune, and even cut one of its arms into one or more pieces – which of course immediately move off in the shape of little barchans. The target dune may quite simply lose a large part of itself in this way, before the impacting dune joins it. So even if the impacting dune merges well with the target dune, the latter may have lost so much sand that the net result is negative, in other words the target dune ends up smaller than it was before the collision (Fig. 6.5).

Although field observations are possible (thanks to aerial monitoring over several years, or even decades), these phenomena are above all studied by simulating dunes and their dynamics, either digitally or analogically, creating miniature dunes under water. By using water instead of air, the saturation length can be reduced by a factor of 1,000 (the ratio of water density to air density), and so the dunes (and their dynamics) can be reproduced at a scale of 1 to 1,000 (Fig. 6.5)) [7, 9, 14].

Ultimately, we can understand why all the dunes remain roughly the same size, if we suppose that they are big enough to keep growing under normal wind conditions

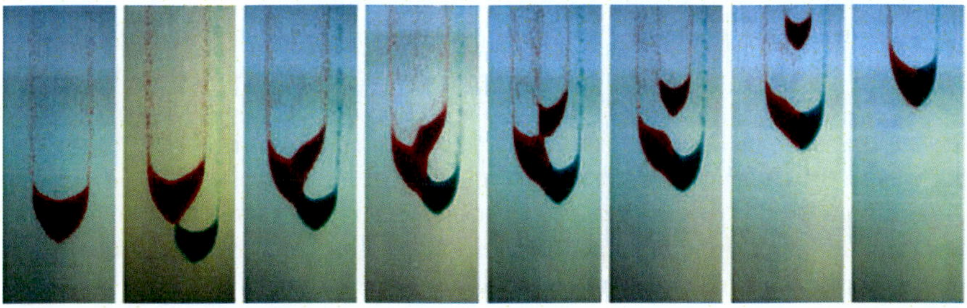

Fig. 6.5 Dune collision. Here we see the sequence of successive events (the wind direction is towards the top of the pictures). The small dune cuts an arm off the large dune. This arm moves off to form a small new dune while the impacting dune joins the remainder of the target dune to recreate a dune of normal appearance [8]. © P. Hersen

but they are then cut down to size by collisions and winter storms. The average size that we observe is then simply the result of a global equilibrium between periods of slow growth and sudden diminution, and it can only be understood over the long term.

6.10 The Grain of Sand, the Dune and the Corridor of Dunes … What About the Individual, the Flows and the Form?

In the end, how should we understand dunes? At first glance, it seems possible to consider the dune as a coherent collective of grains. Although the dune is made up of a gigantic number of grains, and the sand is moved pretty much grain by grain, overall coherence is obtained because the isolated movement of grains is largely governed by the very result of that coherence, i.e. the shape of the dune. It is the shape of the dune that diverts the movement of grains according to its humps and hollows. The shape of the dune is therefore both the result and the controlling factor of the movement of grains. The wind clearly does not control the form, it simply provides the driving force and the external constraint under which the form evolves. There is no control of the form by the grains of sand themselves: the form is self-controlled, or self-organised.

By treating the dune as an individual, we are immediately drawn into the dialectic entailed by the separation between the individual and its environment: survival being entirely dependent on input from the environment, calculating the balance of gains and losses … Fundamentally, the creation of a boundary directly raises the problem of flows across that boundary. Why define an individual, something isolated and indivisible, if it can only survive through exchange with the external environment and if we can only understand its dynamics in terms of these exchanges, or even by the breaking up of the individual into several new individuals?

The same problems arise at the level of dune corridors: how can we calculate the balance of inflows and outflows? How can we understand the very existence of well-defined and delineated corridors hundreds of kilometres long, their appearance and disappearance, their shared characteristics and differences?

At the level of the grain, on the contrary, one might think that there is no boundary and that the problems disappear. There are only movements that are faster on hard ground than on large deposits of sand, rapid movement with the wind and long periods trapped inside those deposits. But this description is also based on individuals: the grains. And as one might imagine, the same questions of flow arise. The grains are eroded by the impacts they sustain during transport, and we must take into account the resulting dust that escapes; they might be covered by immovable deposits inside the dune, or again they might tear up new grains of sand from stony ground …

Above all, with this small-scale description, we lose the whole idea of organisation, of coherence, that appears at larger scales, and so we lose the notion of form – that of "the dune", that of the "dune corridor"... Yet the form is the outward manifestation of the organisation of flows, and at the same time, as we have seen, the thing that organises them. The mobile form provides the solution for avoiding the problems of individual delineation without losing the small-scale information. One can seek to describe and even explain these fascinating temporary coherences in terms of evolving, self-organised forms. This also avoids the problem of control of the form (how and by what?). In the end, the form should perhaps not be considered as the mysterious privilege of the individual, the definition of which is both practical and problematic, but as the result of and the driving force behind the coherence of flows (see Figs. 6.6 and 6.7).

In the light of their behaviour, some of our colleague biologists have suggested that dunes have all the characteristics of living beings: coherence, well-defined forms, dynamic equilibrium, dependence on the external environment, meeting to generate new individuals. And indeed, not only does the description appear to be the same, but also the problems it raises! Does that mean that the solution could also be the same?

Fig. 6.6 Landscape in the Rub Al Khali (Arabia). Where is the individual, between the grains of sand, the small isolated barchans, which gather into waves of transverse dunes, and finally form the megadune? There is just a continuity of forms, on ever larger scales, with the flow of sand creating the forms and the forms regulating the flow. © S. Douady

Fig. 6.7 Barchan field. Although the dunes are individual, they appear to mix up easily. © S. Douady

References

1. Anderson R.S., Sorensen M., and Willetts B.B. (1991) A review of recent progress in our understanding of aeolian transport, *Acta Mech. [Suppl]* **1**, 1–19.
2. Andreotti B., Claudin P., and Douady S. (2002) Selection of dune shapes and velocities. Part 1: Dynamics of sand, wind and barchans, *Eur. Phys. J. B* **38**, 341–352.
3. Andreotti B., Claudin P., and Douady S. (2002) Selection of dune shapes and velocities. Part 2: A two-dimensional modelling, *Eur. Phys. J. B* **28**, 321–339.
4. Bagnold R.A. (1954) *The physics of blown sand and desert dunes*, 2nd edition, Chapman (New York).
5. Cooke R., Warren A., and Goudie A. (1993) *Desert Geomorphology*, UCL Press (London).
6. Hersen P. (2004) On the crescentic shape of barchan dune, *Eur. Phys. Jour. B* **37**, 507–514.
7. Hersen P. (2005) Flow effect on the morphology and dynamics of aeolian and subaqueous barchan dunes, *J. Geophys. Res. Earth-Surface* **110**, F04S07.
8. Hersen P. and Douady S. (2005) Collisions of barchan dunes as a mechanism of regulation, *Geophys. Res. Lett.* **32**, L21403.
9. Hersen P., Douady S., and Andreotti B. (2002) Relevant length scale for barchan dunes, *Phys. Rev. Lett.* **89**, 264301.
10. Hersen P., Andersen K.H., Elbelrhiti H., Andreotti B., Claudin P., and Douady S. (2004) Corridors of barchan dunes: stability and size selection, *Phys. Rev. E.* **69** 011304.
11. Kroy K., Sauermann G., and Herrmann H.J. (2002) A minimal model for sand dunes, *Phys. Rev. Lett.* **88**, 054301.
12. Pye K. and Tsoar H. (1990) *Aeolian sand and sand dunes*, Unwin Hyman (London).
13. Schwammle V. and Herrmann H.J. (2003) Solitary wave behavior of sand dunes, *Nature* **426**, 619.
14. Web site: http://www.lps.ens.fr/ hersen/

Chapter 7
Morphodynamics of Secretory Endomembranes

François Képès

7.1 Some Preliminary Reminders

Cellular living beings are divided into two groups, the prokaryotes and the eukaryotes. Prokaryotes are the bacteria, whose little cells contain no real nucleus to enclose their genetic material. Eukaryotes, covering all non-bacterial cellular beings, have cells with nuclei. Among the eukaryotes, we distinguish between organisms that spend most or all of their existence as isolated cells and those that possess a number of cells. The former are unicellular protists, like the amoeba, for example. The latter are multicellular organisms: plants and animals.

For a cell to exist, an interior space must be defined, which means establishing a boundary with the outside environment. The role of boundary is played by the cell's envelope. In bacteria, this envelope comprises two concentric layers. On the outside, a wall that allows the cell to resist explosion from the high internal osmotic pressure. This pressure should give the bacteria a spherical form. But the outer wall also gives the cell a certain rigidity, enabling some bacteria to be rod-shaped. On the inside there is a cell membrane, the plasma membrane, which is mainly responsible for the selectivity of exchanges between the cell and its environment, making it essential to cell viability. Between the outer wall and the cell membrane, the aqueous phase is called the "periplasm". The cytoplasm is the aqueous phase enclosed by the cell membrane. The central part of the cell, the nucleoid, contains the genetic material, in the form of a double helix of DNA combined with proteins. In general, a bacterium possesses four main compartments: the two layers of its envelope and the two aqueous compartments they delineate, the periplasm and the cytoplasm.

As a first approximation, with spherical cells, the size ratio (one dimension) between eukaryotes and prokaryotes is roughly 20 to 1. The surface area ratio (two dimensions) is therefore 400, and the volume ratio (three dimensions) is 8,000. As metabolic activity is proportional to volume and exchanges are proportional to surface area, eukaryotes have a deficit compared to prokaryotes of 8,000/400 i.e. a

F. Képès (✉)
Research director at CNRS; Founding director of the Epigenomics Project, Genopole®;
Associate member of the Research Centre for Applied Epistemology, École Polytechnique
e-mail: Francois.Kepes@epigenomique.genopole.fr

P. Bourgine, A. Lesne (eds.), *Morphogenesis*, DOI 10.1007/978-3-642-13174-5_7,

factor of 20 in their ratio of metabolic activity to membrane exchange area. And yet all metabolic activity is accompanied by a proportionate exchange of matter with the exterior. This apparent paradox is resolved by a considerable increase of membrane exchange areas inside eukaryotic cells. Membrane surfaces of much greater area than the cell membrane form compartments within these cells: they are known as endomembranes.

The envelope of a eukaryotic cell also has two layers, a cell membrane enclosed by a wall. In unicellular eukaryotes, this wall plays a similar role to the outer wall of bacteria. In multicellular beings, the problem of osmotic pressure is less crucial, because most of the cells have other cells as neighbours. However, the aerial part of plants has to raise the sap from the roots, which can represent considerable hydrostatic pressures. In keeping with these needs, the plant cells have highly-structured walls. In animal cells, as far as cell walls are concerned, the cells are surrounded by extracellular matrices, the mechanical characteristics of which depend on the tissue in question, from the fluidity of the blood to the rigidity of bones.

7.2 Introduction

7.2.1 Cell Membrane and Translocation

In today's organisms, the cell membrane simultaneously possesses two characteristics that are essential for protecting and concentrating the cell contents and for allowing the exchanges necessary to all metabolic activity. The membrane must be generally impermeable to most compounds and at the same time specifically permeable to certain compounds. It is reasonable to postulate that a primitive form of membrane would have possessed these characteristics to a lesser degree than today: it would have been less generally impermeable, and less specifically permeable. While these characteristics gradually developed during evolution, one particular mechanism became established very early on, playing a role in specific permeability. This is the export of proteins to outside the cell, proteins that participate in the cell's relations with other cells, its defence or its nutrition. All proteins are synthesised in the cytoplasm by the cellular machinery. Proteins that are exported must therefore pass through at least one membrane – and in practice they never pass through more than one. Given the physical characteristics of membranes, this passage is hardly likely to be a spontaneous phenomenon. For just as the proteins are usually hydrophilic, so the heart of the membrane is hydrophobic. A molecular machinery with ancestral characteristics, that has been preserved throughout the living world, is responsible for this passage through the membrane, which is known as translocation. Among bacteria, this machinery is associated with the cell membrane. The transported protein thus passes directly into the extracellular space, the cell wall providing no real barrier to its diffusion. In eukaryotic cells, the machinery of translocation is associated with the abundant endomembrane known as the "endoplasmic reticulum" (ER). The synthesised protein is not dispatched outside the cell, but into the aqueous phase (the "lumen") of the ER, which is itself distributed through the cell.

7.2.2 Eukaryotic Secretory Pathway

Other steps will then be necessary for the protein to reach the outside of the eukaryotic cell. The protein no longer crosses a biomembrane, but undergoes a sequence of packing and unpacking steps in compartments bounded by endomembranes. These steps leading to the exit of the protein (exocytosis) follow the eukaryotic "secretory pathway"[1], the successive compartments of which constitute the greater part of the endomembranes: ER, Golgi apparatus, secretory granules, and vacuoles or lysosomes (Figs. 7.1 and 7.2). Along the secretory pathway, the molecules in transit arrive in various different destinations and are matured. In particular, many of the proteins are glycosylated (the addition of sugars) and subject to specific cleavages. In the opposite direction, endocytosis is a means of internalising external elements. Endocytosis uses specific compartments, and then possibly compartments that are also used for exocytosis.

The ER is a network of membranous tubes and flattened cisternae, all in one piece. On the one hand, it constitutes the nuclear envelope of eukaryotic cells. On the other, most of the ER adjoins the cell membrane in the form of a meshed network of tubes. ER cisternae run through the cell, linking up these two assemblies [26, 28, 52, 58, 76]. In places, the ER differs by producing portions of compartment

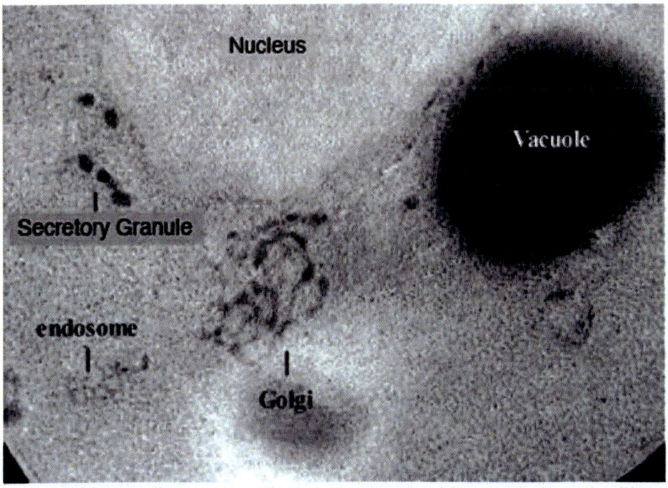

Fig. 7.1 Thick section of a yeast cell. A yeast cell has been fixed with glutaraldehyde and the membranes impregnated with reduced osmium and then lead citrate. The nucleus, devoid of ribosomes, is a lighter shade, while the vacuole is very dark. The Golgi apparatus appears as a network of interconnected tubules with nodal swellings. The cell wall is visible in the two lower corners of the image. Enlargement approximately x 60,000. © A. Rambourg

[1] We use the term secretory pathway in the general case, but etymologically, this term should be limited to the case of proteins secreted in a medium that is indeed extracellular, but nevertheless enclosed within an organ or tissue. In fact, it is equally applied to excreted proteins, or constituent proteins of intracellular secretory compartments.

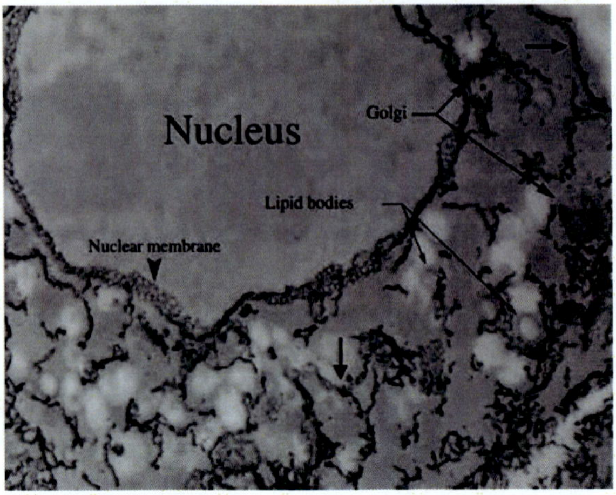

Fig. 7.2 Thick section of a plant cell. A maize root cell has been impregnated with zinc iodide-osmium (ZIO). The endoplasmic reticulum impregnated with osmium (ER, *vertical arrow*) forms a continuous network between the nuclear membrane, perforated with pores, and the ER subjacent to the cell membrane (*horizontal arrow*). The dictyosomes (Golgi), visible in side view (*above*) or in face view (*below*), are often connected to the ER. Lipid bodies, and vacuoles or light plastids, can also be observed. Enlargement approximately x 10,000. © F. Képès

that contain proteins and lipids destined to travel further along the secretory pathway. These differentiations give rise to the Golgi apparatus, named after its discoverer [21]. The morphology of this apparatus varies considerably from one cell type to another. In yeasts, it takes the form of a few dozen apparently distinct elements. Each element is in one piece and of a mainly tubular nature [45, 53, 59]. In plants, the Golgi apparatus comprises a few hundred distinct elements that move along the network formed by the ER. Each element is polarised and in one piece. It is of a mixed nature, being tubular on the periphery and cisternal in the centre [10, 37, 43]. In animals, the Golgi apparatus is generally single and situated close to the nucleus. It often takes the form of a long, interconnected ribbon, along which perforated and compact regions alternate. A cross-section of the ribbon brings to light a remarkable polarised organisation [57]. Among all the eukaryotes, differentiations appear in the Golgi apparatus, giving rise to secretory granules. These secretory granules are subsequently released, and then they fuse with the cell membrane, either as soon as they reach maturity or all simultaneously, in a regulated way (for example the axonal tips of animal neurons). Certain elements travelling through the Golgi apparatus do not go towards the granules, but constitute the vacuole, or its analogue in animals, the lysosomes. These compartments are involved in the destruction and recycling of macromolecules and lipids, and in maintaining the osmotic equilibrium of the cell. This is how they receive their constituent parts, often hydrolytic enzymes (catalysts to break down molecules). Many of the molecules that are broken down in the vacuole or lysosomes by these hydrolytic enzymes arrive there through endocytosis.

7.2.3 Other Eukaryotic Compartments

Apart from its nucleus and secretory compartments, the eukaryotic cell also contains other membrane-bounded compartments, in particular peroxysomes and mitochondria, and the plastids of plant cells. We shall not dwell any longer here on these compartments, the evolutionary origin of which is distinct from that of the secretory endomembranes. Incidentally, given the dynamic interrelations between eukaryotic compartments, counting them would be a perfectly arbitrary operation.

7.2.4 Cytoplasm, Cytoskeleton and Compartmentalisation

The cytoplasm has extremely high viscosity and molecular density, constituting a considerable hindrance to the free diffusion of cell components. This can result in a slowing down that is incompatible with survival for the cell functions that rely either on great rapidity or on the meeting of large molecules or endomembrane compartments. What would our nervous reactivity be like if diffusion was the only means of transporting our neuromediators in membranous vesicles along narrow axons up to a metre long? In this light, we should expect to discover optimising mechanisms to counter the effects of the slowness of free diffusion: microcompartmentalisation of certain functions, and systems of guided, motor-driven transport. Logically, these mechanisms should be more elaborate in large cells than in small ones. And indeed, the intracellular skeleton or cytoskeleton, which allows to actively guide transport, is far more developed in large eukaryotic cells than in bacteria. In addition, in the cytoplasm, and sometimes inside endomembrane compartments of eukaryotic cells, there exist microcompartments that are not bounded by membranes. It is generally accepted that some of these microcompartments emerge locally as a result of the dynamics of a cell function. These are non-equilibrium stationary structures, and they only maintain themselves during their period of activity, when they are crossed by a substantial flow of matter and energy [47]. This can be the case for a multienzyme complex, for example. We shall not go into these microcompartments in any greater detail, but it should be noted that we will encounter endomembrane compartments that also constitute non-equilibrium stationary structures [45].

7.3 Morphodynamics of Membranes

7.3.1 Biological Membranes

All biomembranes contain diverse lipids, usually arranged in a bilayer, which is why we talk of the two "leaflets" of the membrane. They are of heterogeneous composition. In addition to the lipids (phospholipids, sphingolipids and sterols), they also contain non-lipid components: proteins, carbohydrates and peripheral polymers, which are often reticulated (coats, fibres). It is useful to differentiate the three types of heterogeneity that make up the distinguishing characteristic of biomembranes.

Firstly, different membranes have very diverse compositions. Secondly, the two leaflets of a natural lipid bilayer generally have different compositions. Lastly, the chemical composition is heterogeneous along the plane of the membrane. Despite these heterogeneities, biomembranes adopt a limited number of basic morphologies that have been conserved during evolution. Electronic transmission microscopy is the most widely-used method of observing membranes. However, this technique can generate ambiguities, because it produces 2-D projections of 3-D endomembranes. It is nevertheless important to identify these basic morphologies. A significant step in this direction was made when the ubiquity of membrane morphologies with cubic symmetry – possessing the same symmetries as a cube – was recognised [33, 34]. These cubic membrane morphologies had been described in thousands of microscopy studies over the previous thirty-five years, under various names. Representative examples had been drawn from all cell types and all kingdoms, for most types of membranes. Landh [33, 34] showed that they corresponded to 2-D projections of various families of cubic membranes. What is remarkable is that this cubic morphology provides, for a given volume, both a maximum exchange area and a minimum free energy of stabilisation. In addition, the cubic membranes of a given compartment often have the same structure in a wide variety of different species.

In the living cell, biomembranes undergo segregation, fission and fusion. These events are necessary to ensure the differential routing of secretory cargos and containers. We shall now examine the morphogenetic dimension of these three types of event.

7.3.2 Segregation

An endomembrane compartment is defined by its composition, its location and its morphology. These three aspects are obviously interrelated. As far as the composition is concerned, the problem is how to draw the boundaries in a dynamic compartment, complex in both structure and function [64].

Two processes play a role in this context. Firstly, it is necessary to establish molecular gradients all along the secretory pathway, which serve as positional signalling. Secondly, it must be possible to segregate molecular components that are initially mixed together.

7.3.2.1 Positional Gradient

To discuss the protein gradient, we should start by recalling that the proteins travelling through the secretory pathway follow a sequential path from the ER. What distinguishes them is therefore their possible retention in an intermediate compartment or a different routing during the later stages. As their initial path is identical, how does each species of protein "know" where to go or when to stop? Indeed, we can observe a protein gradient from the ER through to the cell membrane. To understand its mechanism, it is important to remember that the boundaries are blurred, in other words the distributions of different proteins along the secretory pathway partly overlap [55, 64]. The protein gradient is generally understood to be the consequence of a fundamental gradient, providing information that is used by the proteins. In function of this fundamental gradient, the proteins position themselves spontaneously, or separate through a process comparable to distillation [2, 8, 79]. The pH of the lumen in each compartment or the thickness of the lipid bilayer are two parameters that have been proposed to explain the fundamental gradient. The degree of protein glycosylation is another possible parameter that has not yet been explored. In any case, *the protein gradient is both a cause and a consequence of the information gradient.* As they both necessarily have a molecular substrate, distinguishing between cause and effect is rather an arbitrary operation. Let us suppose, for example that the pH gradient is informational. The pH decreases along the secretory pathway because proton pumps are active in certain compartments, and not in the first one. But these pumps are membranous proteins preferentially located in these compartments by means of the informational gradient based on the pH. In addition to its location, it is possible that even the activity of a pump may depend on the parameter that it helps to modify. For example, when it is put in place, the proton pump passes through all the compartments, including the first one, but it will only be activated below a certain pH, therefore in a later compartment. In brief, the pH depends on the activity or preferential location of a membranous protein that is activated or positioned according to the pH.

7.3.2.2 Molecular Segregation

Lateral segregation – along the plane of the membrane – of molecular components that are initially mixed up is an elementary process in endomembrane dynamics. One of its effects is to direct the molecules towards different destinations. It is effected by membranous microdomains, like the "sphingolipid rafts" in the "caveolae", which are invaginations in the membrane of animal cells [25, 54]. To give a brief idea of this, let us take the case of sphingolipids that can establish hydrogen bonds with each other, enabling them to cluster together. Phospholipids, which cannot establish such bonds between themselves or with sphingolipids, are excluded from these sphingolipid clusters, leading them to group together by exclusion [70]. These sphingolipid clusters and groups of phospholipids recruit different proteins because of their differential chemical affinities or the different thicknesses of their hydrophobic sections. This is the physical basis of the differential recruitment of proteins based on the thickness of the lipid bilayer. An "intrinsic" membranous protein comprises hydrophilic regions and at least one hydrophobic region that crosses the membrane. To adapt locally to the form of the protein while minimising the changes required, lipids change their length and cross-section, and the membrane changes its form or content (Fig. 7.3a). This physical constraint should not be underestimated. The 3-D tension at the surface of the protein is of the order of the local compression module of the membrane multiplied by the ratio between the thicknesses of the hydrophobic segment of the protein and the membrane. As the local compression module of a membrane is about 300 bars, we can see that a difference of 10% between the two thicknesses generates enormous pressure on the surface of

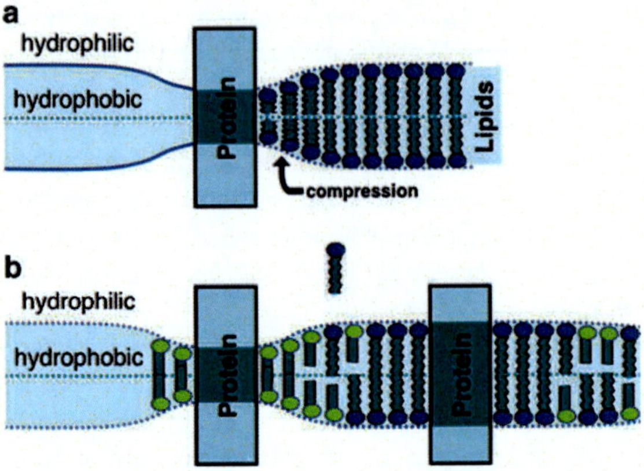

Fig. 7.3 Cross-section of a model membrane. (**a**) Constraints exerted by a protein on a membrane containing one sole type of lipid. Near the protein, the lipids change length and cross-section, and the membrane is deformed. (**b**) A membrane containing several types of lipids adapts more easily to the insertion of diverse proteins. The protein on the left has a short transmembrane hydrophobic domain, to the contrary of the protein on the right. Long lipids tend to segregate around the protein with the long hydrophobic domain, and vice-versa. This is the physical origin of the differential affinity of a protein for certain lipids, an important phenomenon in the lateral segregation of constituents along the plane of the membrane. Light color: hydrophilic; dark color: hydrophobic. © F. Képès

the protein: about 30 bars. This is enough to affect the function and the lifespan of the protein, as has been demonstrated experimentally. It is clear that a biological membrane containing several types of lipids can better adapt to the insertion of diverse proteins that a membrane that only contains one type of lipid (Fig. 7.3b).

Another mode of differentiation along the plane of the membrane exploits the interactions between proteins in a membrane. These interactions can be indirect (independent of the membrane) or direct (dependent on the membrane). Direct interactions, the only ones to be discussed here, can be repulsive [5, 22] or attractive. An interesting particular case is the one where the protein induces an anisotropic inclusion on a flexible membrane [11, 18]. Collectively, these inclusions produce a long-range attractive force between proteins, leading to their aggregation. Digital simulations show a transition between compact clusters and line or rings [11]. The rings are reminiscent of the initial stage in the formation of vesicles or tubules [19]. In passing, it is interesting to note that anisotropic inclusions favour the formation of cubic membranes [18].

7.3.3 Fission

Fission is the segmentation of reticular, tubular or vesicular differentiations, obtained by imposing a bend in an initially flat membrane or cisterna. Auxiliary proteins can induce membrane curvature, by provoking asymmetries between the two leaflets of the membrane. The protein coats that surround the cytoplasmic side of membranes play a role in the process of curvature, fenestration and fission of the membrane. If, in addition, the site of differentiation is fixed, then the cisternae produced consecutively will tend to stack up, because their size and the ambient viscosity will prevent them from diffusing very far from the fixed production site. When stacks are created, it is possible for them to be stabilised by what are called "matrix" proteins, the role of which is too controversial for us to discuss here. Mechano-enzymes (proteins that carry out mechanical work by consuming chemical energy) are required for the final constriction of the membranes, leading to the fission proper. Lastly, in the case of animal cells, the scattered pre-Golgi elements gather into one sole body by attaching themselves to cytoskeletal filaments, by means of a molecular motor (a protein that travels along the filament by consuming chemical energy).

7.3.3.1 Membrane Curvature

Let us start by examining the way that an asymmetry can be created between the two leaflets. The idea that lipidic asymmetries between the two layers could facilitate membrane curvature was proposed back in 1974 [69]. This hypothesis has inspired a large number of experimental works. From a physical point of view [67], when the lipids have the shape of a truncated cone, this favours the formation of vesicles. If they are more cylindrical, then flat bilayers are favoured. Inverted, truncated cone-shapes would favour the formation of "bud necks". The lipid composition is affected by various enzymatic activities. In particular, certain phospholipid transfer proteins play an essential role in secretion, with morphogenetic consequences [60]. "Flippases" favour asymmetry by moving specific lipids from one leaflet to the other, but in a skewed way. Certain lipids like the sterols exert a stabilising counter-effect on these asymmetries. In their absence, it only requires an excess of 1% of phospholipids in a leaflet to trigger membrane budding. The introduction of detergents (molecules that partition into the bilayer but do not self-assemble) such as geraniol is sufficient to cause the growth of long, tangled tubules, enclosed by a bilayer [9].

7.3.3.2 Fenestration and Fission with Coat Proteins

Coat proteins often play a crucial role in the fission of membranes, of which they coat the cytoplasmic side. We can illustrate this fact with the example of COP II ("COat Proteins II"), a regular, two-dimensional polymer complex [3, 38]. COP II is involved in the first morphogenetic event in the secretory pathway [4]. This is the fenestration[2] of ER cisternae to form interconnected tubules (Fig. 7.4) and the fission of these tubules [44, 45]. These authors demonstrated directly on living

---

[2] Fenestration is the gradual perforation of a cisterna, producing pores and ultimately leading to the formation of a tubular network.

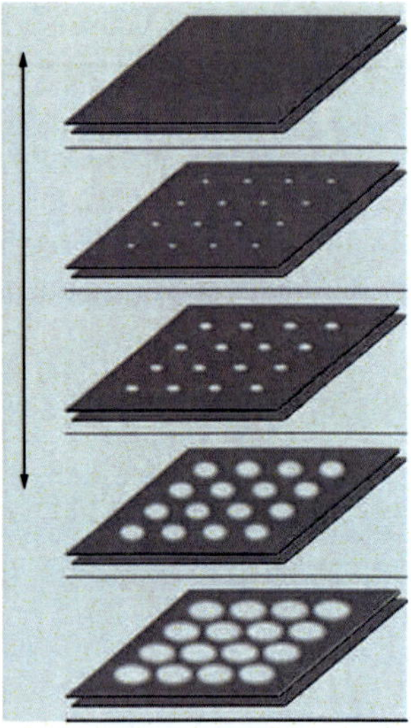

Fig. 7.4 Fenestration and filling-in (saccularization) of a model membrane cisterna. The two flat membranes enclosing a cisterna are shown at different stages of fenestration, from a simple cisterna to a network of interconnected tubules. From top to bottom, fenestration. From bottom to top, saccularization

material the close interrelationship between internal dynamics and global form: in the absence of upstream cargo, there is no transport function to perform, and the container disappears. A model of the interrelationship has been proposed on this basis [30]. The molecules considered in this model are a soluble secretory protein (alpha-factor), a resident ER protein (BiP), the membrane nucleator Sec12p, the regulator Sar1p and the coat protein COP II. The compartments are the cytoplasm, and the lumens and membranes of the ER and the post-ER compartment. The aim is to segregate the alpha-factor and the BiP, directing the former into the post-ER compartment – which must be formed at the same time – and leaving the latter in the ER. The model stipulates that random fluctuations will, at a certain moment, gather a critical concentration of the nucleator Sec12p at a certain spot on the ER membrane. This allows for the recruitment of a corresponding amount of the inactive regulator Sar1p, moved from the cytoplasm towards the Sec12p (Fig. 7.5). Sec12p then catalyses the activation of Sar1p. The activated Sar1p leads to the formation of ER-derived tubular domains, as illustrated in Fig. 7.4 [1]. Then Sar1p [1] or COP II [32] selectively mobilises secretory proteins like alpha-factor within this tubular ER. The resident ER protein, BiP, is not selectively enriched during this process. The model is incomplete without the addition of one of the two following hypotheses.

- *Hypothesis 1*: additional Sar1p is selectively recruited on the cytoplasmic side of the ER membrane at the same spot where alpha-factor is enriched on the other

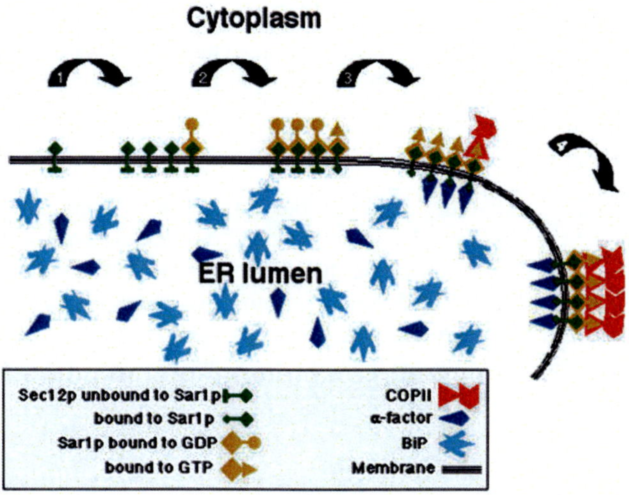

Fig. 7.5 Successive molecular events in the segregation of secretory material into a coated area of endomembrane. The secretory protein alpha-factor and the resident protein BiP are initially mixed in the ER lumen. *Step 1*: Sec12p molecules cluster, thus nucleating an ER export site. *Step 2*: these Sec12p molecules recruit Sar1p-GDP (inactive) and start to exchange their GDP for GTP, producing Sar1p-GTP (activated). *Step 3*: the Sec12p/Sar1p complexes recruit the secretory protein alpha-factor and start to recruit COPII. They induce membrane tubulation, symbolised by slight curvature. *Step 4*: COPII has been fully recruited and self-polymerises into a membrane coat. This coat induces fission of the membrane tubules, symbolised by strong curvature. © F. Képès

side. Activated Sar1p then recruits cytoplasmic COP II on the tubular ER membrane, thus forming a coat on this surface.

- *Hypothesis 2*: COP II self-polymerises into a curved coat, giving the membrane a local curvature.

Whichever the hypothesis, this molecular dynamics results in the membranous tubes of the tubular ER being sectioned into a set of short, disconnected tubes. At this stage, alpha-factor has been enriched in the new post-ER compartment, unlike the resident protein BiP. Both hypotheses relate, in different ways, to mutual autocatalytic recruitment between the secretory cargo (alpha-factor) and the coat (Fig. 7.5). The relevance of this autostimulation loop to the interrelationship between content sorting and container formation is very hard to test in the laboratory, and would be much more easily explored through simulation [30].

7.3.3.3 Stacking of Golgi Elements by Sequestration of the Nucleator

In the baker's yeast *Saccharomyces cerevisiae*, the ER export sites, defined by the position of the Sec12p nucleator, are distributed throughout the ER and change from one export event to the next. Consequently, the Golgi elements that form from these sites are remoter from each other. In contrast, another budding yeast, *Pichia pastoris*, has relatively fixed export sites, in which Sec12p appears to be sequestered

[6, 62]. As a consequence, each Golgi element is pushed away by the next one to come out of the site. As the high viscosity of the cytoplasm slows down their diffusion, these Golgi elements tend to stack up near to where they are formed [6, 59, 62].

7.3.3.4 Constricting with Dynamin

Dynamin, together with other proteins like actin, exerts direct morphogenetic effects. Dynamin can self-assemble into rings or spirals. It can also assemble around membranes to form helical tubes, the diameters of which are similar to those of endocytotic vesicles (Fig. 7.6) [72, 73]. Dynamin is recruited on the membrane by its interactions with lipid derivatives and with an auxiliary protein. If dynamin is supplied with chemical energy, the tubes undergo constriction and fragmentation, a clear indication that dynamin is a mechano-enzyme.

7.3.3.5 Aggregating with Microtubules

While plant cells contain a few hundred small Golgi elements, animal cells typically contain only one (or very few) giant Golgi apparatus. This difference can be attributed chiefly to the link between the animal Golgi apparatus and a type of oriented intracellular "skeleton", composed of filaments known as microtubules (see

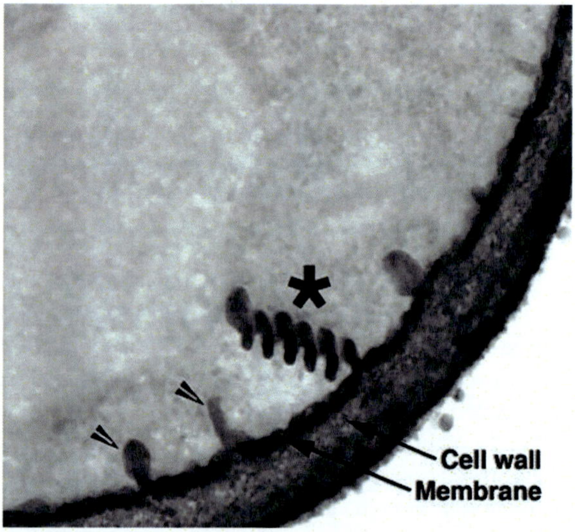

Fig. 7.6 Membrane constriction and endocytosis. A yeast cell has been fixed with glutaraldehyde and the membranes impregnated with reduced osmium and then lead citrate. The specimen was photographed by A. Rambourg under two angles separated by 10°. The two photos were then combined to make an analglyph image that can be used by placing a cyan lens in front of the right eye and a red lens in front of the left eye. Enlargement approximately x 40,000. The thick wall of the yeast is delineated on the cell side by the plasma membrane. This membrane displays several figures of endocytosis, two of which have been arrowed in the picture. One is in face view, the other in side view. The same figure of endocytosis is visible at the end of the corkscrew-shaped membranous tube, indicated by a star. Apparently, this must be a wine yeast. © A. Rambourg

Chap. 5). This link is caused by a microtubule motor, a protein capable of moving along the oriented microtubule by consuming chemical energy. The motor pulls the Golgi membrane towards the extremities of the microtubules, which converge towards their main organising centre, near to one of the poles of the nucleus [36]. All the small pre-Golgi elements gather there, facilitating their fusion into one giant Golgi apparatus. The unique Golgi apparatus is therefore usually located near a pole of the nucleus. If we de-polymerise the microtubules by means of a chemical treatment, then the small pre-Golgi elements no longer converge towards the microtubule organising centre, and hundreds of small distinct Golgi stacks remain scattered throughout the cell.

7.3.4 Fusion

Membrane fusion is a very common process in endomembrane dynamics. In particular, it occurs during the coalescence of two membrane-bounded compartments and during the fenestration of cisternae (Fig. 7.4). The mechanism of fusion starts with a tethering event, particularly necessary if the two membranes to be fused are initially remote from each other. The next step is apposition between the membranes, or docking. To trigger the docking, it may be necessary to provoke movement in the membranes to be fused. An event of molecular recognition ensures the specificity of interactions between membranes. Finally, the fusion proper occurs. Although these different stages will be described separately in what follows, the interactions between the mechanisms responsible for these stages will become obvious.

7.3.4.1 Tethering

The tethering of two endomembrane compartments is defined as the establishment of a low-energy link at distances greater than the thickness of a membrane (> 7 nm). Such attachments can be observed during intra-Golgi transports and between the ER and the Golgi apparatus [51]. Many tethering proteins form very long, thin structures, suggesting that they function as chemical poles hooking up two remote objects. The most well-characterised tethering factor appears in the yeast cell as a protein 150 nm long, Uso1p, involved in transport between the ER and the Golgi apparatus[3] [46, 81]. Its analogue in mammals, p115, is a protein 45 nm long that stimulates intra-Golgi transport [68, 77].

7.3.4.2 Docking

Stable docking is defined as the stage where the membranes are maintained at a small distance (< 7 nm). There exists a family of protein complexes involved in membrane docking at various stages in the secretory pathway [80]. The archetype

[3] By way of comparison, the yeast cell has a diameter of about 5 000 nm, and a secretory granule is about 100 nm.

is the complex called Exocyst. In the case of polarised secretion, these complexes localise preferentially in the main site of exocytosis [17, 23, 29, 74, 75].

7.3.4.3 Priming

Regulatory proteins are involved in various segments of the secretory pathway. In yeast, Ypt1p regulates the step between the ER and the Golgi apparatus in conjunction with the tethering factor Uso1p [66]. Sec4p regulates a later step, the fusion of the secretory granule with the cell membrane, through interaction with Exocyst. In mammalian cells, Rab1 regulates an intra-Golgi step with the tethering factor p115. It seems likely that proteins from the Rab1 or Ypt1p families play an additional role by linking the endomembranes to the cytoskeleton motors, thereby facilitating long-range movements of membrane compartments [12].

7.3.4.4 Recognition and Pairing

It is the recognition between two SNAREs (Soluble NSF Attachment REceptor) that is believed to be responsible for the specificity of each particular interaction between two membranes.[4] SNAREs are very evolutionarily conserved proteins that are found on membranes [16, 65]. They exist as pairs, made up of one Q-SNARE and one R-SNARE. The compartments of the secretory pathway each have their own distinct combination of SNAREs, although any given SNARE may be localised in several compartments. It has been suggested that protein pores between apposed membranes may facilitate fusion [35]. According to this theory, the partial dissociation of sub-units of the pore leads to radial opening of the pore, allowing lipids to enter into the amphiphilic clefts between sub-units. These lipids then enlarge the pore, which serves as the starting point for the fusion. It is believed that pairs of SNAREs are involved in this physiological phenomenon.

7.3.4.5 Fusion

Fusion occurs between a secretory granule and an acceptor membrane, or between the apposed membranes of a cisterna in the process of fenestration. Membrane fusion is thermodynamically unfavourable, requiring physical and biochemical resources. The Q- and R-SNAREs form complexes in the same membrane. These complexes can be destroyed with the consumption of chemical energy by the fusion protein NSF (N-ethylmaleimide-Sensitive Fusion protein) and its co-factor α-SNAP (Soluble NSF Attachment Protein). This destruction enables the SNAREs to pair, no longer with their partner in the same membrane, but with that of the other membrane [24, 48, 50, 71].

[4] On this basis, genomics allows us to draw some conclusions about the maximum number of steps of pairing and fusion in the secretory pathway of an organism, by examining the number of different SNAREs coded by its genome. In yeast, for example, Sed5p and Tlg2p are the only SNAREs involved in the exocytic process [50], consistent with the in situ observation of only two processes of fusion along the pathway [45].

7.4 Functional Models

Without dwelling too long on the controversies of the past, this section will examine the three main models of secretory morphodynamics (Fig. 7.7).

During the 1960s, Mollenhauer and Morré (cf. their review published in 1991 [42]) studied Golgi stacks in thin sections of plant cells and partly purified fractions. They discovered that in addition to flattened cisternae (saccules) and vesicles, the Golgi stacks contain tubular projections forming networks of tubules on the edges of the cisternae. They observed a polarisation of the stack, with the secretory granules increasing in size as they progress from one side of the stack to the other. These granules reach their maximum size on the "exit" (*trans*) face of the stack, where they are released by rupture of the tubular network. The authors postulated that the loss of cisternae at the *trans* face is compensated for by the formation of new cisternae at the "entry" (*cis*) face of the stack. These new cisternae are formed by the fusion of precursors from the ER. In sum, in this *saccular migration* model, the membrane and its cargo are believed to progress through the Golgi stack while maturing at the same time. This model fails to explain the inhomogeneous distribution of enzymes through the Golgi stack, which was observed subsequently. For example, how does an early-acting enzyme remain in the *cis* region, when it should be transported through the stack by the migration of its cisterna? However, this model enjoyed a revival of interest from 1993 on, when a molecular explanation for this distribution of Golgi enzymes was added to it [7, 40, 41]. In this *cisternal maturation-progression* model, the appropriate enzymatic gradient in the Golgi apparatus is maintained by the selective retrieval of early-acting enzymes, which are transported from the *trans* compartments to the *cis* compartments by vesicles that circulate in the opposite direction to the dominant movement of cisternae and their cargo (Fig. 7.7a).

In the meantime, another model came to prominence, before losing ground to the cisternal maturation-progression model around 1998. This was the *vesicular transport* model, based on the same observation of an enzymatic gradient. For this observation led to the development of cell-free assays, reconstituting in vitro the transport of proteins through the Golgi apparatus [20]. This has proved to be a powerful technique for identifying the cytoplasmic factors involved in the transport of macromolecules to and through the Golgi apparatus. The results obtained are generally held to be compatible with the vesicular transport model. This model postulates that transport through the Golgi apparatus is effected by the budding, anterograde targeting (in the direction of maturation), and fusion of transport vesicles from one compartment to the next [14, 15, 27, 61, 65, 78] (Fig. 7.7b). To simplify, we could therefore say that the two dominant models are based on the sequential movement of vesicular shuttles between specific compartments. The vesicular transport model postulates the anterograde movement of vesicles carrying the secretory cargo (in the direction going from the ER towards the granules). The cisternal maturation-progression model proposes the retrograde movement of vesicles containing Golgi enzymes to be recycled.

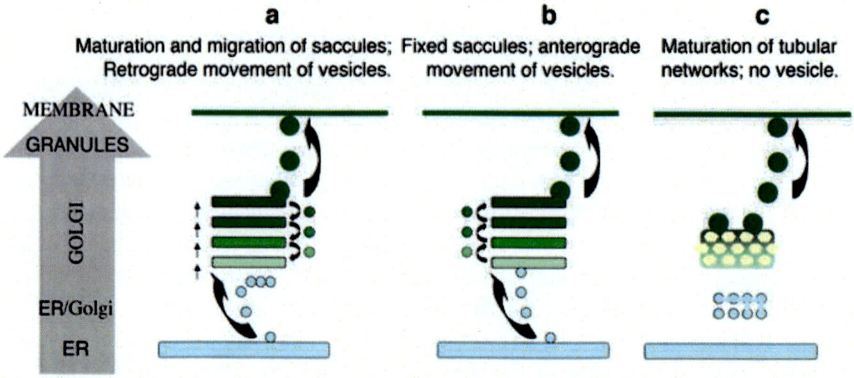

Fig. 7.7 Functional models of secretory morphodynamics. The secretory material enters the ER (*bottom*), reaches the Golgi apparatus via an intermediary compartment, passes through the Golgi apparatus, and then finds itself in the secretory granules, which are liberated outside the cell by fusion with the cell membrane (*top*). Apart from the material's entry into the ER, which involves crossing a membrane, all the other segments of this secretory pathway correspond to membrane fissions or fusions during which the material simply undergoes packing and unpacking events. Several models have been proposed to explain this process. (**a**) *Cisternal maturation-progression model*. This model stipulates that new cisternae are formed by the fusion of precursors coming from the ER. The membrane and its cargo are believed to migrate in the anterograde direction through the Golgi stack, maturing along the way. The loss of cisternae at the top of the Golgi stack is compensated for by the formation of new cisternae at the bottom. The appropriate enzymatic gradient in the Golgi apparatus is maintained by the selective retrieval of early-acting enzymes, which are shuttled back from later compartments to earlier ones in vesicles that circulate in the opposite direction to the net movement of the secretory cargo and cisternae. (**b**) *Vesicular transport model*. The proteins are transported from the ER to the Golgi apparatus, and from one cisterna to the next within the Golgi apparatus, by means of vesicles that successively bud from one compartment and fuse with the next one. Here the cisternae in the Golgi stack are considered to be fixed, and the vesicles migrate in the anterograde direction, thereby transporting the secretory cargo. (**c**) *Continuous flow model*. The ER cisternae fenestrate and then tubulate. The tubular ER network thus formed then transforms into a pre-Golgi tubular compartment, by the breaking of the tubular network, all the more significant as the secretory transit is slow. The junctions of these membrane tubules gradually swell and mature into post-Golgi compartments, in particular secretory granules in certain cell types. These granules are finally released and join the cell membrane. This model predicts that there should be few vesicles, as is observed in yeasts and plants. © F. Képès

On the morphological level, all these models are based on the electronic microscopy study of ultrathin sections of cells (typically 0.5 μm). These ultrathin sections only allow the observation of small fragments of the Golgi apparatus. And yet this apparatus is often very widely spread through the cytoplasm, and its form is complex. In particular, it is easy to confuse vesicles and tubes in a thin section where one can only observe the two-dimensional projection (Fig. 7.8). To avoid this trap, it is necessary to photograph the object from different angles, then use techniques of stereoscopy [56] or virtual reality [31] applied to morphological data as raw as possible. These methods are based on the brain's capacity to reconstruct the volume of objects on the basis of partial information. These morphological studies have revealed the low presence of vesicles in yeast and plant cells, thereby contradicting

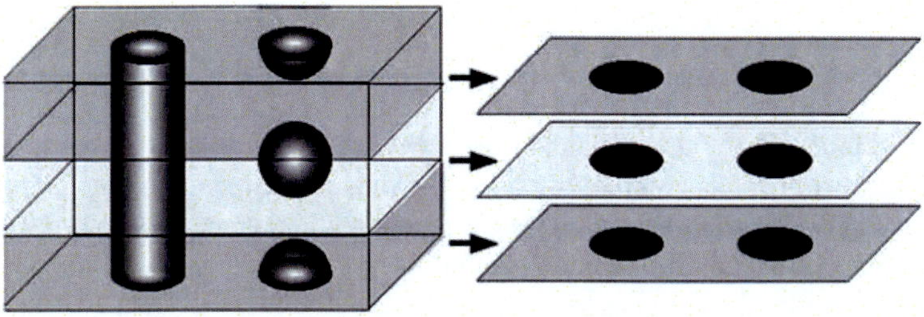

Fig. 7.8 Three-dimensional view or reconstruction from serial sections. The biological objects are represented on the left, as they would be perceived by stereoscopic analysis of a thick section. They are portions of a cylinder (tube) and three spheres (vesicles) of the same diameter. This thick section can be cut into a series of three sections symbolised by the alternating grey/white/grey backgrounds. Each of these serial sections are then photographed separately, providing the three micrographs shown on the *right* (2-D projections of 3-D objects being visible in each section). Together, these micrographs produce ambiguous results. In trying to reconstruct the original objects, one might conclude that they were two cylinders, or six spheres, or one cylinder and three spheres

the predictions of the vesicular models. In addition, they have brought to light the strong anatomical continuity of the Golgi apparatus, including along the *cis-trans* axis. This continuity is incompatible with a process of migration of independent cisternae within the stack. Another, secondary argument is that it allows us to dispense with the idea of cisternal migration to explain the movement of molecules in the anterograde direction. Vesicular models have another weakness, namely their inability to explain the molecular details of the specific localisation of proteins resident in each compartment.

Some further observations allow to propose a model, the *continuous flow* model, which does not have these disadvantages. As an illustration, here is one key observation. In yeast, each Golgi element is a tubular network [58]. When the thermosensitive *sec7* mutant is transported at a temperature where the mutation is expressed, the secretory granules gradually disappear. The Golgi networks increase in size, then the networks fill in, becoming flat cisternae that stack up, resembling the Golgi stacks of plants, or the compact zones of animal Golgi stacks. The cisternae are connected to each other, and very few vesicles can be found associated with these stacks. If the cells are then transported at a temperature where the mutation is not expressed, the formation of secretory granules is resumed and the Golgi stacks gradually return to the initial state of tubular networks [58]. These results can be interpreted directly by supposing that the *sec7* mutation blocks the segregation and release of secretory granules on the *trans* side of the Golgi unit. It can be observed that this blockade leads to the accumulation of Golgi membranes, as if there was a continuous flow of membranes in the anterograde direction that had been blocked. The conclusion drawn from this type of observation is that in eukaryotic cells, the anterograde transport of molecules via the Golgi apparatus appears to result from a continuous flow of membrane and does not require the presence of small vesicles (Fig. 7.7c).

7.5 Conclusions

7.5.1 Themes

Careful examination of the dynamics of secretory compartments throughout the eukaryotic world reveals common themes. The most fundamental theme is that the secretory cargo and container specify each other, in a less general but similar sense as the animal and its environment do, according to Maturana and Varela [39]. This leads us to consider the Golgi apparatus and related structures as compartments that are formed by the very action of transporting a secretory cargo, of functioning. Consequently, stopping the function leads to the disappearance of the structures associated with it, as observed in vivo. In comparison, only the most ancestral secretory compartments are less dissipative: the cell membrane, and to a lesser degree the ER. So the perspectives of evolution (see below for the details) and stability combine to suggest that the Golgi apparatus and related structures are differentiations of the ER, the main secretory endomembrane in terms of quantity.[5]

The second common theme is the crucial importance for endomembrane morphogenesis of the dynamic transition between cisterna and tubular network, by gradual perforation in one direction, and by filling-in in the other.

The third theme is the role of the continuous flow of membrane in the anterograde transport of molecules from the ER to and through the Golgi apparatus.

The last common theme is the self-regulating equilibrium between anatomical continuities and discontinuities in the endomembrane system. As this equilibrium depends on the secretory activity, it provides a source of morphological variability according to cell type or, for a given cell type, according to environmental conditions.

7.5.2 Evolutionary Perspectives

Beyond the source of variability just mentioned, it appears that divergent strategies pave the evolutionary road trodden by the different eukaryotic kingdoms.

As described earlier, only animal cells use microtubules to assemble an almost-unique Golgi apparatus; animal and plant cells display cohesive Golgi elements; animal cells, plant cells and some yeasts possess a Golgi apparatus presenting stacks, and this is not the case for other types of yeast. These characteristics could therefore be considered as three "inventions" produced over the course of evolution. The first invention of a basic way to stack Golgi precursors opened the way to the second invention, which partially stabilises these stacks with what are called "matrix" proteins. In turn, this second invention opened the way to the third: the bringing together of small stacks to form an almost-unique Golgi apparatus [30]. This grouping together could be symbolised by the connection (in anatomical continuity) between the peripheral tubular parts of the small stacks to form non-compact

[5] In evolutionary terms, the ER itself probably derives from the cell membrane.

areas of the Golgi ribbon in animal cells. From the same point of view, the inner cisternal parts of the small stacks would produce the compact areas of the same ribbon. From this simplifying perspective, *S. cerevisiae*, *P. pastoris*, the plants and the animals could be considered as representatives of each of these evolutionary levels.

We do not know exactly which process is optimised by these three inventions. For example, if the unique Golgi apparatus of animal cells is broken up by the destruction of microtubules, the secretory kinetics does not seem to be affected. A simple optimisation of secretory kinetics does not, therefore, seem to be the correct response. However, applying the evolutionary perspective to the second "invention" brings to light the notion of compromise between versatility and stability. For this invention can be interpreted as the appearance of structural proteins that stabilise the type of informal stack present in *P. pastoris* into the type that can be found in plants and animals. Stacking in yeast represents a case of non-equilibrium structure [30, 47] with a purely dynamic organisation, only lasting for as long as the cell secretes. This characteristic provides the micro-organism with the versatility necessary to adapt to eminently variable secretory loads, themselves a consequence of a changing environment. For example, when little secretion is needed, the biosynthetic capacity of the yeast can be transferred to tasks other than the making and moving of endomembranes, secretory cargos and molecular machineries. Conversely, in the more stable environment of cells existing within a multicellular organism, the Golgi apparatus is closer to an equilibrium structure, because its stability now includes a static component. This increased stability could optimise the secretory process in the context of a relatively constant secretory load. It could also allow a faster resumption of secretion after an accidental halt.

In sum, these divergent strategies affect above all the levels of stacking, stabilisation and assembly of the Golgi apparatus. They probably testify to a compromise between versatility and stability to adapt the secretory function to the *degree* of environmental variability.

7.5.3 Questions

Certain facts presented here, particularly the non-equilibrium characteristic, suggest that we should turn on its head the classic question: "How is the compartment maintained in the face of a constant flow of secretory material?" and ask instead: How could the compartment be maintained if there were no flow of secretory material?"

Another classic question is: "How are these proteins, which appear to define the identity of a given compartment, recruited by the compartment?" This question raises the prospect of an endless cascade of specific molecules recruiting other molecules. However, an end is in fact imposed on this cascade by the limited number of proteins and their genes within an organism. It would therefore be more fruitful to rephrase the question as: "How does the identity of each compartment emerge dynamically from first principles and from molecular interaction and feedback?"

7.5.4 Prospects

Many of the major controversies in cell biology, and especially in the field of endomembrane dynamics, would be settled if live cell observation could attain sufficient spatial *and* temporal resolution. So far, the dialectic has mainly been between a fantastic spatial resolution applied to dead objects (electron microscopy), and kinetic observations of live cells (light microscopy). Kinetic observations using electron microscopy [45] are no more than substitutes, requiring statistical evaluation of the raw data and synchronisation of the secretory process in a cell population. This synchronisation in turn requires the use of chemical or genetic treatment. One of the problems with this approach is that the history of the individual cell is not always faithfully reflected by averaged observations on a population.

In the fairly near future this gap will be filled by the appearance of new photonic microscopies that it will be possible to apply to live matter, which will allow an unprecedented level of spatial resolution, breaking the wavelength barrier [13]. In the meantime, the molecular definition of secretory compartments and the tools available can only get better [49, 63].

Acknowledgments Discussions with Alain Rambourg have been essential to the writing of this chapter.

References

1. Aridor M., Fish K.N., Bannykh S., Weissman J., Roberts T.H., Lippincott-Schwartz J., and Balch W.E. (2001) The Sar1 GTPase coordinates biosynthetic cargo selection with endoplasmic reticulum export site assembly, *J. Cell Biol.* **152**, 213–229.
2. Axelsson M.A., Karlsson N.G., Steel D.M., Ouwendijk J., Nilsson T., and Hansson G.C. (2001) Neutralization of pH in the Golgi apparatus causes redistribution of glycosyltransferases and changes in the O-glycosylation of mucins, *Glycobiology* **11**, 633–644.
3. Barlowe C. (2002) Three-dimensional structure of a COPII prebudding complex, *Dev. Cell* **3**, 467–468.
4. Barlowe C., Orci L., Yeung T., Hosobuchi M., Hamamoto S., Salama N., Rexach M.F., Ravazzola M., Amherdt M., and Schekman R. (1994) COPII: a membrane coat formed by Sec proteins that drive vesicle budding from the endoplasmic reticulum, *Cell* **77**, 895–907.
5. Bartolo D. and Fournier J.B. (2003) Elastic interaction between hard or soft pointwise inclusions on biological membranes, *Eur. Phys. J. E* **11**, 141–146.
6. Bevis B.J., Hammond A.T., Reinke C.A., and Glick B.S. (2002) De novo formation of transitional ER sites and Golgi structures in Pichia pastoris, *Nat. Cell Biol.* **4**, 750–756.
7. Bonfanti L., Mironov A.A. Jr, Martinez-Menarguez J.A., Martella O., Fusella A., Baldassarre M., Buccione R., Geuze H.J., Mironov A.A., and Luini A. (1998) Procollagen traverses the Golgi stack without leaving the lumen of cisternae: evidence for cisternal maturation, *Cell* **95**, 993–1003.
8. Bretscher M.S. and Munro S. (1993) Cholesterol and the Golgi apparatus, *Science* **261**, 1280–1281.
9. Chiruvolu S., Walker S., Israelachvili J., Schmitt F.J., Leckband D., and Zasadzinski J.A. (1994) Higher order self-assembly of vesicles by site-specific binding, *Science* **264**, 1753–1756.
10. Cunningham W.P., Morré D.J., and Mollenhauer H.H. (1966) Structure of isolated plant Golgi apparatus revealed by negative staining, *J. Cell Biol.* **28**, 169–179.

11. Dommersnes P.G. and Fournier J.B. (1999) N-body study of anisotropic membrane inclusions: membrane mediated interactions and ordered aggregation, *Eur. Phys. J. B* **12**, 9–12.
12. Echard A., Jollivert F., Martinez O., Lacapere J.J., Rousselet A., Janoueux-Lerosey I., and Goud B. (1998) Interaction of a Golgi-associated kinesin-like protein with Rab6, *Science* **279**, 580–585.
13. Failla A.V., Albrecht B., Spöri U., Schweitzer A., Kroll A., Hildenbrand G., Bach M., and Cremer C. (2003) Nanostructure Analysis using Spatially Modulated Illumination, *ComPlexUs* **1**, 77–88.
14. Farquhar M.G. (1985) Progress in unraveling pathways of Golgi traffic, *Ann. Rev. Cell Biol.* **1**, 447–488.
15. Farquhar M.G. and Palade G.E. (1981) The Golgi apparatus (complex) - 1954-1981 - from artefact to center stage, *J. Cell Biol.* **91**, 77s–103s.
16. Ferro-Novick S. and Jahn R. (1994) Vesicle fusion from yeast to man, *Nature* **370**, 191–193.
17. Finger F.P., Hughes T.E., and Novick P. (1998) Sec3p is a spatial landmark for polarized secretion in budding yeast, *Cell* **92**, 559–571.
18. Fournier J.B. (1996) Nontopological saddle-splay and curvature instabilities from anisotropic membrane inclusions, *Phys. Rev. Lett.* **76**, 4436–4439.
19. Fournier J.B, Dommersnes P.G., and Galatola P. (2003) Dynamin recruitment by clathrin coats: a physical step? *C. R. Biol.* **326**, 467–476.
20. Fries E. and Rothman J.E. (1980) Transport of vesicular stomatitis viral glycoprotein in a cell-free extract, *Proc. Natl. Acad. Sci. USA* **77**, 3870–3874.
21. Golgi C. (1898) Sur la structure des cellules nerveuses des ganglions spinaux, *Arch. Ital. Biologie* **30**, 278–286, in French.
22. Goulian M., Bruinsma R., and Pincus P. (1993) Long-range forces in heterogeneous fluid membranes, *Europhys. Lett.* **22**, 145–155.
23. Grindstaff K.K., Yeaman C., Anandasabapathy N., Hsu S.C., Rodriguez-Boulan E., Scheller R.H., and Nelson W.J. (1998) Sec6/8 complex is recruited to cell-cell contacts and specifies transport vesicle delivery to the basal-lateral membrane in epithelial cells, *Cell* **93**, 731–740.
24. Guo W., Sacher M., Barrowman J., Ferro-Novick S., and Novick P. (2000) Protein complexes in transport vesicle targeting, *Trends Cell Biol.* **10**, 251–255.
25. Harder T. and Simons K. (1997) Caveolae, DIGs, and the dynamics of sphingolipid-cholesterol microdomains, *Curr. Opin. Cell Biol.* **9**, 534–442.
26. Hawes C.R., Juniper B., and Horne J.C. (1981) Low and high voltage electron microscopy of mitosis and cytokinesis in maize roots, *Planta* **152**, 397–407.
27. Hawes C.R., Brandizzi F., and Andreeva A.V. (1999) Endomembranes and vesicles trafficking, *Curr. Opin. in Plant Biol.* **2**, 454–461.
28. Hepler P.K., Palevitz B.A., Lancelle S.A., McCauley M.M., and Lichtscheidl I. (1990) Cortical endoplasmic reticulum in plants, *J. Cell Sci.* **96**, 355–373.
29. Hsu S.C., Hazuka C.D., Foletti D.L., and Scheller R.H. (1999) Targeting vesicles to specific sites on the plasma membrane: the role of the sec6/8 complex, *Trends Cell Biol.* **9**, 150–153.
30. Képès F. (2002) Secretory compartments as instances of dynamic self-evolving structures, *Acta Biotheoretica* **50**, 209–221.
31. Képès F. (2003) Membrane traffic in virtual reality, *Traffic* **4**, 13–17.
32. Kuehn M.J., Herrmann J.M., and Schekman R. (1998) COPII-cargo interactions direct protein sorting into ER-derived transport vesicles, *Nature* **391**, 187–190.
33. Landh T. (1995) From entangled membranes to eclectic morphologies: cubic membranes as subcellular space organizers, *FEBS Lett.* **369**, 13–17.
34. Landh T. (1995) In *Fundamentals of Medical Cell Biology*, edited by E. Bittar, JAI Press (Greenwich).
35. Lindau M. and Almers W. (1995) Structure and function of fusion pores in exocytosis and ectoplasmic membrane fusion, *Curr. Opin. Cell Biol.* **7**, 509–517.
36. Lippincott-Schwartz J. (1998) Cytoskeletal proteins and Golgi dynamics, *Curr. Opin. Cell Biol.* **10**, 52–59.

37. Manton I. (1960) On a reticular derivative from Golgi bodies in the meristem of Anthoceros, *J. Biophys. Biochem. Cytol.* **8**, 221–231.
38. Matsuoka K., Schekman R., Orci L., and Heuser J.E. (2001) Surface structure of the COPII-coated vesicle, *Proc. Natl. Acad. Sci. USA* **98**, 13705–13709.
39. Maturana H.R. and Varela F.J. (1987) *The tree of knowledge*, Shambhala (Boston).
40. Mironov A.A., Weidman P., and Luini A. (1997) Variations on the intracellular transport theme: maturing cisternae and trafficking tubules, *J. Cell Biol.* **138**, 481–484.
41. Mironov A.A Jr., Luini A., and Mironov A.A. (1998) Synthetic model of intra-Golgi traffic, *FASEB J.* **12**, 249–252.
42. Mollenhauer H.H. and Morré D.J. (1991) Perspectives on Golgi apparatus and function, *J. Electr. Microsc. Technique* **17**, 2–14.
43. Mollenhauer H.H. and Morré D.J. (1994) Structure of Golgi apparatus, *Protoplasma* **180**, 14–28.
44. Morin-Ganet M.N., Rambourg A., Clermont Y., and Képès F. (1998) Role of endoplasmic reticulum derived vesicles in the formation of Golgi elements in sec23 and sec18 Saccharomyces cerevisiae mutants, *Anat. Rec.* **251**, 256–264.
45. Morin-Ganet M.N., Rambourg A., Deitz S.B., Franzusoff A., and Képès F. (2000) Morphogenesis and dynamics of the yeast Golgi apparatus, *Traffic* **1**, 56–68.
46. Nakajima H., Hirata A., Ogawa Y., Yonehara T., Yoda K., and Yamasaki M. (1991) A cytoskeleton-related gene, uso1, is required for intracellular protein transport in Saccharomyces cerevisiae, *J. Cell Biol.* **113**, 245–260.
47. Norris V. and Fishov I. (2001) Hypothesis: membrane domains and hyperstructures control bacterial division, *Biochimie* **83**, 91–97.
48. Owen D.J. and Schiavo G. (1999) A handle on NSF, *Nat. Cell Biol.* **1**, E127–E128.
49. Patterson G.H. and Lippincott-Schwartz J.A. (2002) Photoactivatable GFP for selective photolabeling of proteins and cells, *Science* **297**, 1873–1877.
50. Pelham H.R. (2001) SNAREs and the specificity of membrane fusion, *Trends Cell Biol.* **11**, 99–101.
51. Pfeffer S.R. (1999) Transport-vesicle targeting: tethers before SNAREs. *Nat. Cell Biol.* **1**, E17–E22.
52. Preuss D., Mulholland J., Kaiser C.A., Orlean P., Albright C., Rose M.D., Robbins P.W., and Botstein D. (1991) Structure of the yeast endoplasmic reticulum : localization of ER proteins using immunofluorescence and immunoelectron microscopy, *Yeast* **7**, 891–911.
53. Preuss D., Mulholland J., Franzusoff A., Segev N., and Botstein D. (1992) Characterization of the Saccharomyces Golgi complex through the cell cycle by immunoelectron microscopy, *Mol. Biol. Cell.* **3**, 789–803.
54. Puertollano R. and Alonso M.A. (1999) MAL, an integral element of the apical sorting machinery, is an itinerant protein that cycles between the trans-Golgi network and the plasma membrane, *Mol. Biol. Cell.* **10**, 3435–3447.
55. Rabouille C. and Nilsson T. (1995) Redrawing compartmental boundaries in the exocytic pathway, *FEBS Lett.* **369**, 97–100.
56. Rambourg A. (1969) L'appareil de Golgi : examen en microscopie électronique de coupes épaisses (0,5-1μ), colorées par le mélange chlorhydrique-phosphotungstique, *C. R. Acad. Sci. Paris* **269**, 2125–2127.
57. Rambourg A., Clermont Y., and Hermo L. (1979) Three-dimensional architecture of the Golgi apparatus in the Sertoli cell of the rat, *Am. J. Anat.* **154**, 455–476.
58. Rambourg A., Clermont Y., and Képès F. (1993) Modulation of the Golgi apparatus in *Saccharomyces cerevisiae sec7* mutants as seen by three-dimensional electron microscopy, *Anat. Rec.* **237**, 441–452.
59. Rambourg A., Clermont Y., Ovtracht L., and Képès F. (1995) Three-dimensional structure of tubular networks, presumably Golgi in nature, in various yeast strains : a comparative study, *Anat. Rec.* **243**, 283–293.

60. Rambourg A., Clermont Y., Nicaud J.-M., Gaillardin C., and Képès F. (1996) Transformations of membrane-bound organelles in sec14 mutants of the yeasts *Saccharomyces cerevisiae* and *Yarrowia lipolytica*, *Anat. Rec.* **245**, 447–458.
61. Robinson D.G., Hinz G., and Holstein S.E.H. (1998) The molecular characterization of transport vesicles, *Plant Mol. Biol.* **38**, 49–76.
62. Rossanese O.W., Soderholm J., Bevis B.J., Sears I.B., O'Connor J., Williamson E.K., and Glick B.S. (1999) Golgi structure correlates with transitional endoplasmic reticulum organization in *Pichia pastoris* and *Saccharomyces cerevisiae*, *J. Cell Biol.* **145**, 69–81.
63. Rossanese O.W., Reinke C.A., Bevis B.J., Hammond A.T., Sears I.B., O'Connor J., and Glick B.S. (2001) A role for actin, Cdc1p, and Myo2p in the inheritance of late Golgi elements in *Saccharomyces cerevisiae*, *J. Cell Biol.* **153**, 47–62.
64. Roth J. (1997) Topology of glycosylation in the Golgi apparatus, in *The Golgi apparatus*, edited by E.G. Berger and J. Roth, Birkhäuser (Basel), pp. 131–162.
65. Rothman J.E. (1994) Mechanisms of intracellular protein transport, *Nature* **372**, 55–63.
66. Sacher M., Jiang Y., Barrowman J., Scarpa A., Burston J., Zhang L., Schieltz D., Yates Jr 3rd., Abeliovitch H., and Ferro-Novick S. (1998) TRAPP, a highly conserved novel complex, *EMBO J.* **17**, 2494–2503.
67. Sackmann E. (1990) Molecular and global structure and dynamics of membrane and lipid bilayers, *Can. J. Phys.* **68**, 999–1012.
68. Sapperstein S.K., Walter D.M., Grosvenor A.R., Heuser J.E., and Waters M.G. (1995) p115 is a general vesicular transport factor related to the yeast endoplasmic reticulum to Golgi transport factor Uso1p, *Proc. Natl. Acad. Sci. USA* **92**, 522–526.
69. Sheetz M.P. and Singer S.J. (1974) Biological membranes as bilayer couples. A molecular mechanism of drug-erythrocyte interactions, *Proc. Natl. Acad. Sci. USA* **71**, 4457–4461.
70. Simons K. and Van Meer G. (1988) Lipid sorting in epithelial cells, *Biochemistry* **27**, 6197–6202.
71. Söllner T., Whiteheart S.W., Brunner M., Erdjument-Bromage H., Geromanos S., Tempst P., and Rothman J.E. (1993) SNAP receptors implicated in vesicle targeting and fusion, *Nature* **362**, 318–324.
72. Sweitzer S.M. and Hinshaw J.E. (1998) Dynamin undergoes a GTP-dependent conformational change causing vesiculation, *Cell* **93**, 1021–1029.
73. Takei K., Haucke V., Slepnev V., Farsad K., Salazar M., Chen H., and De Camilli P. (1998) Generation of coated intermediates of clathrin-mediated endocytosis on protein-free liposomes, *Cell* **94**, 131–141.
74. TerBush D.R. and Novick P. (1995) Sec6, Sec8, and Sec15 are components of a multisubunit complex which localizes to small bud tips in Saccharomyces cerevisiae, *J. Cell Biol.* **130**, 299–312.
75. TerBush D.R., Maurice T., Roth D., and Novick P. (1996) The Exocyst is a multiprotein complex required for exocytosis in Saccharomyces cerevisiae, *EMBO J.* **15**, 6483–6494.
76. Thiéry G. and Rambourg A. (1976) A new staining technique for studying thick sections in the electron microscope, *J. Microscopie Biol. Cell.* **26**, 103–106.
77. Waters M.G., Clary D.O., and Rothman J.E. (1992) A novel 115-kD peripheral membrane protein is required for intercisternal transport in the Golgi stack, *J. Cell Biol.* **118**, 1015–1026.
78. Wattenberg B.W. (1991) Analysis of protein transport through the Golgi in a reconstituted cell-free system, *J. Electr. Microsc. Technique* **17**, 150–131.
79. Weiss M. and Nilsson T. (2000) Protein sorting in the Golgi apparatus: a consequence of maturation and triggered sorting, *FEBS Lett.* **486**, 2–9.
80. Whyte J.R. and Munro S. (2001) The Sec34/35 Golgi transport complex is related to the exocyst, defining a family of complexes involved in multiple steps of membrane traffic, *Dev. Cell* **1**, 527–537.
81. Yamakawa H., Seog D.H., Yoda K., Yamasaki M., and Wakabayashi T. (1996) Uso1 protein is a dimer with two globular heads and a long coiled-coil tail, *J. Struct. Biol.* **116**, 356–365.

Chapter 8
From Epigenomic to Morphogenetic Emergence

Caroline Smet-Nocca, Andràs Paldi, and Arndt Benecke

The biological *forms* that surround us or that are the physical manifestation of our own existence deserve particular attention. Not just because of the transcendental beauty that we ascribe to them emotionally, but also for fundamental scientific reasons. The *primus inter pares* of these reasons is the concept of *functional form*. The sunflower, the structure of a seashell or the form of our organs are not solely the result of a process optimising the emergence, robustness and reproducibility of these forms; they also include a notion of *two-scale functionality*: the functionality of the form itself, and its functionality in the context of a living organism composed of several forms. The ultimate functionality that can be attributed to any living organism is its reproduction. Consequently, the different functions of the constituent forms of an organism must obey this same global *raison d'être*. The conditions for fulfilling this double functionality entail massive constraints, in addition to the constraints of the physico-chemical world, on the emergence of forms – on biological morphogenesis – which is not the case for the morphogenesis of non-living assemblies. It is very interesting to note that each living organism is composed of several functional forms, not one of which can, on its own, accomplish the ultimate functionality of reproduction. The forms of the living being are therefore interdependent, and the ultimate function of these forms is only accomplished by the whole. This distribution of the reproductive function over different forms is at the very origin of the emergence of the simplest organisms, and it constitutes the second stage of evolution. After the emergence of forms capable of self-reproduction – RNAs forming catalytic three-dimensional structures – the emergence of protected micro-environments around these first forms thus marks the appearance of a new concept: that of the *organism*, associated with a distribution of the functionalities of the whole over different individual forms. Consequently, functional biological forms are not only a physical manifestation of the living being, but also at the origin of life itself. This idea leads us directly to the second particularity of biological morphogenesis: during each cycle of reproduction of an organism, every one of these

A. Benecke (✉)
Institut des Hautes Etudes Scientifiques, Bures-sur-Yvette, France
e-mail: arndt@ihes.fr

P. Bourgine, A. Lesne (eds.), *Morphogenesis*, DOI 10.1007/978-3-642-13174-5_8,

sub-functionalities, and therefore of these forms, must be reproduced. This entails a reduction of local entropy ("a local creation of order") that is not instantaneous but *lasts over time*: think for example of embryogenesis. This temporal extension required for the production of the biological mass of the new generation introduces another huge constraint on morphogenesis. Although the individual functionality of each form often only comes into operation at the end of its morphogenesis, yet the survival of the whole – the survival of the organism – must be assured at every moment. In other words, the morphogenesis of the individual components cannot at any time produce an intermediate form of the organism that is not viable. It is not, therefore, so surprising to observe that the reproduction of an organism follows a very similar course to the evolutionary path that has led to its current existence, and under the same constraint of constant viability.[1] The *information* about the viability of the set of forms during their reproduction is therefore conserved and inherited by the new generation at the same time as the information about the functionality of the individual forms. The reproduction of this information about the *morphogenetic dynamics* in the context of living forms is the third particularity of biological systems. The fourth and final particularity is connected to the processes of transmission of the information between successive generations of the organism – the encoding of the information. The issues of how the functionality of individual forms is encoded, and how the morphogenetic dynamics at the scale of the organism and/or at the scale of organism evolution is encoded, can be summed up in one question: *how is morphogenetic emergence encoded*? This last particularity is the least well-understood, and this chapter seeks to bring together the first views and hypotheses about the way it is achieved in living organisms.

Our discussion of these views and hypotheses about the encoding of the information necessary for morphogenetic emergence will propose two main messages: (i) the encoding of this information implies *local and global functional forms* of the organism's genome, and (ii) in the case of multicellular organisms, morphogenetic emergence is achieved through cell differentiation, which is itself a function of the morphodynamics of the genome mentioned in point (i).

The first section of this chapter will remind the reader of the concepts of DNA encoding and the dynamic three-dimensional structure of the genome. The second section will discuss the known mechanisms of epigenetic encoding and how they play an essential role in the encoding of morphogenetic information. The third section explores in a more explicit, and therefore more technical way, a new connection between DNA encoding and encoding by epigenesis; this connection creates epigenomics, a term that we shall use instead of *"local and global functional forms of the genome"* or *"morphodynamics of the genome"*: this forms the basis for morphogenetic emergence, the subject of the fourth and final section.

[1] See Chap. 11, devoted to the evolution of living forms.

8.1 Genetic Inheritance, Regulation of Gene Expression, and Chromatin Dynamics

There are two types of living organisms: unicellular and multicellular. In this chapter we shall almost exclusively be discussing multicellular organisms, because for the phenomena that interest us here, unicellular organisms can be regarded as special cases of the former. This concept of uni-/multi-cellularity should not be confused with the distinction between prokaryotes and eukaryotes, the former of which are always unicellular, while the latter are mainly but not always multicellular. Our discussion will be focused on eukaryotic organisms because the local and global functional forms of their genome are richer, and their complexity is more exemplary for the messages we wish to propose to the reader. There are many examples of the spatial and dynamic organisation of the prokaryotic genome, and the conclusions drawn from them with regard to information encoding are very similar to those obtained for eukaryotes. For a particularly promising and inspiring example, the reader could consult the works on the solenoid model [12]. The concept of biological form also exists at two scales: the macromolecular forms, and the forms that are visible to the naked eye. The latter are either composed of a set of cells presenting a physical appearance by which it stands out from the rest – the sunflower, for example – or generated by the products or physical manifestations of a cellular activity – as in the case of seashells. The concept of *macromolecular form* is particularly important in the context of our discussion, because we shall argue that the basis of encoding of a form at the multicellular scale is always associated with the functional form of a molecular hyperstructure [20, 16] – the genome in its active state [3].

Morphogenesis at the multicellular scale depends on the state of the constituent cells, either directly through their physical presence, or indirectly through their activity. The activity or identity of a cell is established through the activity of all the genes expressed in that cell. To present a well-defined activity or identity, the cell expresses a *genetic programme* – a functional genetic network – which, in response to intra- and extra-cellular stimuli, supplies the cell with a set of proteins and RNA – the products of the expression of its genes – at the required moment and in the required quantities. An organism possesses many different genetic programmes, transmitted by its ancestors; the choice of one or another by the cell therefore represents the first step in multicellular morphogenesis. This choice is primarily determined by the regulation of gene expression at the scale of the whole genome, following complex and coordinated signalling between the different constituent cells of the organism. The greater part of this gene expression regulation takes place at the level of the gene transcription stage [2].

8.1.1 Gene Transcription and the Regulation of Gene Expression

The expression of the vast majority of genes is finely controlled and changes according to the needs of the cell. Intra- and extra-cellular signals directly affect the levels of expression of the target genes, either positively or negatively, and specific programmes will be established according to the state of the cell and all the signals received. All the activity of each gene can be regulated by a multitude of different signals, the composition and activity of a genetic programme is specific to a given signal and the two are consequently mutually specified. The process of gene transcription – in other words the production of RNA from a primary sequence of DNA – is performed by a machinery of great complexity, comprising a DNA-dependent RNA polymerase and associated factors, the general transcription factors. The composition of this machinery and its molecular sub-complexes is dynamic and the variations in its composition (involving some 50 to 120 different molecules) result from variations in its activity. The assembly of the machinery is regulated, and it therefore already represents a first level of gene transcription regulation. The second, predominant level of transcriptional regulation is based on the use of specific transcription factors for sequences of regulation specific to each gene. The use of transcription factors provides a multitude of degrees of freedom in transcriptional regulation. Taken as a whole, the transcription factors, their binding sequences, their cofactors and their capacity to respond to signals constitute the *genomic regulon* of the organism. The activity of this genomic regulon depends on the three-dimensional structure of its substrate, the DNA [2].

8.1.2 Genomic Structure and its Impact on Transcriptional Regulation

In eukaryotes, particularly, the DNA is not present in naked form in the cell, but on the contrary combined with proteins and other small molecules. This assembly, known as *chromatin*, is an important structural element for the organisation of the genome within the cell. It is involved in multiple dynamics at local and global levels – the condensation of chromosomes during mitosis being an emblematic example of global dynamics. As these structural dynamics are regulated processes, chromatin plays a major role in transcriptional regulation [2]. The structural dynamics of chromatin is mainly regulated by post-translational modifications of its protein components, the histones [2, 4, 8, 13, 14].

One of the most important points has been the demonstration that modifications of chromatin and its basic component histones are not only necessary to the stage of derepression prior to the activation of a gene, but also necessary to the process of activation itself. Consequently, it is not surprising that transcriptional regulation entails a modification of the chromatin structure of the target genes and neighbouring areas. These modifications, which take place at the same time at the level of the chromosome, the nucleosome chain, the individual nucleosomes and the histones (see Fig. 8.1), occur in great number and are only partly understood. They are orchestrated by the transcription factors and performed by transcriptional coregulators, only recently discovered. Just like transcription factors, coregulators can act either positively (coactivators) or negatively (corepressors) on the activity of a given transcription factor (or group of such factors). Coregulators do not interact specifically with DNA (unlike specific transcription factors, which bind to a given genome

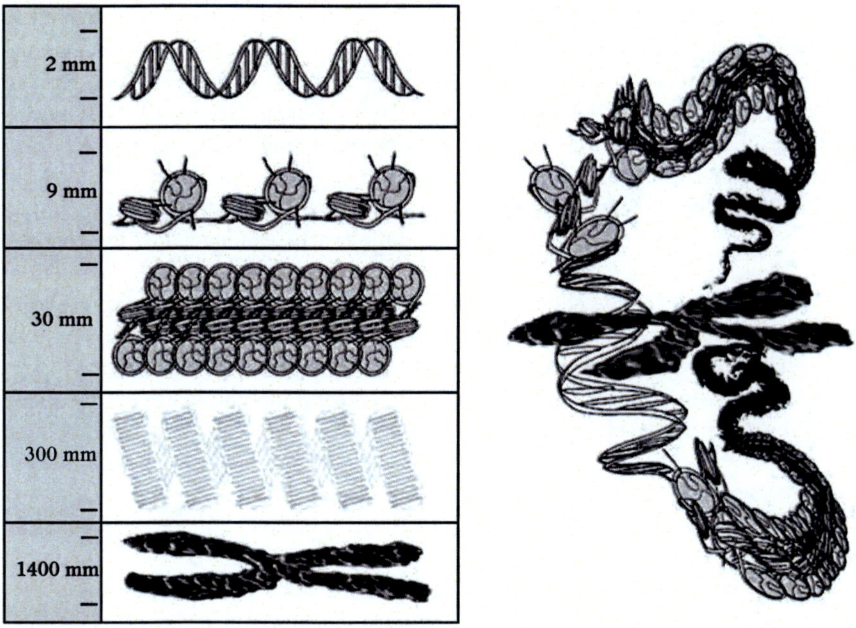

Fig. 8.1 Chromatin form of the genome. On the *left*, the major forms of chromatin at different scales, from naked DNA to condensed chromosome, via fibres of 9 nm, 30 nm and 300 nm. NB: these disconnected levels of organisation are abstractions; in the cell there is continuity between all these levels, as illustrated on the *right*

sequence), but they nevertheless present a certain degree of specificity with regard to the genes of which they coregulate the expression (unlike general transcription factors). They provide an essential link between gene-specific regulators and cell signalling, and thereby contribute to genomic plasticity [2, 4, 8, 13, 14].

The discovery of these coregulators has resulted in the development of a new model, to explain the way the transcriptional machinery of any cell is capable of correctly interpreting the information about transcriptional regulation contained in each genome sequence, and incorporating it into the intra- and extra-cellular signals that control its activity [2, 4, 8, 13, 14].

More broadly, this view has led to the introduction of a new concept of code, at the level of the *chromatin* structure, which is now thought to be a key player in both negative (repression) and positive regulation (activation). In current models of eukaryotic transcriptional regulation, the transcription factors start by binding to their cognate DNA binding site and then, once activated, they orchestrate the remodelling of the chromatin to enable the transcription to take place. But in a completely contradictory manner, there is experimental evidence indicating that the activation of transcription factors can happen before they bind to the DNA; other experiments have shown that the binding of transcription factors to DNA can, at least in certain cases, require a "pre-open" state of the chromatin (see below) and can, moreover, be independent of the transcription process. These facts suggest strongly that the binding of transcription factors to DNA is not necessarily the first step in the sequence of events responsible for transcriptional regulation. These discoveries have led to a model being proposed to explain how the transcriptional machinery reads, interprets and executes the information about the regulation of gene expression. This model introduces the concept of a *second code of gene regulation*, based on modifications of histones (the chromatin sequence). The existence of these modifications and their role in transcriptional regulation are very well documented experimentally, and it is now recognised that they play a major role in the processes induced by the coregulators. The idea of a code operating at the level of the composition and combination of these chromatin modifications is, on the contrary, completely new. We propose that coregulators displaying a particular affinity for a histone tail[2] (modified or not) will specifically recognise certain "chromatin sequences", as a result of which their concentration around these sites will increase, causing the chromatin to "breathe" (a local conformational change of alternate opening and closing, under the combined influence of corepressors and coactivators). This chromatin breathing has indeed been observed, for example in relation to the ATP-dependent remodelling complex SWI-SNF. The transcription factors can then reach the DNA in these sites where the chromatin has been "pre-opened", and they shift the equilibrium between "closed" and "open" states of the chromatin in favour of the latter, allowing the transcriptional machinery to access its target. Interestingly, this model also allows to take into account the compact stacking of nucleosomes and its role in epigenetic activation and repression in a similar fashion to the scenario of transcriptional activation just described: the epigenetic regulation factors (for example DNA-methyltransferases) remove the nucleosomes to be repressed from the hypercycle[3] maintained by the coregulators, inducing a condensed state for the chromatin in their neighbourhood [2, 4, 8, 13, 14].

Thus, the model suggests the emergence and evolution of a second level of coding and regulation, above the first level of coding based on the primary DNA sequence. Over the course of evolution, the appearance of chromatin and its modifications has gone more or less hand-in-hand with that of the coregulators. We propose that the coregulators are to the chromatin sequence what the transcription factors are to the DNA sequence. Histone modifications have been observed in large numbers, simultaneous with transcriptional regulation events, of activation as much as repression. For the moment, it is the acetylation and deacetylation of the histones that has attracted the most attention, but it is clear that histone methylation provides an alternative path of transcriptional activation [2, 4].

As chromatin can be seen as a superstructure enveloping the DNA, in which the DNA is wound around protein complexes, it is perfectly possible that chromatin modifications are the initial step in gene activation, and not simultaneous or subsequent, as had been thought. It seems obvious that the "chromatin sequence" will

[2] Histone proteins have a highly structured central section, the "body", and more flexible end parts on each side, the "tails".

[3] For the theoretical foundations of hypercycles, see [20].

depend on the code associated with the underlying DNA sequences, as well as on the activity of the DNA sequence during the different processes involving this sequence specifically. Nevertheless, the chromatin structure will also contribute to epigenetic regulation (for example in the centromeric and telomeric regions of chromosomes) and provide a direct link between epigenetic regulation and the physiology of cancer or of development, as we shall see in Sect. 8.3. For example, the chromatin structure depends on the cell type and varies over the course of the cell cycle, which supports the model proposed. Thus, the second level of coding proposed here is neither independent nor completely dependent on the underlying DNA sequence [2, 4, 8, 13, 14].

Lastly, according to this model, it will be possible to predict certain characteristics of DNA sequences, on the basis of experimental observations and taking into account the constraints imposed on the architecture of regulatory sequences to obtain reliable and univocal action of the coregulators. Furthermore, it offers the possibility to deduce, from the action of the coregulators, certain architectural properties of the transcription factor binding sites [2]. We shall now see how the structure and dynamics of chromatin are also at the root of the transmission of information through the mechanisms of epigenesis, which play a central role in the establishment of a multicellular morphogenesis.

8.2 Epigenetic Mechanisms, Epigenetic Inheritance and Cell Differentiation

Epigenetics concerns all the heritable modifications occurring during mitosis or meiosis that result in a modulation in the function of a gene, and which cannot be explained by changes in the nucleotide sequence. From a biological point of view, epigenesis is expressed in phenotype variations linked to modifications of cell properties, which are perpetuated in cell lines but are not due to genotype modifications. To sum up, all the cells in the same organism possess the same genetic information in terms of nucleotide sequence, whereas epigenetic modifications vary according to the cell type or the state of differentiation. Epigenetic modifications play a role in the regulation of gene expression over time and space. They include the methylation of DNA, post-translational modifications of histones, the local structure of DNA linked to the position of nucleosomes along the nucleotide sequence, its global structure (for example its organisation into domains), and its nuclear location.

8.2.1 DNA Methylation: Epigenetic Marker of Transcriptional Repression

8.2.1.1 DNA Methylation and Post-Translational Modifications of Histones

The methylation of DNA is a factor controlling gene expression by inhibiting the function of regulatory elements such as promoters, enhancers, isolators and silencers. Methylation profiles vary according to cell type or state of differentiation, and they are responsible for the differential gene expression required to acquire a phenotype. Methylation profiles are transmitted from a mother cell to a daughter cell during mitosis, and also to the gametes during meiosis, and then from the gametes to the embryo. This information is therefore heritable in the same way as the nucleotide sequence of DNA. Methylation has a repressive effect on gene expression: there is a strong correlation between the degree of methylation of promoters and enhancers and the scale of transcriptional repression in the associated gene. More generally, a high degree of methylation coincides with transcriptionally inactive regions of chromatin, also called "heterochromatin". Conversely, a low degree of methylation coincides with transcriptionally active regions called "euchromatin". There is, for example, a clear correlation between the expression of genes governing morphogenesis during the development of vertebrates and the absence of methylation of the associated DNA sequences [7].

On the molecular level, the methylation of cytosines in the 5 position (called 5-methylcytosines) occurs essentially within CpG pairs (about 70% of CpG pairs are methylated in somatic cells) through enzymes, the Dnmts (*DNA-methyltransferases*). These pairs are not randomly distributed through the genome; they are concentrated in or near to transcription regulatory regions. Cytosine methylation stabilises the nucleosome and it is associated with the recruitment of large macromolecular complexes capable of specific binding to methylation sites, the MeCP1/2 (*methyl-CpG-binding proteins*), comprising sub-units that bind to the methylated DNA, the MBD (*methyl-binding domains*). The binding of these complexes covers several dozen kilobases (kb) on each side of the methylation site, at the level of the promoter or enhancer, and it has a repressive effect on transcription by preventing the transcription factors from accessing the nucleotide sequence. Moreover, DNA methylation is closely linked to post-translational histone modifications via the enzymes involved in these processes, such as the HMTs (*histone-methyltransferases*) or the HDACs (*histone-deacetylases*), which are recruited notably by the MeCP1/2 complexes and reinforce the repressive effect on transcription [7].

The signature of long-term transcriptional repression (*gene silencing*) involves DNA methylation, the hypoacetylation of N-terminal histone tails and the methylation of lysine residues of histone protein, all at the same time. For example, the HDACs and HMTs can interact directly with the Dnmts to direct the DNA methylation to regions of hypoacetylated chromatin. At the same time, these modifications reinforce the "silence" of the chromatin, although the sequence of events has yet to be determined. The best-characterised example is that of lysine 9 of histone H3 (H3-K9) [8]. Likewise, MeCP2 binds selectively to methylated regions of DNA and simultaneously to HDAC, leading to stabilisation of the nucleosome. Conversely, active genes are characterised, at the level of their control regions, by the absence of DNA methylation, hyperacetylation of histones and a particular profile of histone methylation. More generally, methylation is associated with a stable nucleosome state, whereas acetylation provokes destabilisation and the unwinding of nucleosomal DNA, thereby making it more locally accessible (which is revealed, for example, by an increase in its sensitivity to ADNase I). In addition, the histone H1, responsible (through its interaction with the internucleosomal DNA) for the condensation, at a higher level, of nucleosomes into organised fibres, binds preferentially to methylated DNA and thus propagates the inactivity along the chromatin fibre. Globally, DNA methylation is therefore closely linked to the histone code, at least in part. DNA methylation therefore plays a predominant role in the processes of differentiation, development and morphogenesis [7].

8.2.1.2 The Role of DNA Methylation in Differentiation and Development

Although all cells possess the same genetic material, they have the capacity to acquire a different phenotype over space and time. This process, *differentiation*, is partly characterised by the differential methylation of genes according to their usefulness in a given tissue. The transmission of the methylation profile through cell divisions allows the same phenotype to be conserved throughout a cell line. This process is accomplished, during replication, by the recognition of methyl-CpG complexes located on the coding strand of the DNA and the methylation of the corresponding cytosines on the non-coding strand.

The development of a mammal embryo requires the presence of the two parental genomes. Although most of the genes are expressed biallelically, some of them, essential for development, are only expressed from one sole allele (monoallelic expression), from either the paternal or the maternal genome. Such genes can be subject to *genomic imprinting*, when the alleles inherited from each parent do not function in the same way, due to different epigenetic information, essentially at the level of the DNA methylation profile. There are more than sixty genes, spread over eleven chromosome regions, that are subject to this genomic imprinting. The differential methylation observed in these genes derives from the different methylation of the gametes, according to whether the gene comes from the oocyte or from the spermatozoon. This is because in the primordial germ cells – the precursors of the gametes – DNA methylation is almost nonexistent; it appears during maturation, and then in a different way in oocytes than in spermatozoa. In these cells, the parental imprinting is effaced by massive demethylation of the DNA, and new imprints are then constructed during gametogenesis according to the sex of the new embryo [7].

During development of the embryo, the first phases are characterised by complete demethylation of the genome, except for the genes subject to imprinting, which appear to be protected. The purpose of the demethylation of the other genes is to efface the regulation induced by the methylation profile of the parental genes. The genes are remethylated before implantation (at the blastocyst stage), resulting in the establishment of a new epigenetic identity specific to the embryo. The genes subject to genomic imprinting contribute additional information at the genome level, and they play a crucial role in development, because they modulate the expression of the genomes over space (cell type) and time [7].

8.2.1.3 The Role of DNA Methylation in the Inactivation of the X Chromosome

Unlike the Y chromosome, the X chromosome contains many genes that are essential to the cell. Now, females possess two copies of the X chromosome, while males only have one, and yet we cannot observe any greater expression of the products of these genes in females. This phenomenon, known as dosage compensation, is produced in female mammals by the inactivation of one of the X chromosomes, at random and at a very early stage in embryonic development. As a result, each individual, male or female, only possesses one active X chromosome. The inactive X chromosome in the female X takes the form of heterochromatin, localised to the periphery of the nucleus, and it is known as the "Barr body".

For each cell (in the female), the inactivation of one of the two X chromosomes takes place during the early stages of development, affecting either the X chromosome from the paternal genome or the one from the maternal genome, and it is transmitted to the whole of its cell line. Thus, whole regions of the organism will have the same inactivated X chromosome. Consequently, the tissues of female individuals are mosaics of two cell types. In the primordial germ cells, the two X chromosomes X are totally demethylated and therefore active before entering meiosis. They are then inactivated again by remethylation [11].

The mechanisms of X chromosome inactivation are as yet poorly understood. A gene coded by the inactive X chromosome, *Xist*, is involved in the process of inactivation through a *cis* mechanism, that is to say on the chromosome to which it belongs. This gene does not code for a protein, but only for an RNA that binds to the inactive X chromosome at the level of the Barr body. DNA methylation plays a role in maintaining the state of inactivation. In particular, one can observe methylation of the *Xist* locus of the active X chromosome and, conversely, an absence of methylation of the *Xist* locus in the inactive chromosome. Moreover, one can observe methylation of the histone H3 and deacetylation of H4 on the nucleosomes upstream of *Xist*. These histone modifications, which probably contribute to the formation of the heterochromatin, appear just after the binding of *Xist* to the Barr body and before transcription ends. More generally, the genes situated on the inactive X chromosome are largely methylated, whereas their counterparts on the active chromosome are not [11].

8.2.2 Structural and Functional Organisation of Chromatin: Spatio-Temporal Regulation

The regulation of gene expression also comprises a spatial and temporal component, because it depends at the same time on the structure and the dynamics of the chromatin, on the formation of specialised domains at the level of the chromosome and on the location of the chromosomes in the different nuclear compartments.

8.2.2.1 Chromatin Domains: the Functional Units of Gene Expression

Genes are expressed at different levels, in specific cells and at a precise moment during development. Whether a gene will be transcribed or not depends on a combination of the state of the chromatin and the availability of transcription factors. Chromatin plays a central role in this process: the hyperacetylation of histones maintains a transcriptionally active state and, at a higher level, the functional domains into which chromatin is ordered are responsible for regulating transcription. Distinct chromatin domains cover several kilobases on either side of a gene (locus). They are delimited by genetic elements such as enhancers, LCR (locus control regions), S/MAR (scaffold-matrix attachment regions) and isolators. The enhancers and LCR are the targets of transcription factors, and therefore directly involved in the level of expression of a gene, by increasing the speed of transcription. They also play a role in ordering chromatin into a suitable spatial configuration for transcription (active chromatin hub) and in the targeting of chromatin to specific nuclear sites (for example, the "transcription factories"). The S/MAR are domains of about 1 kb, localised to the boundaries of chromatin domains, which act to anchor the nuclear matrix and regulate the nuclear architecture. The S/MAR are made up of regions of unpaired DNA, which coincide with regions rich in AT, the origins of replication, the recognition sites of topoisomerase II and the 3'-UTR. Interactions between the proteins of the nuclear matrix and single-stranded S/MAR serve to stabilise the transcriptionally active state of the chromatin. Thus, the S/MAR facilitate transcription by reducing the mechanical constraints imposed on DNA during this process, and by localising chromatin domains to transcription centres associated with the nuclear matrix (see Sect. 8.2.2.3 on nuclear compartments). Isolators are elements that demarcate the boundaries between chromatin domains with distinct condensation states. They function by isolating a domain from the influence of neighbouring domains or by preventing the regulatory elements of one domain from activating a promoter situated in another domain. Epigenetic modifications at the level of the isolators can also influence the structure of domains and control gene expression. For example, isolators can recruit diverse acetyltransferases to prevent chromatin inactivation in an active locus, by propagating the methyltransferase activity of a neighbouring locus. In a particular cell type, an active gene may find itself surrounded by genes that are not useful in this context and are therefore not expressed; the isolator can protect the active gene from the neighbouring chromatin, which is present in a highly condensed, methylated and hypoacetylated form and which would otherwise have a distant inhibiting effect on the gene's transcription [1, 9].

8.2.2.2 Chromosome Domains

The chromosome is the ultimate state of condensation of chromatin. It is the most compact chromatin structure, of about 1 μm, and it is found in mitosing cells. Indeed, this form is more favourable to the transmission of genetic material from mother cell to daughter cell during mitosis. In interphase cells, chromatin is found in a more diffuse form, favourable to the processes of replication and transcription. The chromatin of an interphase cell is constituted of nucleosomes that are compacted together by interactions with histone H1 to form fibres of 30 nm diameter. These fibres are themselves rolled up in the form of loops containing between 10 and 150 kb, held together at the base by a protein complex containing topoisomerase II, which also controls the degree of DNA supercoiling in these loops. The function of the loops is to demarcate the DNA into domains (their number coincides with the number of genes in the genome), each domain comprising a group of genes regulated by one sole mechanism (see Sect. 8.2.2.1 above). However, even in an interphase cell, containing mainly euchromatin (the diffuse form of transcriptionally active chromatin) we can observe that about 10% of the genome remains in the form of heterochromatin, that is to say in a condensed form, usually localised on the periphery of the nucleus [1, 9].

Certain regions of the genome are made of *facultative heterochromatin*, that is to say regions that are specifically inactivated during certain stages of development or differentiation. This is the case for the Barr body, the inactive form of the X chromosome in females, described above in Sect. 8.2.1.3. Other regions of DNA are preferentially in the form of *constitutive heterochromatin*, like the highly repetitive sequences (satellite DNA) such as the centromeres and telomeres: they are devoid of protein-coding genes and always transcriptionally silent. These regions play a major role in maintaining the integrity of the genome and in the transmission of genetic information. The *centromeres*, which contain most of the constitutive heterochromatin, are specialised chromosome regions, involved, during replication, in the binding of each chromatid to the microtubules constituting the mitotic spindle, via a protein complex, the *kinetochore*. In particular, in mammals, genes are observed to associate differentially with the centromeric heterochromatin according to their level of expression. Consequently, the centromere plays a distant role in the inactivation of genes. *Telomeres* are repetitive sequences of DNA, identical in all vertebrates and localised at each end of the chromosomes. The 5'-3' strand has one end that is longer and rich in G, while the complementary strand has a telomeric sequence that is shorter and rich in C. A reverse transcriptase, telomerase, which contains an RNA complementary to the 3' end, adds new repetitive sequences in 3', allowing to maintain a short single-stranded tail in 3' during cell division. The telomeres play an important role in the complete replication of the chromosome; they protect the DNA from nucleases, prevent the fusion of chromosome ends and favour the interaction of chromosomes with the nuclear envelope [1, 9].

The ordering of chromosomes into functional domains is required for various biological processes. Epigenetic effects come into play, alongside regulation processes dependent on the nucleotide sequence, in the establishment of the architecture and function of the chromosomes. For example, epigenetic factors play a role in the assembly of centromeric elements and their transmission during replication. Although both the DNA sequence and the size of centromeres vary greatly from one species to another, the inclusion in the centromeric chromatin of an H3 histone variant, the centromere protein A (CENP-A), is a common characteristic. Recent studies have also demonstrated the presence of dimethylated H3 histone on lysine 4. The centromeres are flanked by large regions of *pericentric* heterochromatin, containing trimethylated H3 histones on lysine 9 combined with HP1 (*heterochromatin protein 1*), an adaptor protein that binds and stabilises the methylated lysines and also combines with Dnmt to maintain the chromatin in a transcriptionally repressed state [1, 9].

The localisation of a gene can have marked effects on its expression during development. For example, heterochromatin can have a distant effect on gene expression. If, during a transposition or a chromosomal translocation, a normally active gene should enter a region near the heterochromatin, it can become inactive. This phenomenon is known as position effect variegation. In parallel, the state of expression of a gene can also orient its chromosomal and nuclear localisation. An inactive gene will be relocalised to the periphery of the nucleus, within the heterochromatin. Likewise, a gene that is to be expressed will be moved from an inactive locus associated with the nuclear envelope (see below, Sect. 8.2.2.3) towards an active euchromatin site in the centre of the nucleus, before transcription [1, 9].

8.2.2.3 Nuclear Compartments and Chromosome Territories

In eukaryotes, the genetic material is isolated in the nucleus and separated from the cytoplasm by the nuclear envelope, which acts as a permeable membrane allowing the controlled entry of cytoplasmic compounds and proteins into the nucleus and the release into the cytoplasm of nuclear elements such as messenger RNAs. These exchanges take place notably through the nuclear pores. The nuclear envelope is made up of two membranes, fused in places to form circular pores containing a protein complex called the nuclear pore complex (NPC). These complexes act like plugs, blocking the pore and regulating the transport of molecules between the nucleus and the cytoplasm. The inner surface of the envelope is composed of nuclear *lamina*, a fibrillar network that serves as a support to the nuclear envelope and for the anchoring of heterochromatin localised at the periphery of the nucleus (for example, the Barr body). The nuclear envelope is less impermeable than the cytoplasmic membrane. The NPC allows the entry of small molecules through diffusion, whereas proteins enter or leave the nucleus via receptors, provided they possess the appropriate short sequence of basic amino acids: a nuclear localisation signal or a nuclear export signal respectively. The NPC is composed of an assembly of basic proteins found in all NPCs, combined in a dynamic and adaptable way with different proteins that modify the transport properties of the NPC and thereby modulate the response of the nucleus according to physiological needs [1, 9].

The cell nucleus is organised and compartmentalised into several regions, which are not, however, bounded by membranes. These sub-organelles contain very high concentrations of nuclear constituents, which suffices to demarcate distinct, specialised compartments capable of performing specific functions. This compartmentalisation is therefore a dynamic process that allows the formation of certain nuclear sub-organelles according to the needs of a particular cell at a given moment in its life. These nuclear compartments include the nucleolus, various nuclear bodies such as Cajal bodies or PML (promyelocytic leukemia bodies), transcription factories, OPT domains (Oct1/PTF/Transcription) and interchromatin granules.[4] The nucleolus is one of the major nuclear domains of interphase cells, because it is the site of

[4] See Chap. 7 on intracellular organisation and morphogenesis.

ribosome biosynthesis. It is also involved in the synthesis and treatment of ribosomal RNA. The nuclear bodies are highly specialised units, defined morphologically and functionally by the proteins concentrated in them. For example, the PML are composed of a protein of the same name and involved in different processes such as transcription, DNA repair, regulation of the cell cycle, apoptosis, etc. but their real biological function has not yet been clearly defined. The OPT domains and transcription factories are regions with high transcriptional activity, because they are rich in transcription factors and RNA-polymerase II/III. Certain chromosomes are preferentially associated with these compartments (the chromosome 6, for example, is closely associated with OPT domains). However, the concentration of transcription factors in certain compartments is not necessarily synonymous with a high level of transcriptional activity: they may simply be storage zones [1, 9, 18].

Chromatin does not take up the whole volume inside the nucleus, but precise zones, with heterochromatin on the periphery and euchromatin in the middle. The interchromosomal space contains the machineries for the transcription and maturation of transcribed RNAs (pre-RNA messengers), such as the interchromatin granules. Each chromosome occupies a very specific territory, the chromosome territory, and this spatial organisation is related to functional needs. This is testified by the architectural reorganisation of the nucleus that takes place during the various stages of differentiation. Thus, it appears that the spatio-temporal organisation of chromatin in the nucleus is closely connected to a transcriptional need with regard to a given group of genes and vice versa. An organisation is also established within each chromosome territory, with active genes on the periphery and inactive ones in the centre. In particular, we can observe substantial reorganisation of these territories during phases of DNA replication and mitosis [1, 9].

A body of evidence now clearly indicates the existence of a structural and functional organisation of the constituents of the nucleus to perform its various functions, following mechanisms that remain obscure. The organisation of nuclear compartments is a dynamic process, in space and time, in the same way as the organisation of the genes and of chromatin. The mechanisms of nuclear compartmentalisation are as yet poorly understood, but they appear to be subjected to multiple topological constraints imposed by the nucleus itself. We can, however, underline the role of the nuclear matrix, which appears to be equivalent to the cytoskeleton, serving as a support to the organisation of the nucleus. The nuclear matrix is an insoluble fibrillar network made of proteins (including the nuclear lamina), which maintain the architecture of the nucleus and serve as scaffolding for the organisation of the different elements. For example, chromatin loops occupy a limited space in the nucleus, and we can also observe that the pre-RNA messengers newly-synthesised during transcription are concentrated in well-defined zones, around the gene from which they originate, at the periphery of the chromosome territory: the new RNA messengers are in fact associated with the nuclear matrix, which subsequently directs them from the centre to the periphery of the nucleus (during which process the introns are removed by alternative splicing), ultimately to be exported to the cytosol via the NPC [1, 9].

8.2.2.4 Cell Memory: Epigenetic Regulation by Polycomb and Trithorax Proteins

During the development of a multicellular organism, the cells adapt their genetic programme to acquire different phenotypes in response to a series of stimuli. In a cell line, the expression profile of genes involved in the process of differentiation is conserved over the numerous mitotic divisions. This cell memory is encoded by genetic and epigenetic factors that enable the cell to transmit its programme of expression to its descendants. A group of proteins and genes store a memory of the transcriptional programmes of genes playing an essential role in development. Among these proteins, two sets of antagonistic regulators intervene to maintain the genes in a transcriptionally repressed or active state: the Polycomb group (PcG) and the Trithorax group (TrxG) respectively. PcG and TrxG are necessary notably to maintain the gene expression profile *Hox*, a class of regulators highly evolutionarily conserved and involved in many development-related cell processes: the target genes of Hox transcription factors play a role in organogenesis, cell differentiation, cell cycle regulation and apoptosis, and cell adhesion and migration. Most of the proteins coded by the genes[5] *TrxG* and *PcG* play a part in large macromolecular complexes that modify the local structure of the chromatin around the target genes to induce transcriptional repression or activation. It also appears that the Polycomb group is capable of organising interactions between different chromosome regions. Furthermore, the composition and properties of the PcG and TrxG complexes differ from one target locus to another, and the various PcG proteins can have different expression profiles according to the tissue or stage of development [10].

To sum up, epigenetic factors intervene to modify the local structure of the chromatin through biochemical modifications of the DNA and the histones. These modifications can spread along the DNA over several nucleosomes and act, at a larger scale, on the chromatin dynamics. We can thus clearly observe regions of euchromatin, structurally more diffuse, favourable to transcription, and regions of heterochromatin, condensed and transcriptionally silent. In addition, the epigenetic coding also has a topological dimension, which plays an essential role in the gene expression profile: the localisation of a gene in an active or inactive chromatin region has an impact on its state of expression. Conversely, the activity or inactivity of a gene leads to the transfer of that gene to a transcriptionally active or inactive region of the chromatin. At a larger scale, the position of a gene in the chromosome territory and in the nucleus affects its accessibility to the transcriptional machinery. An active gene will generally be localised on the periphery of the chromosome territory and in the centre of the nucleus, associated with highly transcriptionally active compartments, while inactive genes are more likely to be buried inside the chromosome, sometimes at the interface with regions of constitutive heterochromatin like the centromeres or telomeres, and on the periphery of the nucleus, in contact with the nuclear lamina [1, 9, 10, 18].

---

[5] Following convention, we denote in italics the gene, here *PcG*, associated with the protein of the same name, here PcG.

8.3 The Link Between Epigenetic Information and the Regulation of Gene Expression

Each cell contains more information than it actually uses. Thus, from the first embryonic stages in the development of a multicellular organism, the cells begin to differentiate so as to acquire different phenotypes. Genetic and epigenetic information is processed to allow the expression of the appropriate genes in each cell type. In parallel, all this information, genetic and epigenetic, is transmitted from a mother cell to its descendants in a cell line, during mitosis or meiosis. By playing a central role in the encoding and transmission of genetic information, DNA is at the forefront of this process. Genes code proteins that will perform many different functions in the cell. Some of these proteins will, in return, have an effect on gene expression. They include the RNA-polymerases, the transcription factors and their coregulators, and other elements playing a part in cell replication and DNA repair, which will ensure the integrity and transmission of genomic information. The epigenetic information, described in Sect. 8.2, adds a second level to the regulation of gene expression and the transmission of expression profiles. All this information, coded in the chromatin, must be integrated by the cell and processed to maintain a correct level of gene expression over space and time.

The regulation of the expression of all the genes must therefore be very fine-tuned. It calls on a combination of different elements, which must all be present together to control the activation of a gene at a given moment in a particular tissue. Gene transcription is therefore a combinatorial operation: a gene is linked to a promoter and regulated in *cis* by several enhancers, themselves regulated in *trans* by different transcription factors associated with a large number of coactivators or corepressors. The enhancers are responsible, together with the transcription factors and their coregulators, for the differential expression of genes over space and time. Thus, a particular combination of promoter, enhancer and transcription factor must be established in a given tissue to enable the expression of a gene of interest. To which we can add the epigenetic modifications that cooperate with the elements involved in transcription to regulate the level of gene expression in a particular context. For example, many transcription factors recruit histone acetyltransferases (HAT) that intervene by remodelling the chromatin to facilitate the access of the transcription factors to the DNA. On this subject, it should be noted that the remodelling of chromatin is a mechanism common to several different DNA-related processes including replication, transcription and repair, and that it can serve as a mediator in the relationships between these different functions [2, 4].

8.3.1 The Link Between DNA Repair and Transcription

Because of spatial proximity, through various different interactions with the chromatin, certain nuclear components do not meet up quite at random. An effect of local concentration around the chromatin can favour the formation of functional complexes operating under particular conditions at a given moment. For example, the DNA distortions caused by specific lesions, due to ultraviolet radiation or other carcinogenic factors, call for repair by means of nucleotide excision repair (NER), in which XP proteins (*Xeroderma Pigmentosum*) play a crucial role. One of the major complexes involved in this process is the transcription factor TFIIH, originally identified as an element in the transcription initiation complex associated with RNA-polymerase II. The TFIIH factor also plays a role in DNA repair through NER: it operates at the level of the incision of the damaged strand via its helicase subunits, XPB and XPD, by intervening in the recruitment of repair proteins by interaction between its subunit p62 and XPG. It has been shown that TFIIH could help to initiate the synthesis of RNA messengers by RNA-polymerase II, once the repair site has been dissociated; it therefore participates not only in the repair of the DNA via its helicase function, but also in the transcriptional regulation of the repaired gene. NER deficiency in humans is linked to severe diseases like Xermoderma Pigmentosum or Cockayne syndrome, resulting from multiple DNA damage affecting numerous genes or regulatory regions. These genetic diseases are associated with heightened sensitivity to sunlight, an increase in the risk of developing skin cancers and neurological dysfunctions [17].

The functioning of TFIIH is the best-known example of the link between transcription and DNA repair. Recently, however, a link between these same processes has been identified at the level of sites of epigenetic regulation, involving a mechanism of base excision repair (BER) and cytosines. Cytosine is an unstable DNA base because it can lead, through deamination, to the formation of uracil and consequently to G : U mismatches. These mismatches can easily be recognised in the DNA and repaired by enzymes called Uracil-DNA Glycosylases (UDG), which are, moreover, highly evolutionarily conserved proteins. These same cytosines are the seat of epigenetic regulation processes through the methylation of DNA and they are liable to lead to G : T mismatches, specifically repaired by Thymine-DNA Glycosylase (TDG). It has recently been shown that TDG combines with an acetyltransferase (HAT), CBP (Creb-binding protein), which acts as a coactivator for many transcription factors in the family of nuclear receptors, including the retinoic acid receptors (RAR) [21]. Furthermore, TDG also intervenes directly as coregulator of the RARs and other nuclear receptors. It has been suggested that CBP might intervene in DNA repair initiated by TDG, remodelling the chromatin through histone tail acetylation, so as to facilitate the access of TDG to the faulty base. Moreover, CBP acetylates TDG and this acetylation induces the dissociation of CBP from the complex and negatively regulates the recruitment of APE, an endonuclease that intervenes in the repair of the abasic site. On the basis of this data, it has been suggested that CBP/TDG, like TFIIH, plays a role in the regulation of transcription after repair. As DNA methylation plays a major role in maintaining the transcriptionally inactive state of a gene at the level of the CpG islands in the regulatory regions, the production of a G : T mismatch followed by its repair by TDG/APE contributes to the restoration of a non-methylated, transcriptionally active site. The dissociation of APE following the acetylation of TDG by CBP and the dissociation of CBP would have the consequence of activating the transcription of the repaired gene by the nuclear receptors via TDG. Thus, the state of acetylation of TDG would direct its activity, according to the target genes, towards a simple DNA repair or towards a possible role in the activation of transcription via the "indirect" demethylation of the DNA, linking TDG to the mechanisms of epigenetic regulation of transcription [6, 15, 21, 22].

8.3.2 CBP/p300, HATs Involved in Cell Growth, Differentiation and Development

CBP and its homologue p300 are very similar in terms of structure and function, and both interact with various elements in the nucleus: transcription factors (Fos, Jun, E1A, CREB, E2F, MyoD, p53, NF-kB, hormone nuclear receptors) and acetyltransferases (PCAF, SRC-1, TIF2, ACTR/SRC-3). CBP/p300 coordinates transcriptional regulation in numerous ways, establishing communication between several transcription factors around the same regulatory region, called an *enhanceosome*, regulating their activity and taking part in the transduction of cell signals towards a transcriptional response. These two acetyltransferases play a part in transcription by modulating the chromatin structure through the acetylation of histone tails, facilitating the access to specific nucleotide sequences for transcription factors, and also for general transcription factors like TFIIB (*RNA Pol II-associated transcription factor B*), TBP (*TATA-binding protein*) and RNA-polymerase II. It should be pointed out that acetyltransferases are also capable of acetylating proteins other than histones, like the transcription factors (p53, MyoD), for example.

In particular, CBP/p300 play a part in the regulation of the hematopoietic system by interaction with numerous transcription factors crucial to the formation of myeloid and lymphoid cells. Interestingly, CBP/p300 can interact just as well with transcription factors involved in differentiation and cell cycle arrest as they can with those involved in cell cycle progression and proliferation. CBP/p300 intervene in the regulation of processes of cell division and differentiation through several different mechanisms. Firstly, CBP/p300 are involved in the expression of genes linked to ligand-dependent nuclear receptors, activated by lipophilic hormones, such as the retinoic acid receptors (RXR/RAR), oestrogen receptors (ER), progesterone receptors (PR), thyroid receptors (TR) and glucocorticoid receptors (GR). These nuclear receptors operate in various cell differentiation pathways and function by recruiting coactivators to perform their transcriptional activity in response to hormonal stimuli. For example, CBP binds to another acetyltransferase, SRC-1 (*steroid receptor coactivator-1*), via its C-terminal region, while the nuclear receptors occupy its N-terminal region. Thus, CBP and SRC-1 cooperatively stimulate both the oestrogen and the progesterone receptors. Secondly, it has been suggested that CBP/p300 plays a role in the suppression of tumours, because the binding of certain viral oncoproteins inhibits their activity, thereby interfering in cell differentiation and favouring cell cycle progression. The binding of the adenovirus E1A protein at the level of one of the protein-protein interaction regions of CBP induces deregulation of the mechanisms involved in control of the cell cycle and differentiation, leading to the entry in S phase and cell proliferation. Conversely, the binding of PCAF (*p300/CBP-associated factor*) to the same region of the CBP is believed to suppress cell growth. PCAF could therefore influence the transcription of genes involved in cell cycle arrest, differentiation and apoptosis [21].

Generally, CBP/p300 offer multiple interfaces that can be occupied simultaneously and combinatorially, to respond to a specific stimulus and trigger the transcription of the associated genes. Moreover, limiting quantities of CBP/p300 in the nucleus could orient their interaction preferentially towards certain partners according to the stimuli received, and deprive other partners of their action. Consequently, the activity of CBP/p300 would only be directed at one sub-group of transcription factors at any given moment: either those involved in cell differentiation, or those involved in proliferation.

Because of their crucial role in the assembly of the transcriptional machinery, CBP/p300, like other coregulators, are at the heart of the transcriptional regulation of specific genes necessary to the processes of cell growth, differentiation and development. However, the mechanisms that enable CBP/p300 to organise the transcriptional machinery in space and time remain largely unknown. And yet this is a question of major interest, especially since functional deregulations of CBP/p300 are involved in several cancers. A malignant transformation of cells, associated with CBP/p300 mutations, is probably caused by the disruption of crucial signalling pathways involving these coactivators. CBP/p300 mutations have been implicated in a wide variety of tumours, such as myeloid leukemia or gastric and colorectal carcinomas. In certain leukemias (AML, acute myeloid leukemia), a chromosome translocation of the genes *MOZ* (monocytic leukemia zinc finger) and *CBP* leads to the expression of a fusion protein MOZ-CBP, where the acetyltransferase domain MOZ is fused with almost all of the CBP. Likewise, the fusion proteins MLL-CBP, where the *MLL* (mixed lineage leukemia) gene is highly homologous to the *trX* gene group in Drosophila, are found in more than 80% of child leukemias. Although the physiological consequences of these processes have not yet been elucidated, it is believed that these chimeric proteins may possess a different specificity than CBP with regard to target genes, and may target a subset of genes regulated by CBP. Moreover, inactivating mutations of a CBP allele in germ lines are the cause of an autosomal dominant genetic disease, Rubinstein-Taybi syndrome, characterised by mental disability, craniofacial anomalies and a susceptibility to neuronal and hematopoietic cancers. These observations indicate firstly, that the level of CBP is limited in the cell and plays a crucial physiological role in embryogenesis, and secondly, that p300 cannot compensate for CBP function [21].

8.3.3 Epigenetics and Oncogenesis

Disruption of the epigenetic equilibrium has a marked impact on chromatin structure and dynamics, and consequently on transcription. To the extent that DNA methylation is used to "turn off" genes, either locally or from a distance, the establishment and maintenance of correct methylation profiles are essential to normal development and functioning in mammals. In particular, profound disruption of DNA methylation profiles have been observed in cancers, generally in the form of a large loss of methylation affecting the whole genome, associated with a local gain in methylation in certain regions. From a clinical point of view, the loss of normal methylation profiles occurs during the early stages of oncogenesis, and it is correlated to the severity of the disease and the metastatic potential of a large number of tumours. On the level of the chromatin, global hypomethylation affects coding regions, introns and repetitive DNA sequences. It contributes to the acquisition of the malignant phenotype by affecting the functional stability of the chromosomes, by reactivating transposable elements and by the loss of normal profiles of the genetic footprint. As for hypermethylation, it mainly affects CpG islands localised in regulatory regions such as promoters. It contributes to the repression of genes, and in particular tumour-suppressing genes. DNA-methyltransferases (Dnmt) are strongly involved in the establishment and maintenance of methylation profiles, thereby playing a crucial role in differentiation. For example, embryonic stem cells deficient in Dnmt are viable up until the induction of differentiation. At this stage, they prove to be incapable of differentiating, and they die. Moreover, Dnmt recruit histone-deacetylases (HDAC) at the level of the promoters to reinforce the repressive effect on transcription. Thus, specific histone acetylation profiles are associated with highly methylated promoters: in the hypermethylated CpG islands of inactivated tumour-suppressing genes, the H3-K9 residue is methylated, whereas in the non-methylated CpG islands of actively transcribed genes, this site is acetylated. Consequently, the equilibrium between methylation and acetylation of H3-K9 could help to regulate the methylation state of the promoter, since these two concurrent modifications precede the hypermethylation of the promoter. Furthermore, the loss of certain epigenetic modifications, such as the acetylation of H4-K16 and the trimethylation of H4-K20, can be considered as an epigenetic marker of malignant transformations and as a factor of prognosis.

New therapeutic strategies are being envisaged, targeting these aberrant gene repressions in cancers. The use of HDAC inhibitors alone is not satisfactory in the re-expression of hypermethylated genes in tumoral cells. On the other hand, their use in combination with demethylating agents has a synergetic effect on the re-expression of genes. Generally, HDAC inhibitors lead to the accumulation of hyperacetylated nucleosomes, favourable to transcription, but they are also capable of inducing the acetylation of transcription factors, particularly the tumour suppressor p53. The action of HDAC probably has an influence on the acetylation state of p53 and on the effectiveness of its binding to the promoter of *p21*, a CDK inhibitor. The transcription of *p21* activated by p53 would have the effect of orienting the tumoral cells towards cell cycle arrest and apoptosis.

The therapeutic prospects for epigenetic modifications are very promising, inasmuch as these modifications are reversible, unlike the genetic factors (i.e. linked to the DNA sequence) involved in cancers. Better knowledge of the mechanisms at work in epigenesis and structural identification of the partners involved would make it possible to define rational approaches for the design of therapeutic compounds [21].

To sum up, the CBP/TDG complex creates a direct link between two fundamental processes: transcriptional regulation and epigenetic regulation, which both play a significant role in the regulation of gene expression. Two facts are particularly interesting: (i) the two mechanisms entail the modification of chromatin structure and dynamics, and in both cases the activity of the histone-acetyltransferase CBP is at the origin of the modifications; (ii) the link between these two mechanisms creates a genomic plasticity, allowing to integrate both information about transcriptional regulation and epigenesis into gene expression to establish specific genetic programmes [5, 15, 21, 22].

8.4 Morphogenomics

In the first three sections of this chapter, we have seen that the genome of an organism cannot be reduced to its DNA; we cannot consider it without the proteins and other associated molecules with which it forms the chromatin structure in eukaryotic cells. Chromatin possesses structural dynamics at several scales, functionally very important, which are themselves an integral part of the encoding and decoding of information about the regulation of gene expression. Even more importantly, we have seen that this chromatin dynamics produces very distinct functional forms at the level of the genome (Fig. 8.1 above), and that these functional forms are directly involved in the coding of information. The plasticity between the different processes related to the genome – an exemplary illustration of which we have discussed here with the role of CBP/TDG interaction in transcriptional regulation, epigenesis and DNA repair by base excision – is achieved by the chromatin, or more precisely by its local and global dynamics, and therefore by its functional forms. We can therefore state not only that chromatin plays a key role in integrating information into the genome, but also that this central function is nothing other than a process of morphogenesis at the macromolecular level. Using the name *"epigenome"* to bring together the constituent molecules of the genome, the chromatin and the epigenetic structures underlines this point, by formulating the morphodynamics of the genome as the foundation of the coding of cell identity. Fig. 8.2 seeks to provide a schematic illustration of this genomic morphodynamics, or "epigenomics". Over time, the different genomic positions along the DNA frequently change their chromatin structure, according to inter-, intra- and extra-cellular metabolism and signalling. This is both the result of local genomic activity and at the same time the cause of that activity – the functional form contains some of the information needed to establish the activity and so to establish itself. Such identity between cause and effect brings strongly to mind the ideas set out by Erwin Schrödinger in his book *What is Life?* [19], and is quite remarkable: it appears to be the first principle used by nature. The first molecules capable of self-reproduction, and therefore of being both the reproductive machine and the very subject of that reproduction, were the first forms to satisfy the definition of the ultimate functionality of the living being. Secondly, the notions of functional form and its morphogenesis are inseparably linked to this

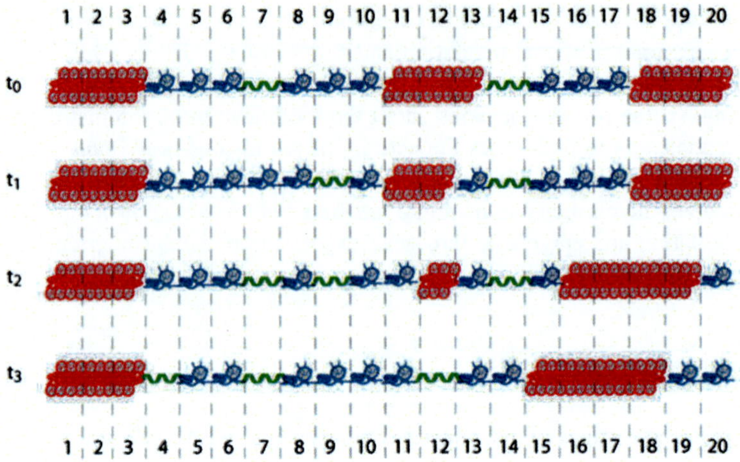

Fig. 8.2 Morphogenomics. At every moment, the genome is present in different local functional forms, from naked DNA (*green*), to condensed fibres (*red*). Over time, following a regulated chromatin dynamics, the different parts of the genome alternate between different functional forms

concept, and have been from the start: catalysis is a process that requires well-defined structures; if these structures are capable of self-reproduction, then these forms emerge anew with every generation. Life itself is therefore the consequence of emergent functional forms; life is morphogenesis.

This logic can be taken even further. In this chapter, we have spoken a lot about the information necessary for the emergence of biological morphogenesis and its encoding. The nature of this information is structural. DNA bases are recognised by transcription factors by their physico-chemical properties, which are themselves a function of the chemical structure of the nucleotide in question. Proteins acquire their function, for example their interaction with DNA sequences, via their three-dimensional structure. Likewise, the concentration of metabolites, or ions, for example, is also perceived by the biological system through specific interactions with the macromolecules present. This specific recognition, which therefore allows to transport information, is only possible thanks to structures – functional forms. Although these structures emerged during evolution, this evolutionary morphogenesis, which allows the encoding and interpretation of the biological information necessary for the reproduction of living forms, is the very manifestation of life.

The functional form emerges following the information coding for its structure, which evolves over time to create even more efficient compound functions to ensure their own reproduction in their dynamic environment. The different local and global functional forms of the genome allow efficient coding of the information necessary to specify the cell programmes and hence the cell identity. In the introduction, we argued that only this regulated cell specification could explain the existence of forms at the scale of organs. This appears to lead to a contradiction. For biological information is most often considered as having a slow dynamics, at the scale of evolutionary time. Our sequence of genes or the elements of their regulation is too highly conserved to explain the multitude of appearances of biological forms. Each human being has a unique face; each zebra has a unique pattern of stripes. That is why the role of the genome in morphogenesis has for so long been considered of secondary importance. Only today has this enigma found an explanation, with a better understanding of the phenomena of chromatin dynamics, epigenetics and their interactions with genomic information. The processes of epigenetic regulation are produced over much shorter time scales than those of evolution – at the level of the generation times of the organism. The processes of the chromatin dynamics are even faster, and their multitude, their combinatorial nature, and the elements of stochasticity that lead to phenomena of stochastic differentiation, allow, in turn, the emergence of biological forms displaying wide variability in detail. The "hard" information encoded in the genes and their mechanisms of regulation ensure the *functionality* of these forms, and therefore *define* the *limits of variability* of a specific biological form. Consequently, cell differentiation, which is a function of the epigenome, is the *support* for morphogenesis. The definition of these *boundaries of the viable morphogenetic space* allow, in particular, to constrain the dynamics of metabolites or other small molecules and their action as morphogenetic agents.

The morphodynamics of the genome (Fig. 8.3) leads to morphogenetic emergence, which is nothing other than the expression of the self-reproduction of a set of functional forms. Isn't life wonderful?

Fig. 8.3 Morphogenomic life cycle

References

1. Bartlett J., Balgojevic J., Carter D., Eskiw C., Fromaget M., Job C., Shamsher M., Trindade I.F., Xu M., and Cook P.R. (2006) Specialized transcription factories, *Biochem. Soc. Symp.* **73**, 67–75.
2. Benecke A. (2003) Genomic plasticity and information processing by transcriptional coregulators, *ComPlexUs* **1**, 65–76
3. Benecke A. (2003) Chromatin dynamics are a hyperstructure of nuclear organization, in *Proceedings Modelling and Simulation of Biological Processes in the Context of Genomics*, edited by P. Amar, F. Képès, V. Norris, and P. Tracqui, Platypus Press (Paris), pp. 31–40.
4. Benecke A. (2006) Chromatin code, local non-equilibrium dynamics, and the emergence of transcription regulatory programs, *Eur. Phys. J. E* **19**, 379–384.
5. Benecke A. and Gronemeyer H. (1998) Nuclear receptor coactivators as potential therapeutical targets: the HATs on the mouse trap, *Gene Ther. & Mol. Biol.* **3**, 379–385.
6. Chen D., Lucey M.J., Phoenix J., Lopez-Garcia J., Hart S.M., Losson R., Buluwela L., Coombes R.C., and Chambon, P. (2003) T : G mismatch-specific thymine-DNA glycosylase potentiates transcription of estrogen-regulated genes through direct interaction with estrogen receptor alpha, *J. Biol. Chem.* **278**, 38586–38592.
7. Esteller M. (2006) CpG island methylation and histone modifications: biology and clinical significance, *Ernst Schering Res. Found. Workshop* **57**, 115–126.
8. Fuks F. (2005) DNA methylation and histone modifications: teaming up to silence genes, *Curr. Op. Genet. Dev.* **15**, 490–495.
9. Gilbert N. and Bickmore W. (2006) The relationship between higher-order chromatin structure and transcription, *Biochem. Soc. Symp.* **73**, 59–66.
10. Grimaud C., Negre N., and Cavalli G. (2006) From genetics to epigenetics: the tale of Polycomb group and trithorax group genes, *Chromosome Res.* **14**, 363–375.
11. Heard E. (2005) Delving into the diversity of facultative heterochromatin: the epigenetics of the inactive X chromosome, *Curr. Opin. Genet. Dev.* **15**, 482–489.
12. Képès F. and Vaillant, C. (2003) Transcription-based solenoidal model of chromosomes, *ComPlexUs* **1**, 171–180.
13. Lesne A. (2006) The chromatin regulatory code: beyond an histone code, *Eur. Phys. J. E* **19**, 375–377.
14. Lesne A. and Victor J.M. (2006) Chromatin fiber functional organization: some plausible models, *Eur. Phys. J. E* **19**, 279–290.
15. Lucey M.J., Chen D., Hart S.M., Phoenix F., Al-Jehani R., Alao J.P., Whiet R., Kindle K.B., Losson R., Chambon P., Parker M.G., Schar P., Heery D.M., and Buluwela L. (2005) T : G mismatch-specific thymine-DNA glycosylase (TDG) as a coregulator of transcription interacts with SRC1 family members through a novel tyrosine repeat motif, *Nucleic Acids Res.* **33**, 6393–6404.
16. Norris V., Fralick J., and Danchin A. (2000) A SeqA hyperstructure and its interaction direct replication and sequestration of DNA, *Mol. Microbiol.* **37**, 696–702.
17. Riedl T., Hanaoka F., and Egly J.M. (2003) The comings and goings of nucleotide excision repair factors on damaged DNA, *EMBO J.* **22**, 5293–5303.
18. Sansam C.G. and Roberts C.W. (2006) Epigenetics and cancer: altered chromatin remodeling via Snf5 loss leads to aberrant cell cycle regulation, *Cell Cycle* **5**, 621–624.
19. Schrödinger E. (1944) *What is Life?*, Cambridge University Press (Cambridge).
20. Schuster P. and Eigen M. (1979) *The hypercycle*, Springer (Berlin).
21. Tini M., Benecke A., Um S.J., Torchia J., Evans R.M., and Chambon P. (2002) Association of CBP/p300 acetylase and thymine DNA glycosylase links DNA repair and transcription, *Mol. Cell* **9**, 265–277.
22. Um S.J., Harbers M., Benecke A., Pierrat B., Losson R. and Chambon P. (1998) Retinoic acid receptors interact physically and functionally with the T : G mismatch-specific thymine-DNA, *J. Biol. Chem.* **273**, 20728–20736.

Chapter 9
Animal Morphogenesis

Nadine Peyriéras

Morphogenesis and cell differentiation are interdependent during the embryonic development of metazoans[1] leading from the fertilised egg to the organism capable of reproduction. Morphogenesis and differentiation also come into play throughout the life of the organism, in the physiological processes of cell renewal or regeneration and in many pathological processes such as cancerogenesis.

In fact, the egg is just one stage in the cycle of life itself, caught up in a spiral that has been constantly turning for nearly four billion years. These two time scales, in which intervene ontogeny and phylogeny respectively, are one of the aspects of the complexity of life. Contrary to the theses of Ernst Haeckel,[2] ontogeny does not recapitulate phylogeny,[3] but the two are interdependent and the one is hardly intelligible without the other. On the one hand, the unique cell of the egg and the processes of morphogenesis are the culmination of a long construction about which we can only speculate. On the other, the emergence of new life forms must pass through the sieve of ontogeny, in the sense that all variations in phenotype have embryological foundations and only those variations that do not compromise the embryonic viability of the organism will be transmitted from generation to generation. Moreover, conceiving the continuity of living beings through the cell prevents us from forgetting that what is transmitted from one generation to the next is not only DNA, but the whole organisation and structuring of the cell and its components, without which the DNA sequence would be nothing.

N. Peyriéras (✉)
Institut Alfred Fessard, Gif-sur-Yvette, France
e-mail: nadine.peyrieras@inaf.cnrs-gif.fr

[1] In the phylogenetic classification of living beings, metazoans are multicellular organisms, see [21].

[2] See the reference work [9] on the state of the art in developmental biology and the historical foundations of the discipline, and also [10].

[3] The reader could compare this current point of view of developmental biology, ruling that knowledge of phylogeny is not sufficient to understand ontogeny, with the apparently contradictory view given in Chap. 11 by a paleontologist, affirming that the ontogeny we can observe today in the laboratory provides precious indications for reconstituting evolutionary history, in other words phylogeny.

P. Bourgine, A. Lesne (eds.), *Morphogenesis*, DOI 10.1007/978-3-642-13174-5_9,

As Richard Lewontin observed[4] [24], the metaphor of development is not the best-chosen for representing ontogeny and the morphogenetic processes underlying the formation of the organism. For morphogenesis is not the simple deployment of a pre-existing form, but rather a succession of disruptions revealing new boundaries between cell populations, new compartments that are deformed and transformed. We can describe certain aspects of these events quite well, but we are still far from understanding them. The task is huge, because ultimately the aim is to understand embryonic development at every scale of observation: molecule, cell, tissue, etc. and integrate these different levels of organisation into a coherent and explanatory model that also takes into account interactions between the organism and its environment.

The cellular aspects of animal morphogenesis have been accessible to observation through microscopy.[5] From the nineteenth century on, naturalists have explored the embryogenesis of a large number of metazoans and described the different levels of cell organisation underlying morphogenesis [32]. Reconstructing the branching process of cell divisions or cell lineage become a key challenge to our understanding of morphogenesis. Up until now, however, complete reconstruction of cell lineage has only been possible for the simplest organisms, in terms of their number of cells. The branching-out of the tree of cell lineage in space and time, the organising of cells into territories and compartments, then into tissues, from tissues to organs, and then the contribution of organs to the functioning of the organism constitute the foundations of our representation of animal morphogenesis. This tradition, based on a description of the cellular and tissue aspects of morphogenesis, has been partly eclipsed by the development of molecular biology[6] in the 1950s–1960s and even more so by the molecular genetics of the 1970s. Since then, preference has been given to describing the levels of molecular and genetic organisation of morphogenesis.[7] However, this description fails to really take into account the parameters of spatial and temporal localisation and event probability, without which the dynamics of processes is almost inaccessible and the integration of the molecular and cellular aspects of ontogeny is impossible. Achieving this integration is equivalent to comprehending the complexity of living organisms, and requires us to rethink our questions and experimental strategies within the context of a new interdisciplinarity.

The present chapter follows this approach, setting out to describe the processes of morphogenesis while highlighting the questions involving complex systems, which are for that reason most likely to stimulate the interest of our colleague theorists.

---

[4] Richard Lewontin sees, in the metaphor of development, the insidious influence of the preformationist theories that prevailed until the nineteenth century.

[5] Identification of the egg as a single cell was only clearly established in the nineteenth century with the development of cell theory recounted in [22].

[6] See [27]. In the 1950s, molecular biology designated the level of molecular organisation of living beings.

[7] Indeed, the period 1970–1990, marking the adoption of strategies of molecular genetics, witnessed a large number of works in developmental biology that were poorly documented in terms of morphological analysis. The current period marks a return to morphological investigations with new imaging tools, notably video-microscopy and time-resolved confocal microscopy, which allow to access the dynamics of biological processes at different levels of organisation.

Some biologists argue that few current biological models are favourable to the implementation of complex system strategies. We believe that the obstacles are not inherent to the biological systems we handle, but are of a conceptual nature, and require us to carry out important theoretical work in order to approach them differently. Can we move beyond this observation, and propose strategies to allow the biology of the twentyfirst century to tackle the reconstruction of multiscale dynamics and the integration of the different levels of organisation in living beings and thereby shift the paradigm?

The dominant paradigm governing biology, and developmental biology in particular, is characterised by the concepts of "genetic switch" [30], and "genetic programme"[8], in which each cell state is determined by a unique, i.e. specific, combination of "on" and "off" genes. Even though the concept of regulation of gene expression is essential to our understanding of ontogeny, we believe that it would be useful to abandon the metaphor of the programme, which gives the false idea that the processes of morphogenesis are comparable to a computer algorithm. But beyond this epistemological debate of the last thirty years, the dominant paradigm needs to be revised to take into account spatio-temporal dynamics at every scale. Apprehending morphogenesis presupposes that we understand the consequences of the expression of any given genetic and biochemical combination, and how the behaviour of a cell is the reflection of its integration into the organism as a whole.[9]

9.1 The Acquisition of Cell Diversity

The question of cell diversity arises before the question of the acquisition of form. It arises with the first cell divisions, whereas the form hardly concerns anything more than the adhesion between embryonic cells. Even when the embryo only has a few pairs of cells, we can already observe a diversification in the biochemical content and even in the morphology of the cells. The processes through which this diversity is generated can only be of three kinds. Firstly, the first cell could itself be heterogeneous. In every instance, and without invoking a process of localisation of one or another RNA or protein in the egg, equal distribution of components, some of which are only present in a small number of copies, is impossible, and cell divisions inevitably generate diversity. Secondly, as soon as the cell divisions produce a cluster of cells that do not all have the same environment, then the cell-environment interaction generates diversification. Thirdly, this diversification is only conceivable if it is relayed by the dynamics of molecular and genetic interactions, where it can be approached in terms of bifurcations.

The processes of embryonic cell diversification lead to the functional differentiation of the cells of the organism. Before the appearance of differentiated functional

[8] The construction of this paradigm and its epistemological consequences are discussed in [8].

[9] In the 1970s, François Jacob based his approach to the development of vertebrates on the concept of levels of integration of biological processes, or "integrons". The cell is the first integron [16].

features, cell diversification is described in terms of the potentialities and morphological and biochemical characteristics that define the cell's identity. Classically, embryologists evaluate cell potentialities by carrying out ectopic or heterochronic transplantations[10] and observing whether the transplanted cell is capable of adopting a future in accordance with its new environment. The restriction of cell potentialities is therefore relative. Likewise, cell differentiation, which is a priori the ultimate step in the restriction of cell potentialities during development, appears as a progressive phenomenon, the reversibility of which is still being debated. Indeed, reversibility only seems to be limited by our ability to bring it to light. It may be that any differentiated cell type can, in a suitable environment, return to a previous state or even acquire properties different to those attributed to its lineage.[11]

9.1.1 Heterogeneity of the Egg: What Is Determined from the Moment of Fertilisation?

Unquestionably, the heterogeneity of the egg is, in many cases, at the origin of embryo cell diversification. The amphibian egg on which much experimental embryology research has been conducted provides a good illustration of this. From the outset, one can distinguish in the egg an animal pole and a vegetal pole, in which are concentrated the food reserves of the embryo, but also certain RNA that have been displaced due to the cortical rotation accompanying the entry of the spermatozoon during fertilisation. The heterogeneity of the egg therefore leads directly to diversification of the embryo cells, which sequester qualitatively and quantitatively different combinations of molecules over the course of their divisions. But the heterogeneity of the egg alone cannot account for the diversification of embryo cells, which must result from the dynamics of cellular, molecular and genetic interactions.

This perspective would allow to explain the behaviour of the system at each moment in terms of its dynamic properties. We could thus escape from the regressive logic that requires us to seek the explanation for the properties of the system at any given moment in a description of its previous state. This line of reasoning leads inescapably to the conclusion that everything has already been determined in the egg. And yet the egg is no more than a relative initial state that is subject, like all the other states of the system, to the laws of physics with which evolution is constantly "tinkering"[12].

---

[10] A transplantation is ectopic when the localisation of a cell transplanted into a recipient embryo is different to its original localisation. It is heterochronic when the ages of the donor and recipient are different.

[11] The lineage of a cell corresponds to its ancestors and descendants. The history of each cell during ontogeny is a combination of its lineage and of the associated morphological and biochemical states. The study of processes of regeneration has played a large part in challenging the dogma of the irreversibility of cell differentiation. See, for example [6].

[12] The concept of "tinkering" (*bricolage*) and its consequences are discussed by François Jacob in [17].

9.1.2 *The Interaction Between Cells and Their Environment and the "Inside-Outside" Hypothesis*

Whatever the modalities, the successive divisions of the single egg cell lead to the emergence of a cell diversity without which there could be no morphogenesis. The causes of this diversification remain speculative. They may be intrinsic to the cell but also involve interactions between the cell and its environment. Indeed, the multiplication of embryo cells rapidly leads to a diversification of the cell environment, and the distinction between cells in the middle of the embryo cell mass and those on the periphery is accompanied by morphological and functional differentiation. In many species, the outer layer of cells is characterised by the acquisition of polarity, where we can distinguish an apical face and a basal face, morphologically and physiologically[13] distinct. This differentiation can be understood as the result of interaction between the embryo cells and their immediate environment, defining an opposition of the "inside-outside" type [33]. On the physiological level, the outer layer of cells protects the embryo, ensuring its integrity and controlling exchanges with the external medium. It most probably also displays biomechanical properties essential to morphogenesis. In mammals, the simple formation of the blastocyst,[14] where one can distinguish an outer cell layer and a cavity partly filled by the inner cell mass that will subsequently form the embryo is already a morphogenesis of which the processes are poorly understood.

9.2 The Anatomical Tradition of Embryology, Identification of Symmetry Breaking and Characterisation of Morphogenetic Fields

Morphogenesis and cell diversification are closely linked in the sense that the acquisition of differential properties of cell adhesion and mobility necessarily lead to the emergence of patterns. Certain morphogenetic events, such as the elongation of the nematode embryo *Caenorhabditis elegans* are brought about essentially through cell deformations [2]. The morphogenesis of vertebrates, on the other hand, is characterised by large-scale cell movements. Firstly, there are the cell movements of gastrulation, which involve all the cell populations of the embryo at the same time and establish the antero-posterior and dorso-ventral axes of the organism and the morphogenetic fields from which the organs originate. Certain cell populations also show large-scale migratory behaviour outside of gastrulation, for example the germ cells that give rise to gametes, or neural crest cells. The embryological origin and behaviour of neural crest cells have been studied systematically since the end of the

---

[13] Polarity in cell metabolism, with the directed transport of proteins or metabolites.

[14] The blastocyst stage designates a mammal embryo before implantation in the uterus. The cells of the outer epithelial layer of the blastocyst do not participate in the formation of the actual embryo, but ensure its implantation in the uterine wall, see [2].

1960s by Nicole Le Douarin. Her work is based on the construction of chimeric embryos by transplanting parts of the neural tube of quails into chicken embryos. The graft cells are identified by histological analysis during the development of the chimeric embryo.[15] These works have demonstrated that neural crest cells give rise to the skull in vertebrates [22]. The neural crest cells that migrate from the dorsal neural tube due to an "epithelium-mesenchyme" type morphological transition[16] are an evolutionary innovation that almost certainly appeared in the chordates before the emergence of the vertebrates.[17] The characterisation of neural crest cells and their contribution to morphogenesis illustrate the paradigm of the anatomical tradition of embryology. If we can identify a cell and follow it so as to know a posteriori what it becomes, we can also identify, at each moment, the cells that share the same future. Whence the concept of "morphogenetic field". It allows to demarcate the populations of cells that prefigure organs. However, this concept raises its own problems. With the operational definition that we have given, the spatio-temporal individuation of a morphogenetic field is not easy to establish.

9.2.1 Symmetry-Breaking in Early Embryogenesis

Starting from the single cell of the egg, embryogenesis draws the antero-posterior and dorso-ventral axes of the organism and covers the functional diversification of the cells, necessarily involving the reorganisation of the symmetrical properties of the system. The ovocyte itself[18] can present spherical, axial or bilateral symmetry, depending on the species. These properties of symmetry can be modified from the moment of fertilisation, as for example in the amphibian ovocyte, which loses its axial symmetry to adopt bilateral symmetry determined by the point of entry of the spermatozoon. However, an apparent symmetry observed at the scale of the overall morphology can be contradicted by the observation of underlying, cellular or molecular levels of organisation. The example of the morphogenesis of the teleost zebrafish *Danio rerio* shown in the figures below illustrates this point. As we can see in the photographs taken during the first five hours of development after egg-laying, Fig. 9.1, the embryo presents axial symmetry. Moreover, systematic experiments of transplantation in a 4-h embryo show that apart from the outer cell layer – already

[15] Histological analysis studies sections of fixed tissue through the staining of cell nuclei, revealing a morphological difference between the nuclei of quail cells and those of chickens.

[16] The detachment of cells from an epithelium (which is a layer of cells that adhere to one another) to produce a loose tissue called mesenchyme (where the cells may display an active behaviour of mobility) is essential to normal morphogenesis. An epithelium-mesenchyme type transition is also involved in cancerogenesis during the formation of metastases.

[17] In which case, neural crest cells would have appeared long before the formation of the skull, [18].

[18] The ovocyte is the female gamete. We speak of the "egg" after fertilisation.

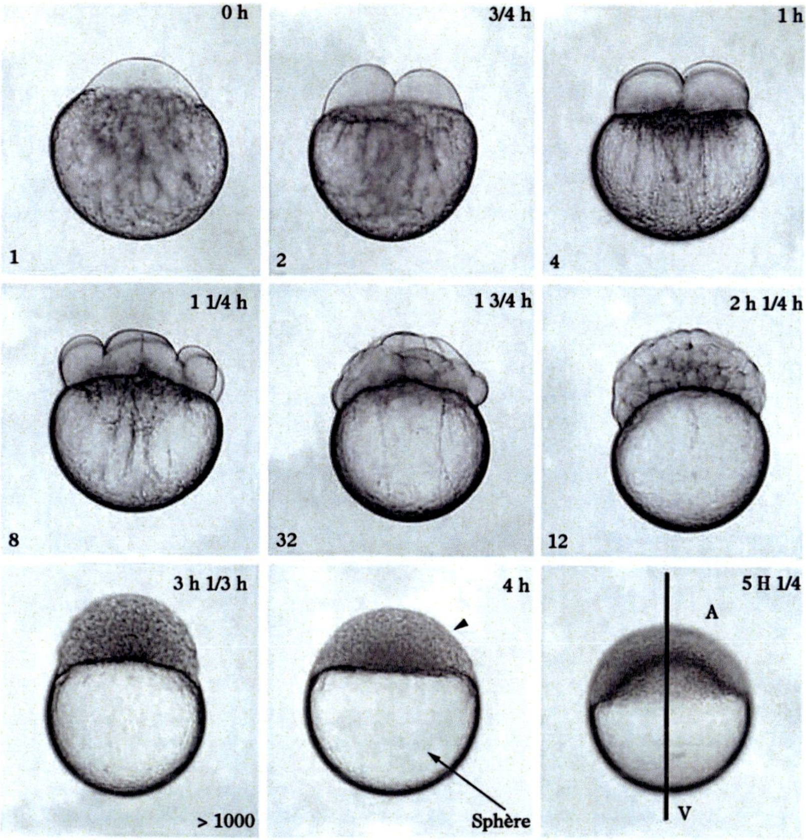

Fig. 9.1 Before gastrulation: an apparent axial symmetry. All the embryos are seen in *side view*, with the animal pole at the *top*. Time elapsed since egg-laying is shown in the *top right-hand* corner of each picture (development at 28°C), and the number of cells or stage of development in the *bottom left-hand* corner. At 4 h, the *arrowhead* indicates the cells and the *long arrow* indicates the sphere of the food reserves or yolk. At 5h1/4 the axial symmetry is oriented along an animal pole (A)-vegetal pole (V) axis, prefiguring the head-tail axis. After [28]

morphologically differentiated – the cells are, at this stage, potentially equivalent. However, a study of the spatio-temporal distribution of different species of RNA in a 4-h embryo shows wide biochemical heterogeneity among the cells. So at this stage, the potentialities of each cell are not fixed by its molecular composition. We must therefore suppose that the dynamic properties of molecular and genetic interaction networks allow a complete restructuring of the biochemical signature of the cells of very young embryos at the moment of transplantation.

Gastrulation, during which the embryo cells cover the food reserves of the yolk and form the antero-posterior and dorso-ventral axes of the embryo, starts 6 h after egg-laying (Fig. 9.2). The dorsal side of the embryo is then morphologically identifiable and the embryo presents a bilateral plane of symmetry. On the biochemical

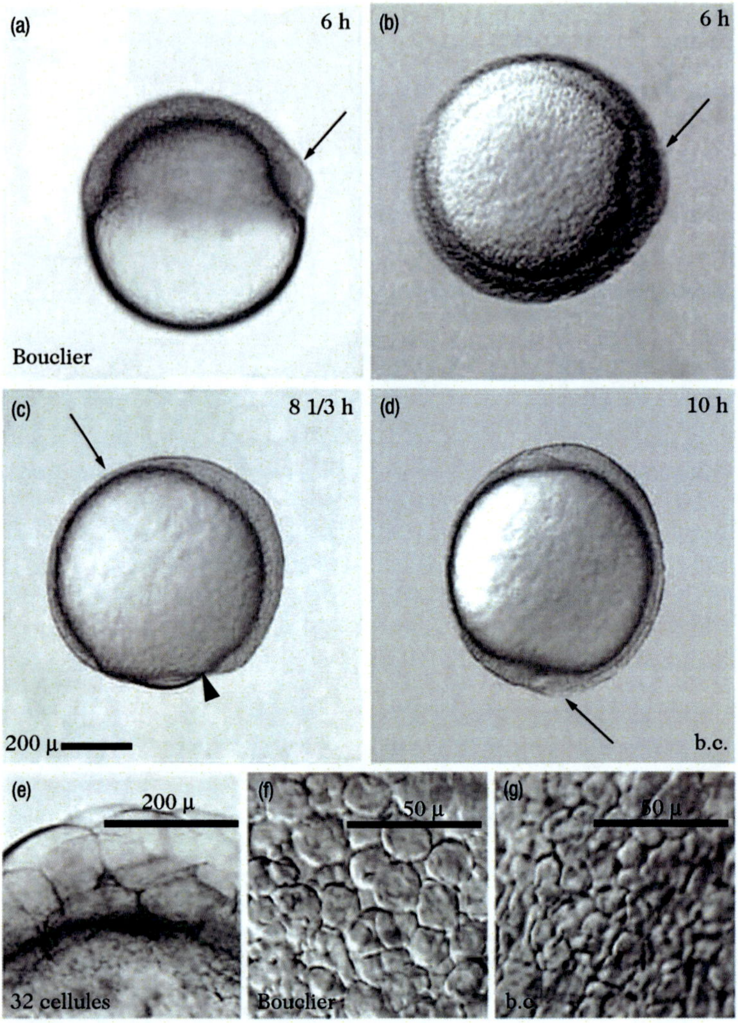

Fig. 9.2 Gastrulation: an apparent bilateral symmetry. (**a**) (*side view*) and (**b**) (*top, animal pole view*) the arrow indicates the bulging of the embryonic shield that marks the dorsal side of the embryo. (**c**) (*side view*) the long arrow indicates the anterior limit of hypoblast migration (inner cell layer that moves forward during gastrulation) and the arrowhead indicates the margin of cells in the process of covering the yolk. (**d**) (*side view*) the arrow indicates the formation of the caudal bud (b.c.). (**e, f, g**): high magnification pictures (measuring bar in microns at the top of each picture) of the embryo cells, of which we can distinguish the membrane contours and the nuclei of some cells. After [28]

level, however, the embryo cells will soon indicate the establishment of left-right asymmetry. It therefore appears that a biochemical signature systematically precedes the morphological manifestation of symmetry breaking.

9.2.2 Formation of Boundaries and Compartments During Organogenesis

Up until the end of gastrulation (10 h of development at 28°C), the morphogenetic fields are hard to distinguish morphologically, and they can only be identified by means of labelling that allows to reconstruct cell lineage. Little by little, however, the morphogenetic fields become individuated, with the establishment of boundaries and compartments. This is illustrated below (Fig. 9.3) by the formation of the notochord and somites, and other, more subtle boundaries would be visible with higher magnification. The notochord is a transitory embryonic structure around which the

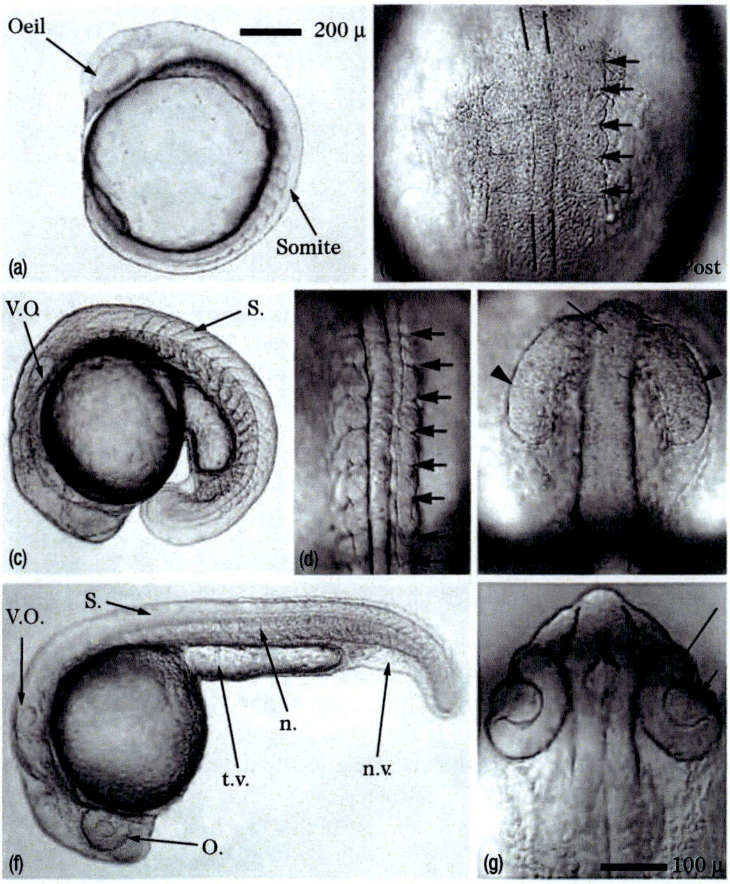

Fig. 9.3 Somite formation: the phylotypic stage. (**a, c, f**): side views. (**b, d, e, g**): top views. (**a**) and (**b**): 4-somite stage. (**c, d, e**): 18-somite stage. (**f**) and (**g**): after 24 h of development. In (**b**) and (**d**), the notochord is delineated by two pairs of lines. The boundary of the somites is indicated by arrows. In (**c**), v.o. otic vesicle, s. somite. In (**e**) the long arrow indicates the anterior brain and the arrowheads indicate the eyes. In (**f**), o. eye, t.v. yolk tube, n. notochord, n.v. ventral fin. In (**g**), the long arrow indicates the retina and the short arrow the lens. After [28]

vertebral column will develop. Somites are blocks of cells that form on each side of the notochord at a rhythm of two pairs per hour, starting from the anterior end and moving back. They will give rise to the vertebrae, the ribs and the skeletal muscles. We shall return to the cyclical process of somite formation later.

It is during somitogenesis that vertebrate embryos resemble each other the most. This is all the more striking as they arrive at this stage having followed very different modes of division and gastrulation. These particular paths are largely linked to the way in which the question of resources is dealt with. The mammal embryo, for example, is dependent on its implantation in the uterus and on the formation of embryonic annexes (the placenta) enabling it to draw nutrients from the mother's blood. In the zebrafish, the mass of reserves is not affected by division and is enveloped by embryo cells during gastrulation, subsequently to be incorporated into the digestive tube.

In 1928, Carl von Baer, studying different vertebrate embryos, set out his "biogenetic" laws, linking morphogenesis and phylogeny[19] [1]. These laws state that within a phylum, the embryogenesis of different species first express the most general features before gradually forming the characteristics that distinguish them. Today, the existence of a phylotypic stage is interpreted as the result of developmental constraints [11]. It is argued that these historical constraints are founded on the impossibility of compromising certain stages of morphogenesis by introducing new variations. Consequently, these constraints will remain speculative as long as we have no explanation for the processes of morphogenesis. Up until now, developmental biology has mainly sought this explanation in an approach to the levels of molecular and genetic organisation in terms of the genetic gain and loss of function.

9.3 The "Bottom-Up" Approach of Developmental Biology

The foundations of developmental biology reside in the encounter between formal genetics and embryology that started at the end of the 1930s. Richard Goldschmidt worked to develop a physiological genetics to focus more on the "activity of genes" rather than their chromosomal localisation, contrary to the concerns of Thomas Morgan.[20] The idea was to interpret the phenotype of mutants, the genotype of which differs from that of the wild type by the presence of a mutation at a given locus. Richard Goldschmidt thus found himself confronted with the difficulty of linking the genotype and the phenotype. He also attempted to phenocopy mutations by means of various treatments of the embryo, consisting in modifying the physical and chemical parameters of its environment. In these types of experiment, it proved

[19] Note that von Baer's laws invalidate the assertion of Ernst Haeckel.

[20] It is important to note that Thomas Morgan had dismissed this question, which he considered to be out of range, when he undertook the study of the transmission of hereditary characters and he and his school drew up the first genetic maps.

to be very difficult to achieve experimental conditions giving results reproducible enough to be interpreted.[21]

In the first half of the twentieth century, Conrad Waddington [38] also pioneered this path of physiological genetics, describing embryogenesis as the route followed by the embryo or its cells in an epigenetic landscape where the bifurcations involve genes, mutations of which affect the development of the embryo. Waddington's epigenetic landscape evokes an energy-related landscape and a dynamic of bifurcations that depends on the spatio-temporal implementation of certain genes.[22] Although it remains heuristic, the scenario proposed by Waddington is far from being an explanatory model of the processes of morphogenesis.

Modern developmental biology has largely revisited, with the tools of formal genetics and molecular genetics, the questions of classic embryology of the German school of *Entwicklungsmechanik* [12], which had imposed its approaches and concepts at the beginning of the twentieth century. At the time, the aim was to understand morphogenesis as a whole, according to the interactions between the different parts of the embryo. The methods applied consisted essentially in the extirpation, transplantation or explantation and recombination of parts of the embryo. Most of the concepts that we use were developed in this context, in particular the concepts of organiser, embryonic induction and tissue competence (the ability of tissue to respond to embryonic induction). Transposed to the context of developmental genetics, the tissue extirpations and transplantations of experimental embryology become genetic losses and gains of function. The phenotypes obtained in the two types of approach are sometimes remarkably similar.[23]

The strategies of functional genetics have allowed to explore the role played in morphogenesis by a large number of "genes". These strategies offer a productive path towards describing the molecular processes underlying cell behaviour. After several decades of stereotypical approaches searching for "genes of development"[24], we have a fragmented picture of genetic and molecular interaction networks, the structure of which remains largely obscure and the dynamics of which has not been envisaged.

The reconstruction of the dynamics of genetic and molecular interaction networks should provide a "bottom-up" type of explanation for the emergence of cell diversity and morphogenesis [5]. We shall now describe three examples of approaches that move in this direction.

---

[21] In a certain way, Emmanuel Farge followed this tradition by interpreting the phenotypic effects of mechanical deformations applied to fly embryos, and this type of experiment remains extremely difficult [7].

[22] See Chap. 16, which presents correspondence between R. Thom and C.H. Waddington on this subject.

[23] The transplantation of the Spemann organiser, for example, is phenocopied (gives the same phenotype) by the localised overexpression of a mutant form of a nodal/activin receptor. See [29].

[24] The concept of genes of development takes us back to the blind alleys of genetic reductionism. See [8].

9.3.1 *Dynamics of Molecular and Genetic Interactions in the Formation of Patterns*

The work achieved by the Garrett Odell's team researching the gene interaction network underlying the selection of neural progenitors by lateral inhibition in *Drosophila* illustrates the interest of a theoretical approach using the appropriate formalisms and working as closely as possible to the descriptions supplied by biologists, who are then able to test the model's predictions experimentally [26] (see Fig. 9.4). The concept of lateral inhibition refers to the capacity of a cell to inhibit its neighbours, leading to the distinction between two cell types through the interaction of the Notch receptor (N) and its Delta ligand (DL). The model allows to establish that the network as represented below leads robustly to a process of lateral inhibition by delaying the process of cell differentiation during a certain time. The stochastic variation in gene expression between the cells is sufficient for the functioning of the network to lead, at a given moment, to the choice of a cell that will be the neural precursor. The model gives support to the scenario of lateral inhibition and therefore competition between neighbouring cells. It also shows the evolutionary potential of this type of network, which is indeed at work in all metazoans, in a large number of morphogenetic processes sharing the characteristics described below. It appears that the evolutionary potential of a genetic network underlying a "morphogenetic pattern" is linked to its robustness, in other words its capacity to control the formation of such a pattern in a wide enough range of initial conditions and parameters.

Here, the molecular and genetic interaction network is modelled in a simplified and idealised context. It is possible that application of the same network in the context of the reconstruction of cell morphodynamics based on 4-D (the three

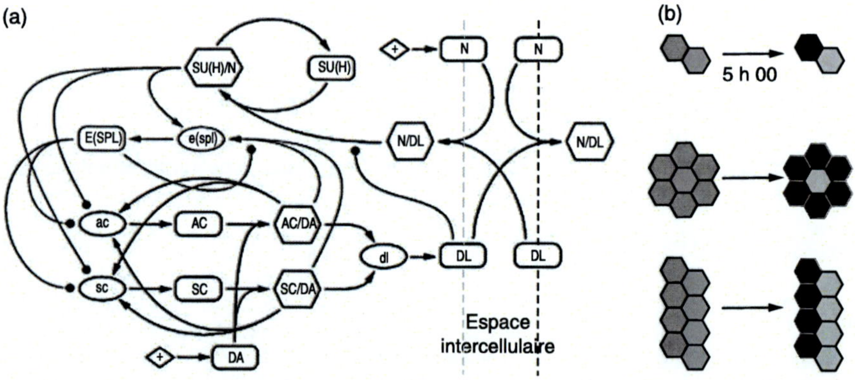

Fig. 9.4 (**a**) A model of the neurogenic network linking the genes and proteins involved: Notch (N), Delta (Dl), achaete (ac), scute (sc), daughterless (da) Suppressor of Hairless (SU(H)), and Enhancer of Split (E(SPL)), which are chosen as the critical nodes of the network. The authors have tested the capacity of this model, and certain variants based on different distributions of the links in grey, using three geometrically different tests of lateral inhibition. (**b**) In all three cases, the light grey cells end up inhibiting their neighbours. After [5]

spatial dimensions and time) in vivo observations may lead to a partial revision of the proposed scenario.[25]

9.3.2 The Concept of Morphogen and Pattern Generation Through the Threshold Effect

Organisation of the antero-posterior axis in *Drosophila* depends on the only clearly-identified morphogen gradient. This is the gradient of the bicoid protein.[26] The formation of *Drosophila* body segments appears as a hierarchical process of molecular and genetic interactions that generates a pattern of stripes, sequentially subdivided to form 14 stripes from which the body segments of the larva are organised simultaneously.[27]

The first stage in this process of segmentation is the establishment of a concentration gradient for the bicoid protein (bcd) in the egg, deriving from the *bcd* messenger RNA anchored at the anterior pole of the egg during ovogenesis. The shape of the gradient depends on the rates of diffusion, translation and degradation of the protein. It is probably also modified by the reaction of the protein with its targets in the nuclei localised on the periphery of the syncytium that the *Drosophila* egg becomes after 14 divisions of its nucleus.

The question debated by several authors over the last four years concerns the variability of the bcd gradient and the processes by which this variability might be filtered to lead to the establishment of a precise posterior expression boundary for the Hunchback protein (hb), which has been described by biologists as a direct target of the bcd protein [37]. This question is original in developmental biology, where quantitative strategies are limited and often unreliable (see Figs. 9.5 and 9.6).

To begin with, the expression boundary of *hunchback* messenger RNA is unclear, becoming well-defined during cycle 14 of nucleus division. If *hb* is directly regulated by bcd through a threshold of its concentration gradient, then the position of the posterior expression boundary of *hb* (at the level of the RNA, or of the protein denoted hb) in relation to the anterior and posterior poles of the embryo should directly reflect the variations in the bcd gradient. And yet in 2002, it was demonstrated that the variability of the gradient has no effect on the specification of the hb expression boundary [13]. The problem is then to identify the processes that

---

[25] Cell dynamics plays an active role in pattern formation. In the *Drosophila* wing imaginal disc, when a cell is beginning to win the competition and inhibit its neighbours, it produces filopodia that increase the surface area of contact with its neighbours so that it can enter into contact with other cells located several cell radii away [19].

[26] Bicoid establishes a concentration gradient in the egg. The diffusion of a potential morphogen secreted into the intercellular space is hindered by interactions with its receptors in the extracellular matrix and on the surface of the cells. To such an extent that in a cellular context, the functional properties of a gradient are almost certainly generated otherwise than by diffusion (filopodia, cell-to-cell transmission).

[27] Pattern generation by threshold effect was initially proposed by Lewis Wolpert, see [39].

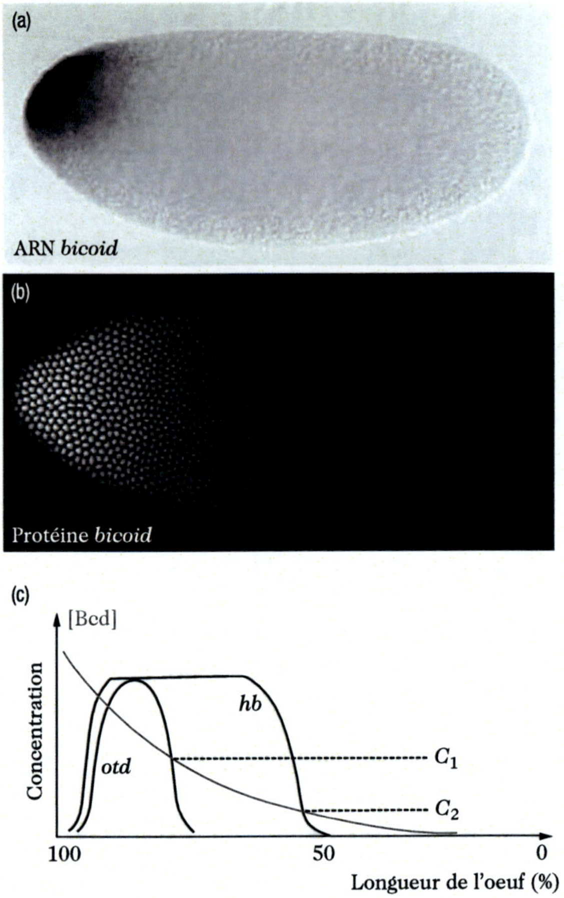

Fig. 9.5 The bicoid gradient. Expression of the *hb* RNA **(a)** and the hb protein **(b)** in the *Drosophila* embryo. **(c)** Graph showing the morphogen activity of bcd and the expression profile of two of its targets, *hb* and *otd*, responding to two different concentration thresholds (abscissa: egg length in %). After [37]

allow for the variability of the gradient to be filtered, and the authors have recently proposed, on the basis of a theoretical model, the intervention of a second gradient emanating from the posterior pole of the embryo, correlated with the bcd gradient by means of the same degradation factor [14].

For other authors [4], there is no need for the variability of the bcd gradient to be filtered. It is the transcriptional process itself that specifies the anterior-posterior localisation of the *hb* expression domain.

The debate remains open, and appears to call for new measurements and greater spatio-temporal precision, particularly in relation to the dynamics of nuclear division cycles. If, as Nathalie Dostatni's team proposes, the bcd gradient on its own is sufficient for domain specification, the processes of refinement of the *hb* boundary

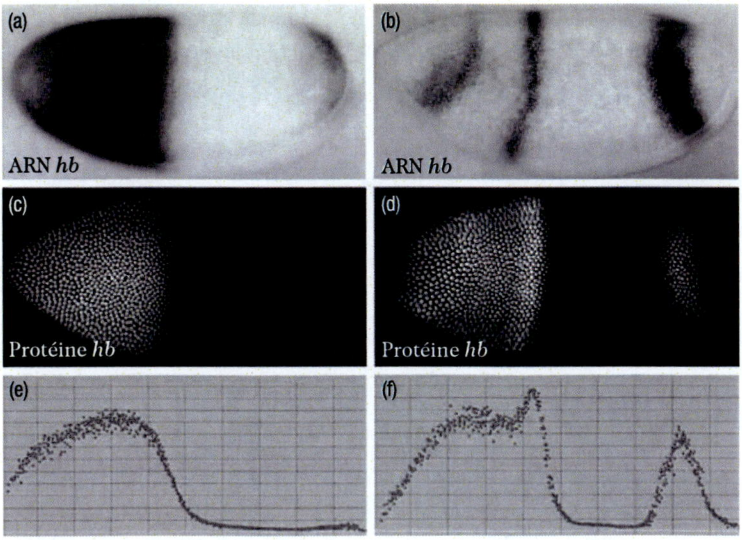

Fig. 9.6 The expression of Hunchback. The pattern of expression of the *hb* RNA **(a, b)** and the hb protein **(c, d)** in the *Drosophila* embryo. The *left-hand column* shows the patterns of expression at the beginning of division cycle 14 and the *right-hand column* in the middle of cycle 14. Below, measurement of fluorescent intensities **(e, f)** corresponding to the patterns of protein (c) and (d) respectively. These figures are inspired by those presented in [37], an article that gives all the references and arguments in the debate

in cycle 14 need to be identified. It may be that these processes bring into play a cooperative link between bcd and DNA or perhaps the reorganisation of chromatin over the course of cell cycles.

9.3.3 The Formation of Somites in Vertebrates: A Model of Coupled Oscillators

The process of segmentation of the *Drosophila* body is a priori very different from the formation of repeated parts in vertebrate embryos. The formation of the somites is a temporal process that gradually extends towards the posterior pole of the organism. A large number of works have shown that the expression of a whole range of genes is cyclically regulated during somite formation, with an oscillation frequency that coincides with the time needed for the formation of a new pair of somites. In *Danio rerio* the process is rapid, as the cycle only takes 30 min. Julian Lewis [23] has shown that the presence of the transcription inhibitors her1 and her7 is essential to the oscillation process, while Notch Delta intercellular signalling is essential to the synchronisation of the oscillations between different cells. Mathematical simulation shows that direct her1/her7 autoinhibition plays a decisive role. The process is robust, despite the stochasticity of gene expression. The predicted oscillation period is close to that observed (30 min) and depends on the time needed for the transcription of her1 and her7. When the coupling parameters are varied, the oscillation period can be much longer; it is dictated by the time delays in the Notch pathway (see Figs. 9.7, 9.8, and 9.9).

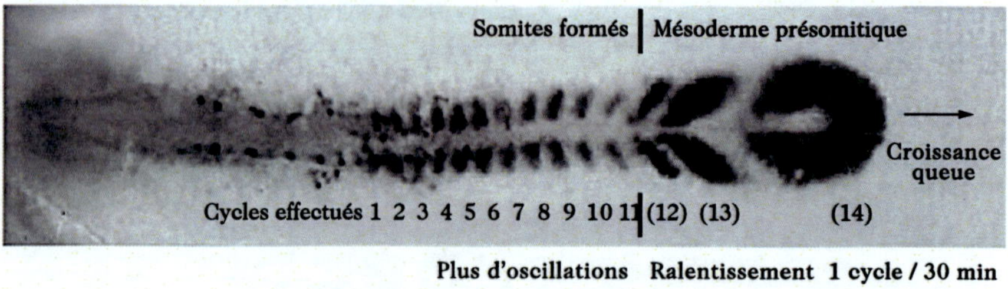

Fig. 9.7 *Danio rerio* embryo at the 10-somite stage, coloured to reveal in situ, here in darker grey, the presence of *deltaC* coding the DeltaC ligand of the Notch receptor. The transcription factors her1 and her7 have similar expression profiles. The oscillation is produced in the region of the presomitic mesoderm (PSM) and stops when the cells are recruited to the newly-formed somites. The oscillations do not suddenly stop, but slow down in the anterior PSM. Consequently, the cells are at different stages in the cycle according to their position in the PSM. The periodicity of the process is defined at the posterior end of the PSM. In each cycle, a new pair of somites is added. The *n*-th pair of somites is made of cells stopped after the *n*-th cycle from the beginning of somitogenesis. After [23]

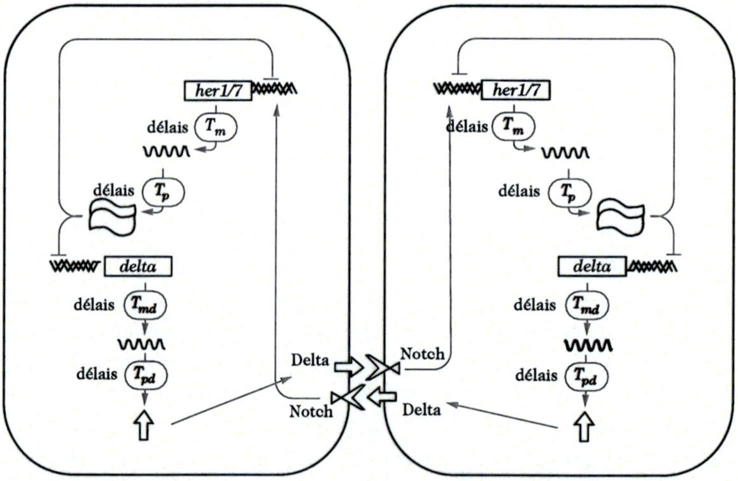

Fig. 9.8 Intercellular Notch Delta signalling synchronises the oscillations between two adjacent cells. Each cell contains an oscillator based on autoinhibition with time delays of the her1/her7 transcription factor dimers. The model shows that the value of the $T_m + T_p$ autoinhibition time delay corresponding to the transcription and translation of *her1* and *her7* is critical to the periodic regime and the oscillation period. The two cells present a variation of 10% in their $T_m + T_p$ values, corresponding to different values of periodicity. After [23]

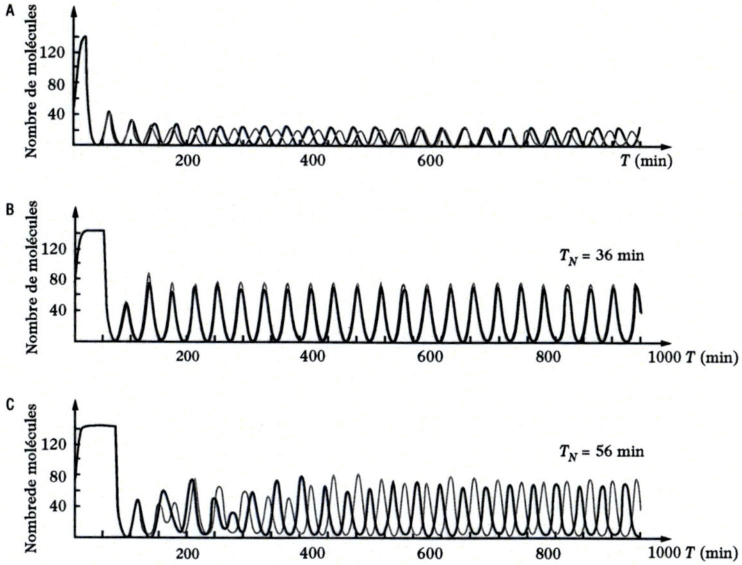

Fig. 9.9 A: Computer simulation shows that in the absence of Notch signalling, the oscillators are not synchronised. B: Notch signalling is active, with a time delay of TN=36 min. It is assumed that the *her1/her7* transcription rate is determined by the product of an increasing function of Delta activity. The cells oscillate synchronously with a period that is the average of the periods of the individual cells. C: Notch signalling is active but with a time delay of 56 min. The cells oscillate asynchronously. These figures are inspired by [23]

Julian Lewis concludes from this study that even with a relatively simple model, the biologist's intuition is not sufficient to make predictions about the behaviour of the system, and that he or she must resort to mathematics.

However, this representation of morphogenesis still lacks any integration of molecular and cellular processes. In fact, in parallel to the oscillation of gene expression, the cells proliferate and leave the presomitic mesoderm to form a pair of somites, and the periodicity of cell divisions and the oscillations of gene expression are the same. The question therefore arises of how the two levels of organisation are linked.

Generally speaking, our understanding of morphogenesis will not be complete without an integration of "bottom up" and "top down" causalities, taking into account the feedback effects of higher levels of organisation on cellular, molecular and genetic dynamics. Such an integration requires the investigation of the individual and collective cell behaviour of the organism. The genomic era had sidelined the anatomical tradition of embryology, which is now staging a return in new strategies of in vivo observation of cell behaviour.

9.4 The Reconstruction of Cell Morphodynamics and the Revival of the Anatomical Tradition of Embryology

The cell behaviours grouped together under the term "cell morphodynamics" are stereotyped. A cell lives and dies. A living cell can divide, change form, modify its interactions with its neighbours, and move. In the multicellular context of the embryo, these individual behaviours can give rise to collective behaviours, and with them the emergence of a pattern or form. Strategies of in vivo observation and the reconstruction of cell morphodynamics allow to extract the parameters that are significant in terms of the emergence of collective movements, pattern formation and eventually the feedback effect on the cell of the ordering of tissues.

9.4.1 Cell Movements and Deformations in Morphogenesis

What are the processes underlying cell movements in the embryo? What role does individual cell mobility play? Do cell movements in the embryo have the characteristics of emerging collective movements? Is the oriented migration of cells or cell populations directed by biochemical gradients? How are the movements of different cell populations coordinated?

These questions, which remain largely unanswered, can be approached through the reconstruction of cell morphodynamics. In the Xenopus, which has opaque tissues, Ray Keller has filmed the outer cell layer of the embryo during the movements of gastrulation. Using electron microscopy, he has also conducted a detailed study of the cell deformations involved in the elongation of the antero-posterior axis of the organism [20]. On the strength of these studies, he has proposed a model of morphogenesis by convergent extension, based on a largely intuitive biomechanical interpretation of the cell behaviours observed. A formal approach could now help to provide a rational foundation to their reconstruction by simulation.

The teleost fish *Danio rerio* is chosen as a model for the study of vertebrates notably because of the transparency of its tissues, allowing in vivo observation of all the cells in the embryo.[28] Cell morphodynamics are then reconstructed by computer, using 4-D imaging of the whole organism. All the characteristic parameters of cell behaviour are thus directly measurable.[29] Through this approach, we hope to be able to reconstruct cell lineage and obtain the measurements needed to interpret cell movements and deformations. It should thus be possible to access indirectly the biomechanical forces at work in the organism. This aspect is essential to our understanding of morphogenesis [15]. Biomechanical forces spread through the tissues at long distance and could determine the coordination of cell population

[28] Films of the development of *Danio rerio* can be seen at http://www.zfin.org; http://depts.washington.edu/fishscop/

[29] The strategies of reconstruction can be consulted on the site of the European Embryomics project: http://www.embryomics.eu

movements. Fast long-range correlation could hardly be achieved by anything other than biomechanical or electrochemical forces. Moreover, when the biomechanical forces at work in the embryo are integrated into the cell, they modulate its metabolic and genetic activity [3]. In brief, the interpretation of biomechanical constraints is an important form of "top-down" causality, hitherto so neglected in our descriptions of the processes of morphogenesis.

9.4.2 Cell Adhesion and Biomechanical Constraints in the Embryo

Biomechanical constraints in the embryo derive from the cohesion of cell layers and the extracellular matrix. The curvature of a cohesive cell layer, like the epithelium enveloping the *Danio rerio* embryo during the first hours of its development, indicates that it confines the other cells of the embryo. This structuring is already the result of cell diversification and of the compartmentalisation that distinguishes between cell populations with very different properties of adhesion. A good number of authors have hypothesised that the segregation of different cell types according to their properties of preferential adhesion is an essential process in morphogenesis. In 1955, Townes and Holtfreter conducted systematic experiments in which they dissociated and mixed cells from amphibian embryos and demonstrated the capacity of cells originating in the same tissue to reassociate and form characteristic patterns [35]. Since then, the molecular foundations of cell adhesion have been largely described by biologists. Membrane glycoproteins of the cadherin family are suitable candidates for conferring differential affinity on cell populations of the organism [25]. The family of ephrins might also play a role in the same type of process. According to the principle proposed by Steinberg, a differential affinity in cells would be enough to lead to their segregation and to the generation of patterns [31]. However, the manipulation of cadherins or ephrins to test for their involvement in the formation of boundaries and compartments remains difficult to interpret, inasmuch as these molecules are connected to the network of molecular and genetic interactions, of which we can predict in a satisfactory manner neither its dynamic behaviour nor its integration by the cell.

9.4.3 The Tensegrity Model

It is clear that we must return to the cell as the site of integration of bottom-up and top-down causalities. In each cell, the cytoskeleton and its molecular motors are connected to the cell surface and, more generally, to the molecular and genetic interaction network. The micromechanical forces of morphogenesis operate at this level. The model of tensegrity [15] proposes that the morphogenesis of tissues depends on the mechanical interaction between the cells and the extracellular matrix that subjects the tissue to isometric tension. Local reorganisation of the extracellular matrix can then lead to the stretching of this matrix and the cells adhering to it. Cell growth and differentiation could depend on local physical distortions or changes in the tension exerted by the cytoskeleton. The formation of a pattern presenting local curvature, buds or branches, could result from the division and migration of one or several cells in a quiescent environment.

An attempt to interpret the generation of forms in terms of a combination of physical forces could lead one to compare biological forms with those generated in inert matter [34]. Do such analogies hold any prospect of furthering our understanding of morphogenesis? Is the explanation of the physical phenomenon useful to our understanding of the biological process? Let us take the example, proposed by D'Arcy Thompson in his book *On Growth and Form*, of the splashes produced by a drop of milk falling into the surface of the liquid and the parallel drawn between the forms produced and the tentacles of a Hydra.[30] Few biologists would see the interest in making such a comparison, because the descriptions that one can make of the two systems and the emergence of their form simply cannot be superimposed. The time scales are very different, as are the components. On the biological level, the fact that the forms are produced from cells, from their multiplication, diversification, deformation and mobility makes any direct transposition impossible.

The example of Turing structures is more difficult to evaluate, and whatever the reasons, it is a real source of inspiration. However, Turing himself, when he wrote "The chemical basis of Morphogenesis" [36], made no claim that a model of reaction-diffusion could explain morphogenesis. Indeed, the pattern stripes observed in *Drosophila* embryos or in *Danio rerio* cannot be explained by the diffusion and reaction of a single activator/inhibitor pair. If we limit ourselves to this description, it is impossible to conceive of a parallel between the stripes of reaction-diffusion and the migration of melanocytes in the fish epidermis. Once again, the cellular context of the biological system makes direct transposition impossible. We could, however, consider cell migration in the plane of the epidermis as diffusion, and introduce parameters of reaction with the adhesion between melanocytes to obtain a description that is satisfactory for the biologist and that conceives the dynamics of the system at the cellular scale.

But a satisfactory explanation of any morphogenetic process certainly requires a combination of top-down and bottom-up approaches, linking the spatio-temporal dynamics at the molecular scale with individual and collective cell behaviour. The descriptions produced by biologists serve as the foundation for a large number of questions that will remain out of our range if we do not improve our ability to take into account the parameters of topology, temporality and even probability that are necessary to a theoretical approach. However, the intention is not to produce exhaustive descriptions. The blind pursuit of understanding through the accumulation of a high flow of systematic measurements could simply divert us from the construction of relevant models. The examples of reconstruction of molecular dynamics mentioned above are remarkable precisely because they raise the question of the elements that must be taken into account to conserve the essential properties of the system.

---

[30] See Chap. 14, devoted to the work and legacy of D'Arcy Thompson.

References

1. von Baer K.E. (1928) *Entwicklungsgeschichte der Thiere: Beobachtung und Reflexion*, Bornträger (Konigsberg), in German.
2. Bard J. (1994) *Embryos*, Wolfe Publishing (London).
3. Brouzes E. and Farge E. (2004) Interplay of mechanical deformation and patterned gene expression in developing embryos, *Curr. Opin. Genet. Dev.* **14**, 367–374.
4. Crauk O. and Dostatni N. (2005) Bicoid Determines Sharp and Precise Target Gene Expression in the Drosophila Embryo, *Curr. Biol.* **15**, 1888–1898.
5. von Dassow G., Munro E., Sunderland B., and Odell G. (2008) *Genetic modules: pattern formation and regulatory dynamics*, complete text at http://raven.zoology.washington.edu/celldynamics/research/genenet/index.html
6. Echeverri K. and Tanaka E.M. (2002) Ectoderm to mesoderm lineage switching during axolotl tail regeneration, *Science* **298**, 1993–1996.
7. Farge E. (2003) Mechanical induction of Twist in the Drosophila foregut/stomodeal primordium, *Curr Biol.* **13**, 1365–1377.
8. Fox Keller E. (2000) *The Century of the gene*, Harvard University Press (Cambridge MA).
9. Gilbert S. (2006) *Developmental Biology*, 8th edition, Sinauer associates (Sunderland MA).
10. Gilbert S. and Raunio A. (eds.) (1997) *Embryology, constructing the organism*, Sinauer associates (Sunderland MA).
11. Gould S. and Lewontin R.C. (1979) The Spandrels of San Marcos and the panglossian paradigm. A critique of the adaptation program, *Proc. R. Soc. London* **205**, 581–598.
12. Hamburger V. (1988) *The heritage of experimental embryology*, Oxford University Press (Oxford).
13. Houchmandzadeh B., Wieschaus E., and Leibler S. (2002) Establishment of developmental precision and proportions in the early Drosophila embryo, *Nature* **415**, 798–802.
14. Houchmandzadeh B., Wieschaus E., and Leibler S. (2005) Precise domain specification in the developing Drosophila embryo, *Phys. Rev. E* **72**, 061920.
15. Ingber D.E. (2006) Mechanical control of tissue morphogenesis during embryological development, *Int. J. Dev. Biol.* **50**, 255–266.
16. Jacob F. (1971) *La Logique du vivant*, Gallimard (Paris), in French. Translation *The logic of life*, Princeton University Press (Princeton), 1993.
17. Jacob F. (1981) *Le jeu des possibles*, Poche (Paris), in French.
18. Jeffery W.R., Strickler A.G., and Yamamoto Y. (2004), Migratory neural crest-like cells form body pigmentation in a urochordate embryo, *Nature* **431**, 696–699.
19. de Joussineau C., Soulé J., Martin M., Anguille C., Montcourrier P., and Alexandre D. (2003) Delta-promoted filopodia mediate long-range lateral inhibition in Drosophila, *Nature* **426**, 555–559.
20. Keller R., Davidson L., Edlund A., Elul T., Ezin M., Shook D., and Skoglund, P. (2000) Mechanisms of convergence and extension by cell intercalation, *Philos. Trans. R. Soc. Lond. B Biol. Sci.* **355**, 897–922.
21. Lecointre G. and Le Guyader H. (2001) *Classification phylogénétique du vivant*, Belin (Paris), in French. Translation *The tree of life: a phylogenetic classification*, Belknap Harvard University Press (Cambridge MA), 2006.
22. Le Douarin N. (2000) *Des chimères, des clones et des gènes*, Odile Jacob (Paris), in French.
23. Lewis J. (2003) Autoinhibition with Transcriptional Delay: A Simple Mechanism for the Zebrafish Somitogenesis Oscillator, *Curr. Biol.* **13**, 1398–1408.
24. Lewontin R.C. (2001) *The Triple Helix*, Harvard University Press (Cambridge MA).
25. Mc Neill H. (2000) Sticking together and sorting things out: adhesion as a force in development, *Nat. Rev. Genet.* **1**, 100–108.
26. Meir E., von Dassow G., Munro E.M., and Odell G.M. (2002) Robustness, Flexibility, and the Role of Lateral Inhibition in the Neurogenic Network, *Curr. Biol.* **12**, 778–786.

27. Morange M. (1994) *Histoire de la biologie moléculaire*, Éditions La Découverte (Paris), in French.
28. Peyriéras N. (2003) *Le Développement des vertébrés*, Éditions Le Pommier (Paris), in French.
29. Peyrieras N., Lu Y., Renucci A., Lemarchandel V., and Rosa F. (1996) Inhibitory interactions controlling organizer activity in fish, *C. R. Acad. Sci. III* **319**, 1107–1112.
30. Ptashne M. (1986) *The Genetic switch*, Cell Press (Palo Alto).
31. Steinberg M.S. (1970) Does differential adhesion govern selfassembly processes in histogenesis? Equilibrium configurations and the emergence of a hierarchy among populations of embryonic cells, *J. Exp. Zool.* **173**, 395–433.
32. Stern C.D. and Fraser S. (2001) Tracing the lineage of tracing cell lineages, *Nat. Cell Biol.* **3**, E216–E218.
33. Tarkowski A.K. and Wroblewska J. (1967) Development of blastomeres of mouse eggs isolated at the 4- and 8-cell stage, *J. Embryol. Exp. Morph.* **18**, 155–180.
34. Thompson, sir d'Arcy W. (1917 and 1942) On Growth and Form, two editions from Cambridge University Press (Cambridge).
35. Townes P.L. and Holtfreter J. (1955) Directed movements and selective adhesion of embryonic amphibian cells, *J. Exp. Zool.* **123**, 53–120.
36. Turing A. (1952) The Chemical Basis of Morphogenesis, *Phil. Trans. R. Soc. London B* **237**, 37–72.
37. Yucel G. and Small S. (2005) Morphogens: Precise Outputs from a Variable Gradient, *Curr. Biol.* **16**, R29–R31.
38. Waddington C.H. (1947) *Organisers & genes*, Cambridge University Press (Cambridge).
39. Wolpert L., Beddington R., Jessell T., Lawrence P., Meyerowitz E., and Smith J. (2002) *Principles of development*, 2nd edition, Oxford University Press (Oxford).

Chapter 10
Phyllotaxis, or How Plants Do Maths When they Grow

Stéphane Douady

10.1 Discovery

Have you ever looked at a fir cone? Or a sunflower, or a pineapple? Almost certainly ... But have you noticed the regular spirals that cover them? Or counted those spirals? Probably not!

We all think we know the things around us [1, 5]. But they can still hold surprises in store for us. In each of the above cases, if you count the number of clockwise and counter-clockwise spirals that cover the whole thing, you get two consecutive numbers in the Fibonacci sequence. This sequence, named after the thirteenth-century monk-mathematician who used it to describe the growth of a hypothetical population of rabbits, is formed by starting with 0 and 1 and defining the next number as the sum of the two preceding numbers. Thus: 0, 1, 1, 2, 3, 5, 8, 13, 21, 34, 55, 89, 144, 233, and so on.

In each of the above plant examples, not only do we obtain numbers from this sequence, but two consecutive numbers. So, for instance, most species of pine cones are (8, 13), small sunflowers are (13, 21) and the biggest, record flowers can go up to (144, 233). What is just as amazing is that we are talking about the exact numbers. There are exactly 144 spirals, for example, and not 143 or 145 (see Fig. 10.1).

And the surprise is even greater when we learn that this property exists in all the botanical elements of plants, not just in pine cone scales or florets (the small, closely-clustered flowers that form the disk of the sunflower), but also in the bracts (the small leaves that protect the flowers) and before that the leaves. And this property can be found from the very origin of plants, in the first fossils, right through to the most highly-evolved species.

You may not believe this. The next time you walk through the woods you may wish to verify it by counting the spirals on all the pine cones you find, as I did the first time I heard about it [12]. But it really is true.

S. Douady (✉)
CNRS Research Director at the Complex Matter and Systems Laboratory (UMR 7057 CNRS-Paris-Diderot), University of Paris-Diderot, Paris, France
e-mail: douady@lps.ens.fr

P. Bourgine, A. Lesne (eds.), *Morphogenesis*, DOI 10.1007/978-3-642-13174-5_10,

Fig. 10.1 *(Left)* Base of a maritime pine cone. The 13 spirals spreading out clockwise from the centre are clearly visible, and the 8 anti-clockwise spirals can also be distinguished. This pine cone is therefore (13, 8). © S. Douady. *(Right)* Sunflower. The florets, each of which is turning into a seed, exhibit 21 clockwise and 34 anticlockwise spirals that can be seen very clearly on the outer edges of the disk. © Y. Couder

10.2 Why?

What is the explanation for this phenomenon? That is the interesting question.

A first type of explanation is geometrical, and linked to the "golden number" τ. As the Bravais brothers demonstrated as early as 1835 [2], if we set out the scales of a pine cone in a regular way, moving up by the same distance each time and turning by a "golden angle" $\phi = 2\pi(1-\tau)$, then the spirals we obtain are necessarily two consecutive numbers of the Fibonacci sequence. This is because of the mathematical relation between the golden number and the Fibonacci sequence: the ratio between two consecutive numbers of the Fibonacci sequence are the successive rational approximations of the golden number written as a continued fraction.[1]

But that only shifts the problem elsewhere, because we now have to explain why the angle between two successive botanical elements should always be the golden angle. One could advance the myth that the golden angle is the foundation of all harmony, and that plants are harmonious. One could also affirm that the angle is genetically coded in the form of a continued fraction, and the golden number is the easiest to encode because its continued fraction only contains 1's, but even the colleague who suggested that idea was only joking.

[1] A continued fraction is a way of expressing any number x as a (possibly infinite) sequence of fractions of integers. We start with the whole approximation n_0 of x: $x = n_0 + r_0$, with a remainder r_0 smaller than 1. It follows that $1/r_0$ is greater than 1, and can be written $1/r_0 = n_1 + r_1$, and so on. By stopping after a given number of terms n_p, we obtain a rational approximation of the number. As the golden number is obtained by solving the equation $\tau = 1/1 + \tau$, we can obtain it by iterating $\tau = 1/1 + 1/1 + 1/1 + 1/\ldots$ As this series only contains 1's, it is the one that converges "the most slowly". If we assume that the approximation of τ to order p is the ratio of two consecutive numbers in the Fibonacci sequence, defined by $u_0 = 0$, $u_1 = 1$ and $u_{q+2} = u_{q+1} + u_q$: $\tau_p = u_q/u_{q+1}$, which is true for the first order, while the approximation to the following order does indeed give $\tau_{p+1} = 1/1 + \tau_p = 1/1 + u_q/u_{q+1} = u_{q+1}/u_{q+1} + u_q = u_{q+1}/u_{q+2}$.

Another type of explanation is finalist: this result occurs because it is advantageous to the plant. The golden number is irrational. Because its expression as a continued fraction only contains 1's, it is the number whose sequence of successive rational approximations converges the most slowly. It is therefore said to be the irrational number the "furthest removed" from being rational. If we now assume that the leaves are regularly separated by the angle ϕ, then as ϕ is also irrational, two leaves will never be superimposed one directly above another. This means that any leaf will never completely overshadow the one below, optimising the reception of light and obviously presenting an advantage for the plant.

The problem with this explanation is that in addition to these "spiral" types of leaf-arrangements, there are also "whorled" arrangements, where several leaves are attached to the stem at the same height, and then at the next level the leafs are positioned over the intervening spaces. Thus, every two levels the leaves are situated vertically above each other, which is clearly not to the advantage of the plant (Fig. 10.2).

This explanation also overlooks the fact that the sun is not always vertically above the plant; it is in the process of rising and setting. Indeed, outside the tropics it is never directly overhead. Even in the most difficult case of the young maple shoot, with whorls of two leaves, growing in the vertical light-well of a forest, we can see that the lower leaves avoid the shadow of the newer ones simply by extending their petioles (the stalks connecting them to the main stem), as shown in Fig. 10.3. And while the leaf remains attached to the same place on the stem, it can change the direction in which it faces. Plants growing in front of windows provide a good

Fig. 10.2 Plant with "whorls of two" leaves: two leaves appear opposite each other at the same height on the stem, followed by another pair at right angles to them, and so on. As the sun is (almost) never vertically overhead, this arrangement does little to reduce the leaves' exposure to sunlight. © S. Douady

Fig. 10.3 Young maple shoot. By growing (especially the length of their stalks), the leaves manage to avoid overshadowing each other, even though they are also "whorls of two". © S. Douady

illustration of this: all the leaves are turned towards the window. When you turn the plant around, all the leaves appear to be "back-to-front". But after a few weeks, the leaves have turned around to face the light. A natural example is provided by the horizontal branches of yews and spruces: although the leaves are perfectly arranged in spirals, they gather horizontally on the left- and right-hand sides of each branch, with the top side of each needle uppermost.

These movements show that the plant's exposure to sunlight really is important. But what these leaf movements – and the widespread existence of whorled arrangements – show is that exposure to sunlight is not dependent on the basic arrangement of the leaves, which is what phyllotaxis deals with.

Another explanation that has been put forward looks at how the bases of the leaves and their attachments occupy the space on the stem. These bases, known as "foliar scales", can be seen clearly on artichokes and pine branches. They also determine the space that is left for each seed in sunflowers, after flowering. This widely-proposed explanation is also finalist: if the elements are stacked up regularly following the golden angle ϕ, then it can be shown that this configuration provides the biggest space possible for each seed. This explanation also relies on the hypothesis of the golden angle, and in fact, it only applies to sunflowers and other composite flowers.

So how can we explain the numbers of spirals? We should abandon the idea of a direct explanation of *why*, and return to the question of *how*. Only when we have answered the *how* can we hope to deduce the *why* from that.

10.3 How?

One point that goes some way towards explaining the universality of these arrangements is that plants all grow in the same way, through their buds. All plants can be

Fig. 10.4 Centre of a pine bud. Clearly visible is the flat, central, embryonic zone around which the primordia develop, turning into needles, scales, etc. The spiral arrangement (in this case (21, 13)) has been determined from the very start. © R. Rutishauser

considered as surfaces on stems that grow through their tip – the bud. Plants form volumes – the stem – by overlaying new surfaces.

There is a zone in the bud that stays permanently in an embryonic state, making cloning easier. When the stem grows, bumps known as *primordia*, develop one after another around this central zone (Fig. 10.4). Recent studies have demonstrated the predominant role of certain hormones like auxine and their transport in the appearance and development of these primordia. The latter then develop into the different botanical elements: leaves, bracts, petals, stamens, florets, etc. A considerable number of genetic studies have investigated how the type of element is determined. What interests us here is how the *position* of this new element is determined.

As early as 1868, Hofmeister was already describing the organisation of the bud and proposing a rule to determine the position of each new primordium [8]. He assumed that the primordia appeared at regular intervals over time (typically one leaf a day during the spring growth), and that the new element was placed in the biggest space left between the previous elements and the central embryonic zone. It was an iterative description of the process, based on purely physical arguments, which was still standard thinking at the time.

But this description only applies to spiral modes, unless we decide to multiply artificially the number of primordia that appear at the same time as a means of describing whorled modes. Mary and Robert Snow showed, in a beautiful series of experiments, well before the emergence of molecular biology in the 1950s, that one plant could present both modes – spiral and whorled – at the same time, and could even change from one mode to the other. A rule had to be found that would be valid in both cases simultaneously. The Snows suggested that it is not the time between the appearance of two primordia that is fixed, but the size of each one: a new primordium appears as soon as there is enough space for it [10]. This rule can be used to describe at the same time the spiral mode and the whorled mode, where two spaces become free at exactly the same time. And it is reasonable to assume that there is a characteristic size for these elements.

The rules proposed by Hofmeister and the Snows are physical rules concerning the occupation of space, and iterative over time. The question is therefore: if we accept these hypotheses, are they sufficient to explain the observations? The answer, in a word, is yes. It simply remains for us to understand how (or why ...)!

10.4 Van Iterson's Tree ... Pruned!

We can look at the whole set of regular spiral positions by stacking up discs on the curved generating surface of a cylinder. In this case, we find one sole parameter: the ratio of the size the discs to the size of the cylinder (known as the Snow parameter, Γ). This parameter is also linked to the parameter that can be constructed from the speed of growth and the average time between the appearance of two primordia (the Hofmeister parameter, G).

In 1904, van Iterson charted all the possible regular arrangements by plotting the angle between one element and the next (the divergence, denoted d) as a function of the Snow parameter Γ. This produces a tree, with one sole possibility with a large parameter – the "opposite" arrangement that can be observed in the lime tree, for example – followed by bifurcations and more and more possibilities as the parameters get smaller and smaller (Figs. 10.5 and 10.6) [11]. Each branch of this tree corresponds to a number of spirals in each direction. If we look at what happens at each bifurcation, when a branch divides in two, we can see that each of the new branches keeps one of the two previous numbers of spirals and takes the sum of the two previous numbers as its new number. It is this summation rule, deriving from the mathematical properties of this periodic network on a cylinder, that generates

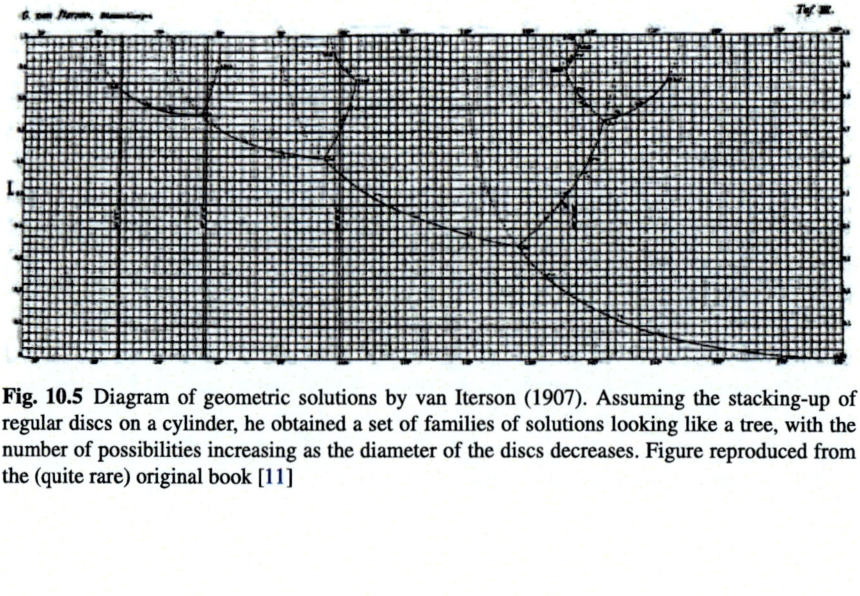

Fig. 10.5 Diagram of geometric solutions by van Iterson (1907). Assuming the stacking-up of regular discs on a cylinder, he obtained a set of families of solutions looking like a tree, with the number of possibilities increasing as the diameter of the discs decreases. Figure reproduced from the (quite rare) original book [11]

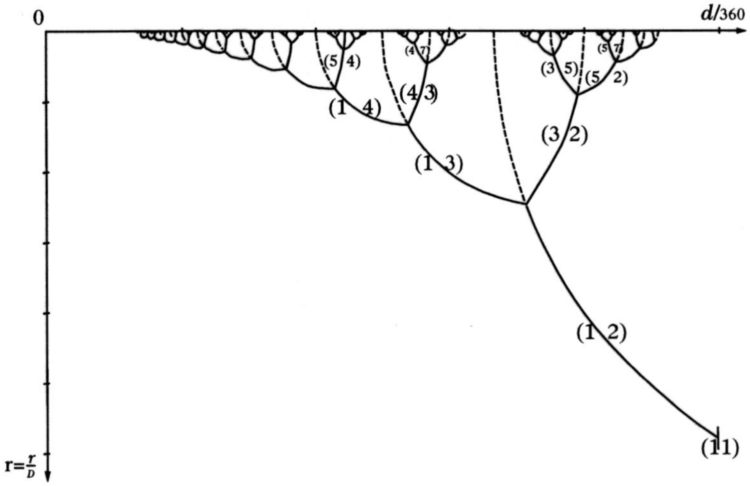

Fig. 10.6 Another van Iterson tree diagram. Along the y-axis, we plot the ratio of the diameter $2r$ of the discs to the diameter D of the cylinder (parameter $\Gamma = r/D$), and along the x-axis we plot the possible angle between one disc and the next (divergence d, divided by 360°). Each section of branch corresponds to a certain number of spirals in each direction (clockwise and then anti-clockwise). The branches mark the boundaries of regions characterised by a unique number of spirals

the Fibonacci rule, in our case. But in fact, in this tree, all pairs of numbers are possible … So why do we only see pairs of numbers from the Fibonacci sequence?

In fact, this tree diagram considers all regular arrangements, without concerning itself about their construction element by element. Now, if we apply the rule of

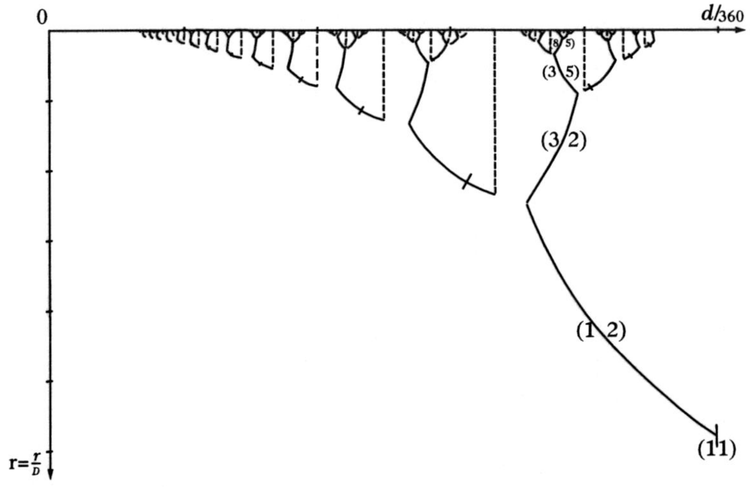

Fig. 10.7 Modified van Iterson tree diagram, taking dynamic constraints into account. This amounts to taking away the base of one out of every two branches. Only one branch remains attached to the ground (connecting all the parameters, from the biggest to the smallest), and that is the branch corresponding to the Fibonacci sequence

Snow or Hofmeister, considering that the elements are positioned one by one, then a fundamental difference appears: all the branches separate [4, 9]. All that remains are single, oscillating branches, and only one of them exists over the whole range of parameters. As the reader may have guessed, this is the branch where the numbers are two consecutive numbers from the Fibonacci sequence (Fig. 10.7).

The separation of the branches simply amounts to the fact that if a geometric stack is possible and if we look at its growth over time, the new position that should correspond to the continuation of the regular stacking up does not correspond to the first possible position, and therefore contradicts the rules of growth.

If we look at the result, we can see that this pruning only spares isolated, unbroken branches, along which we can always observe two consecutive numbers from a sequence constructed using the Fibonacci rule (but with the two initial terms in the sequence varying). So it still remains for us to explain why, in plants, we only observe the "real" Fibonacci sequence, the one starting with (0,1).

10.5 Dynamics

As these rules of growth are physical rules, we can reproduce them in an experiment or, even more simply, perform numerical simulation on them. In this case, we find that the system converges spontaneously towards the regular modes described above, with truncated branches. It is also interesting to be able to vary the parameter (rhythm) of growth. We then observe that the system continuously follows the branch it is on, as long as the branch exists. If that branch disappears, then the system jumps, stabilising again on another branch after a certain, irregular period of time [3, 6, 7].

Now, the Fibonacci sequence corresponds to the longest branch – the only one, in particular, that exists for the biggest parameter of growth. Consequently, to explain the appearance of these numbers in plants, we need only assume that plants always start their growth with the biggest parameter, where only the Fibonacci branch exists, and it is only afterwards that their parameter decreases, to ensure they stay on the branch. And this is indeed the case, as we can see from the botanical data, or in the sunflower, for example. The sunflower starts its growth from two cotyledons (seed leaves) from which the stem and leaves then grow. It can indeed be observed that, starting with these two cotyledons as the first element, the system very quickly converges towards a simple Fibonacci spiral mode. Growth only slows down when the plant is going to produce its flower; the primordia become bracts and large numbers of spirals can be seen on the edges of the flower. In this state, the plant appears to have stopped growing. The parameter of growth only picks up again (and the number of spirals diminishes towards the heart) when the formation of the flower is complete and it bursts into bloom.

Over the course of these transitional regimes, the local compactness of the stacking varies, regularly changing from dense hexagonal stacks to looser, square stacks. These rules of local stacking may be intended to produce compact stacking, but they are only local rules, and the overall result is also constrained by the arrangement of the existing elements, which impose continuity and sometimes, therefore, a less compact mode.

For the sunflower as a whole, however, the stacking is globally very efficient. But this derives simply from the fact that the system continuously follows the branch of possible options, because otherwise there would be jumps between branches, resulting in defects and loss of space.

To be more thorough, we can simulate the Snows' rules, in which case we obtain all the modes, either spiral or whorled. Only the spiral modes allow for the parameter to change continuously, whereas each of the whorled modes only has a limited range of parameters.

With this type of rule, we can even reproduce a number of "abnormal" cases, such as those having numbers of spirals drawn from the "Lucas numbers" (a Fibonacci sequence starting with (1, 3)), or even "bijugate" cases (the number of spirals is the double of a number in the Fibonacci sequence). These abnormal cases can be explained by an initial growth parameter that is lower than normal, enabling the plant to stabilise from the outset on a mode other than that associated with the Fibonacci sequence, for example the mode associated with an initial pair (1, 3), but then keeping the same summation rule.

In this way, we can reproduce in detail the positions of the leaves in cases of abnormal sunflowers, and show how their initial growing conditions were sufficient to cause their "abnormality". In fact, we can consider the final plant as the recording, from the first to the last leaves, of the history of its growth.

10.6 Conclusion

So, to explain the presence of Fibonacci numbers in spiral plants, we must return to the history of the plant. The plant builds itself up piece by piece, and this dynamic construction and its history are the only key to understanding the final result. If the plant follows rules of growth by simple stacking, the result is constrained by the mathematical properties of regular networks. We can therefore consider this growth as a trajectory that unfolds according to circumstances (e.g. external constraints) and within the set of possibilities defined by the mathematical constraints.

So although the result appears to be imposed (the numbers are always from the Fibonacci sequence), it is not imposed directly, by some mysterious kind of control mechanism, but indirectly, by the fact that all the plants share the same geometry of development and, normally, the same history of growth. This point is clearly demonstrated by the abnormal cases that can be explained fundamentally by an external modification of their growing conditions (such as pruning or the application of fertiliser).

More generally, we can therefore look at phyllotaxis as an illustration of the fact that a form is first and foremost the result of the dynamics of growth. If the result appears complex to us, this is mainly because of our own limitations in understanding forms and the logic of their appearance. And if the result appears to us to be imposed, this is mainly because we do not understand the limits to the field of possibilities. To understood the form, we must start by considering it as the unfolding of a history under constraints that remain largely unknown.

References

1. D'Arcy W. Thompson (1917) *On Growth and Form*, Cambridge University Press (Cambridge).
2. Bravais L. and Bravais A. (1837) Essai sur la disposition des feuilles curvisériées, *Ann. Sci. Nat* (seconde série) **7**, 42–110, in French.
3. Couder Y. and Douady S. (2004) La géométrie des plantes: l'art d'empiler, special issue "Les formes de la vie", *Pour la Science*, pp. 51–55, in French.
4. Douady S. (1998) The selection of phyllotactic patterns, in *The symmetry of plants*, edited by R.V. Jean and D. Barabé, Series in Mathematical Biology and Medecine, vol. 4, World Scientific (Singapore).
5. Douady S. (2004) "Forme", *Notionnaire*, Encyclopedia Universalis (Paris), in French.
6. Douady S. and Couder Y. (1993), La physique des spirales végétales, *La Recherche* **24**, 26–35, in French.
7. Douady S. and Couder Y. (1996) Phyllotaxis as a self-organizing process, *J. Theor. Biol.* **178**, 255–312.
8. Hofmeister W. (1868), Allgemeine Morphologie der Gewachse, in *Handbuch der Physiologischen Botanik*, vol. 1, pp. 405–664 Engelman (Leipzig), in German.
9. Levitvo L.S. (1991) Energetic approach to phyllotaxis, *Europhys. Lett.* **14**, 533–539.
10. Snow M. and Snow R. (1962) A theory of the regulation of phyllotaxis based on Lupinus Albus, *Phil. Trans. Roy. Soc. Londres. Ser. B* **244**, 483–513.
11. Van Iterson G. (1907) *Matematische and Microscopisch-Anatomische Studien über Blattstellungen*, Gustav Fisher Verlag (Lena), in German.
12. Web site: http://www.lps.ens.fr/ douady

Chapter 11
The Logic of Forms in the Light of Developmental Biology and Palaeontology

Didier Marchand

11.1 Introduction

If you ask palaeontologists, and indeed anyone interested in the theory of evolution, for the key words that encapsulate it, you will obtain the following results: adaptation, natural selection, speciation, but also ontogeny and phylogeny. The first three key words apply to the future of the individual and by extension to the future of the species: we are therefore dealing with adults of a reproductive age. The two other key words concern (i) the evolution of the morphology from the egg to the adult (individual ontogeny: short timescale) and especially what goes on in the black box called embryogenesis, and (ii) the modification of ontogenetic sequences over time, resulting in changes to adult morphologies (phylogeny: long timescale). After quite a long eclipse, coinciding with the advent of genetics at the beginning of the twentieth century and then the modern evolutionary synthesis in the 1940s, the ontogeny/phylogeny pair has recently enjoyed a steady revival of interest. Firstly under the impetus of S.J. Gould at the end of the 1970s [8], and then with the appearance, at the end of the 1980s, of what we now call developmental biology. This revolution in the biological sciences is comparable in importance to that which shook the earth sciences at the end of the 1960s with the emergence and rapid development of plate tectonics.

The revolution is also under way in palaeontology, in that at the beginning of the 1990s, the results obtained by biologists gave rise to a new approach to evolution, known by the now-classic term of "evo-devo" (for evolutionary developmental biology). As a consequence, palaeontologists are now in a prime position to supply embryologists with reference points to date the appearance of the major developmental innovations, which should help towards a better understanding of the

D. Marchand (✉)
Paleontologist at the Centre des Sciences de la Terre, University of Burgundy, Dijon, France
e-mail: didoux.marchand@gmail.com

P. Bourgine, A. Lesne (eds.), *Morphogenesis*, DOI 10.1007/978-3-642-13174-5_11,

history of the emergence of the major constraints on construction. Without a doubt, this will greatly improve our knowledge of the relationship between ontogeny and phylogeny.

11.2 Palaeontology and Time

It has become a commonplace to say that the time of palaeontology is linear time. And yet in the nineteenth century, this linear time was not clearly taken into account by the catastrophists and it was often neglected in the twentieth century, overshadowed by the "almost" cyclical time of genetics. As H. Tintant, professor of palaeontology at Dijon (University of Burgundy) often remarked [14], fossils and radioactivity are the only absolute proofs of the linear time so dear to S.J. Gould [9]. And indeed, fossils are the only testimony we have of past morphologies, of a reality that has now disappeared. This means that the sequence of fossil forms over time obliges us to adopt an approach that is both historical and biological at the same time, which is equivalent to saying that it prohibits us from using a strictly Cartesian approach. Consequently, the modelling of form should not be taken as anything more than an aid to understanding a biological development that takes place in an environment where, of course, the physical constraints must be respected. But under no circumstances can modelling replace history, or in other words reality, quite simply.

And the biological logic is so constraining, so coercive – as developmental biology has shown – that failure to respect it results in either teratology (malformation) or cessation of the biological process. This means that the morphological changes that occur over the "long timescale" must imperatively respect this biological logic that allows or authorises the construction of an organism.

Two remarks should be made here. Firstly, it is most often the case, over the long timescale, that successive morphological changes bring about a reduction in the number of future possibilities, within a given monophyletic line. H. Tintant liked to recall this phrase of P. Claudel that applies so well to phylogeny [14]:

> The key that unlocks is also the key that locks.

A morphological innovation will prohibit, or at least channel, certain ontogenetic possibilities. For a palaeontologist, studying form, studying changes in form, amounts to studying the possibilities authorised (and realised) by embryogenesis.

Secondly, physical or chemical phenomena are studied by the reproduction of experiments within cyclical time, and the results are only scientifically valid if they are reproducible, or even better, predictable. But if we want to understand the history of the universe, it now appears obvious that a "palaeontological approach" is indispensable for taking stock of the situation. Linear time must take into account not only laws but also what is called, for want of a better term, randomness, which can be considered as a succession of particular events, contingent and often improbable, as A. Lesne put it.

11.3 From the Cell to the Multicellular Organism: An Ever More Complex Game of "Lego"

For the sake of comparison, the 4.6 billion years that have passed since the formation of the Earth are often condensed into one year. From this perspective, life appeared relatively early: sometime around the end of March (3.5 billion years ago). At this stage, it consisted exclusively of unicellular beings. The first piece is therefore simple: the first "Duplo" (that Lego for young children) is in place. Then, at the end of July (about 2 billion years ago), as we can observe in the siliceous cherts in the Gunflint formation in Canada, there were already species with a filamentous structure: the cells divided but often remained attached. Complexification was beginning to take place, but very slowly. For some authors, it was about 1.6 billion years ago (during the famous night of 24 August) that a more elaborate building block appeared, what we might call the first "Lego". This piece, the eukaryote cell, is more complex because it is composed of two or three cellular entities that collaborate: a magnificent success of symbiosis. Retrospectively, we may find it logical to collaborate – to be more efficient – but the essential point is that these cellular entities did not reject one another, as can so often be observed with today's organisms, and there were even effective genetic exchanges. With the eukaryotes, we also witness the appearance of more complex structures in the sense that a number cells stay together to form a multicellular structure, even if it is still very small. And it is reasonable to suppose, as many authors have done, that meiosis and fertilisation were associated with this complexification from which arose what we might call, by comparison with what we observe today, "proto-embryos".

We also know that the first forms that can be described as organisms are not known with certainty to have existed more than 600 million years ago: this takes us up to the middle of November. These organisms originated out of several important innovations. The first of these was almost certainly the extension of the duration of "common growth". The second was the structuring of these cells to form tissues: ectoderm on the outside and endoderm on the inside, followed by specialisations within these tissues: the very first "Lego Technics". The third, and perhaps the most fundamental, was the invention of gastrulation (invagination of tissues) which allows to create spatial relations between the tissues: the invention of 3D structure. For at least the last 600 million years, the gastrula stage has been indispensable to all metazoans, an inescapable step in the construction of organisms.

So there has been a progression from uniqueness: the cell, to the multicellular in one dimension: the filament, and then to the multicellular in two and then three dimensions. One last observation: in organisms without a coelom (like the sponges and coelenterata) the adult structure appears to be fundamentally radial, whereas in coelomates, there is always two-sidedness, at least in the larval stages.

It therefore appears that the establishment of the morphology of organisms is governed by constraints of two orders. Firstly, what has been invented is generally conserved. And each innovation becomes a necessary step on the path towards something more complex. Secondly, the main characteristics of construction are clearly under genetic control, as developmental biology has demonstrated. But as complexification leads to ever more spatial connections between the different parts of the organism, it is highly probable that certain important morphological innovations appeared as secondary consequences, before being taken up by selection.

11.4 The Major Body Plans: In the Early Cambrian, Quite Everything Was Already in Place

The first well-conserved adult animal fauna in the form of imprints (calcified skeletons did not yet exist) were discovered in Australia, near Ediacara, in sediment about 570 million years old (16 November in our condensed year). Since then, many deposits from this Eocambrian period have been found in various parts of the world. What can we learn from this fauna about the main morphological types present at this time, at the very end of the Precambrian? Firstly, forms with a fundamentally radial structure were abundant, in terms of numbers of individuals, and the body plans of some of them are similar to those that exist today.

Secondly, we can see that coelomates were also present and that in certain cases, their bilateral symmetry is well-pronounced. This can be detected in forms believed to be worms or closely connected forms, in future molluscs and of course in future arthropods. On the other hand, it is essential to note that there were no structured, articulated appendages, even in forms resembling arthropods like the *Spriggina*: this morphological innovation, which we now know to be particularly associated with the Hox gene, was not yet definitively established. And the discovery of phosphatized embryos in China, in strata dated at about 600 million years, shows that the main types of development, spiral or radial, were already present.

But fossils from more recent deposits in Chenjiang (Early Cambrian of China; 19 November) and in Burgess (Middle Cambrian of Canada; 20 November) show us a fundamental fact: in these distant eras, all the phyla present on our planet today were already in place, from sponges to vertebrates. But within these phyla, of course, many classes (the class being the taxonomic unit just below the phylum) were not yet present. To give two significant examples:

1. The echinoderms (already present in Ediacara, notably with the genus *Arkarua*) had a calcareous skeleton (stereome) identical to those of present-day echinoderms and, apart from a couple of exceptions (helicoplacoids and carpoids), pentaradial symmetry. However, the five classes known today (crinoids, asteroids, ophiuroids, echinoids and holothuroids) did not yet exist and other classes, now extinct, were present;
2. In the vertebrates, this aspect is even more striking, as the only class present was the *Agnatha*, which is now comprised of lampreys and hagfish.

A third example also offers a wealth of insight. Today, the phylum of arthropods contains three main classes, characterised among other things by very typical numbers of cephalic appendages (apart from very rare exceptions): the *Crustacea* (two pairs of antennae, three pairs of appendages); the *Chelicerata* such as limuli, scorpions and arachnids (one pair of chelicera in front of the mouth, one pair of pedipalps – for example the pincers of the scorpion – and three pairs of appendages); the *Unirama* (one pair of antennae, three pairs of appendages) comprising the myriapods (centipedes) and the insects. In the Burgess deposits, the ancient arthropods (520 million years old) have been remarkably analysed by a good number of researchers and introduced to a much wider public by S.J. Gould in his book *Wonderful Life* [10]. They include the well-known class of trilobites, which became extinct towards the end of the Permian period; this class was characterised, at the level of the cephalon (head), by one pair of antennae and three pairs of legs. But alongside the trilobites, we find a multitude of arthropods possessing a number of cephalic appendages unknown in present-day nature. For S.J. Gould, the conclusion is that in the Middle Cambrian, the number of classes was much higher than it is today. It is on the basis of this observation that he proposed his famous hypothesis of decimation: over the course of time, the number of species falls and the living world becomes more and more stereotyped. Today, we would say that diversity is disappearing.

However, the contributions of developmental biology applied to the study of these Cambrian arthropods allows us to put the problem differently. Indeed, palaeontology has shown us that the appendages of these arthropods, like the members of vertebrates, appear relatively late on in the process of structuring the body. Let us look at two examples:

1. The form known as *Spriggina*, from the Ediacara deposits, presents bilateral symmetry and two sections that are recognisable by their respective positions, a cephalon and a thorax. But ... there are no real articulated appendages;
2. The form found in Chenjiang, called *Haikouichthys*, already possesses a head with its sense organs, a pharynx with its branchial arches, metameric muscles in the shape of chevrons, characteristics observed in present-day vertebrates. But ... without paired appendages. Developmental biology confirms that among modern arthropods, the development of appendages comes after the establishment of the body plan, and in particular after the appearance of the tagmata. The same holds true, of course, for vertebrates.

If we apply these two observations, one of a palaeontological nature and the other embryological, to the arthropods of the Early and Middle Cambrian, we can already recognise certain general morphologies that are easily classified among the *Crustacea*, and others that can be classified just as easily with the *Chelicerata* (merostomes). However, apart from rare exceptions, do not have the appropriate cephalic appendages, if we compare them to the present-day classes. Should we therefore put them into new classes, or simply say that the growth of appendages, under the control of Hox genes, had not yet been developed, or to be more precise, had not yet become stereotyped and fixed.

If this interpretation is correct, it means that in the living world, form is fundamentally dependent on a "genetic programme" that imposes morphological innovations, which appear in fits and starts over the course of linear time. If these innovations are compatible with the existing programme, they can become established and if the environment accepts them, they can even end up by stabilising and becoming characteristics of a given group, whatever the taxonomic level envisaged. Of course, the new forms resulting from this "tinkering" must also submit themselves to the physical and biological constraints imposed by the environment. And when it comes to inter- or intra-species competition, it is essentially the physiological qualities of the organism that make the difference, not its form. But that is quite another kettle of fish.

11.5 The Phylum of Vertebrates: A Fine Example of Peramorphosis

The invertebrates are classified into a large number of phyla, whereas animals with an internal skeleton are all grouped together in the vertebrates phylum. This rather obvious statement nevertheless allows us to affirm that all the new morphological features that appear in this monophyletic group must be based on the "genetic stock" present in the oldest Chinese ancestors. It is therefore this "genetic stock" that is capable of inventing, innovating, or "tinkering", provided that it respects (on pain of death) the rules imposed by the constraints of construction – rules that very probably started to emerge at the end of the Precambrian.

History teaches us that the complexification of an organism is statistically associated with the embryonic and foetal stages. As the oldest embryonic constraints are the first to appear over the course of ontogeny, a certain period of time is necessary for the successive morphological innovations to express themselves: ontogeny will take longer. This peramorphosis (from the Greek *pera*: beyond or across). The description of this fundamental phenomenon is already long-established, dating from the beginning of the nineteenth century and the pioneering work of Haeckel: ontogeny, at least in the early stages of development, cannot do other than to recapitulate phylogeny, because it must obey the oldest rules if the organism is to continue to develop.

What do we observe over the "long timescale" of palaeontology? According to the data collected from the Chinese forms, the fundamental structuring of the bodies of vertebrates, what one could call the hard core, the basic software, is composed of: (i) a head that already has olfactory, optic and probably otic capsules; (ii) a branchial basket made of arches and branchial clefts; (iii) a dorsal nervous axis; (iv) a chevron-shaped muscular system and, of course, (v) digestive and reproductive systems. The locomotive part, on the other hand, is only represented by a poorly defined dorsal fin. This is fundamentally the same structure that characterises present-day *Agnatha*, except that their much longer bodies now enable them to swim more efficiently by undulation, and that original morphologies appeared during the Early Paleozoic, but without deviating from the fundamental *Agnatha* body plan.

With cartilaginous fish, the structuring of this ancestral form was accentuated. The head is more individualised and more efficient: jaws appear, the olfactory apparatus is often enlarged into a pair of nostrils instead of one, and the organ of balance is more highly-developed (three semicircular canals instead of one or two).

The branchial basket is more compact (often with five branchial arches) and the nervous tube is now well-protected by the vertebrae. But the most visible morphological changes concern the appearance of well-individualised pectoral and pelvic fins. Likewise, the appearance of a caudal fin capable of providing propulsion by oscillation radically modifies the form – that of sharks in particular, as the group of rays opted for an original solution in their way of swimming.

In bony fish, the appearance of the dermal skeleton, providing better protection for the brain, allowed the skull to develop much more varied morphologies. Each part of the skull, each bone even, could change shape, sometimes causing considerable changes in the whole head. But whatever the morphological variations observed, and the sometimes drastic changes in form (in certain cases clearly related to developments that could be qualified as teratological), they always remained within the "fish pattern". And even in the *Panderichthys*, a group closely related to the *Tetrapoda*, this morphology endured, although the pectoral and pelvic fins were original insofar as they were almost constructed like the paired limbs of *Tetrapoda*, with a stylopodium (humerus or femur) and a zeugopodium (radius/ulna or tibia/fibula), with the exception of the autopodium (hand/foot) that was still absent. But what it is important to note, combined with these structural modifications of the paired limbs, is the diminution of the other fins and the appearance of a more compact, more solid dermal skull, in which the bones become more and more individualised et correlatively fewer and fewer. And as the formation of dermal bones is related to neural crests, it is reasonable to imagine that the changes occurred very early on in the process of ontogeny, during the fish-tetrapod transition. All these innovations, which we can consider to be coordinated, ended up by profoundly changing the general morphology of these fish.

And then, with the very first tetrapods, whose four limbs ended in hands and feet, the morphology underwent a spectacular transformation, all the more so since the odd fins totally disappeared, apart from the caudal fin, which remained, but in a much reduced form. So at the end of the Devonian, about 370 million years ago, we see the last great morphological innovation of the *Vertebrates*: four-limbedness. But we will have to wait a little longer (a few million years, maybe more?) before these limbs that behave more like fins "full of toes" become truly operational: in this case, the organ clearly came before the function [1].

Since that great morphological innovation, changes in form in vertebrates have often been associated with the restructuring of limbs used to make fins, wings or to shrink, sometimes to the point of disappearing, in vertebrates that became snake-like in form. What the genome has constructed, it can also tinker with or erase.

11.6 The Anomalies of Development: An Opening Towards New Morphologies

We have known since the time of Isidore Geoffroy Saint-Hilaire [3], that the malformations or teratologies (from the Greek *teratos*: monster) which occur spontaneously during ontogeny provide a wealth of information about the way an organism constructs itself: it is often the dysfunctions that lead us to understanding. And this is what Goldschmidt had so remarkably grasped and expounded with great clarity in his celebrated (but little read) article of 1933 [7].

Since the 1980s, we have known that a large number of malformations, particularly those affecting the limbs, are related to (i) what are called the architect genes or Hox genes, or (ii) phenomena of embryo construction that are now correctly identified. In this respect, the data obtained from palaeontology (the old reality) are highly instructive. Three examples appear to be particularly significant.

1. The first is that of snakes. It has long been known that ophidians have lost not only their front legs but also every embryonic trace of these limbs and their associated shoulder girdle (to such a degree that we cannot determine exactly how many cervical vertebrae they have). On the other hand, traces of pelvic girdle and femurs are still visible in forms such as boas and pythons, but even then the tibia/fibula pair (zeugopodium) and the autopodium are absent.
   Recent descriptions of three genera of ophidians from the Cenomanian (about 100 million years ago), fossilised in a marine environment, show that even in that era, the front legs had already totally disappeared. According to [11], the hind limbs were still present, though greatly reduced (measuring from 1 to 3 cm) in relation to the length of the body (between 72 and about 150 cm). The pelvic girdle is easily recognisable (with its three well-individualised bones) and the legs still have a stylopodium (femur) and a zeugopodium (tibia/fibula). The autopodium, on the contrary, is in all three cases incomplete: the tarsus is still present, but the metatarsi are only present in one case. By comparing modern and fossil forms, it would appear that the reduction of the limbs in ophidians started with the autopodium and then continued with the zeugopodium before reaching the stylopodium. Interestingly, the different parts of the hind limbs appear to disappear in the reverse order to their appearance during the fish-tetrapod transition.
2. The second example is that of *Cetacea*. The works of Gingerich et al. [4–6] and of Thewissen et al. [13] have shown that cetaceans are rooted in the artiodactyls, the earliest representatives of which (*Pakicetus* from the Early Eocene, about 50 million years ago) were tetrapods with perfectly functional limbs. Subsequently, with forms like *Basilosaurus* from the height of the Middle Eocene (about 38 million years ago), the limbs had diminished considerably. In these very elongated forms (some more than 15 metres long), rather snake-like in appearance, the front limbs had already changed into paddles, while the hind limbs were engaged in a process of considerable reduction. As in the ophidians, the pelvic girdle and femurs were still easily recognisable, although reduced in size, while the zeugopodium had shrunk much more and the tibia and fibula were partly fused; in the autopodium, the bones of the tarsus were welded together and the metatarsals were reduced in length and number. In the Late Eocene, the autopodium of cetaceans had disappeared and the zeugopodium was hardly recognisable. Today, what remains of the pelvic girdle is no longer attached to the vertebral column and the femur, often present, can only be identified by its position in the organism. It is interesting to note the similarity to the ophidians.

3. The third example is that of the *Sirenia*. The ancestral form, *Pezosiren* [2] dates from the beginning of the Middle Eocene. It was a form with a well-developed pelvis and short but complete limbs, apparently enabling it to move on land. From the height of the Middle Eocene, however, the limbs started to shrink, and in a form like *Halitherium taulannense* [12] the autopodium had already disappeared and the zeugopodium had shrunk. As in the two previous examples, the reduction of the hind limbs started with the autopodium and moved towards the stylopodium; in parallel, the pelvic girdle became more and more vestigial, with the reduction and then the disappearance of the ischiatic foramen (an opening between two bones of the pelvis, the ischium and the pubis) then the ever more marked reduction of the acetabulum (the hollow where the head of the femur fits into the pelvis), thus reflecting in the pelvis the gradual reduction of the femur.

In the three cases described, modification of the form of the limbs is strictly dependent on the "genetic programme", and no "morphing" type model can be envisaged: in these cases the method of D'Arcy Thompson and derivative methods prove to be unusable. And the same is true for the evolution of the toes in *Equidae*, which shrank slightly before completely disappearing, or almost. This was the case for the front toe V, which suddenly disappeared in the transition from *Epihippus* (Late Eocene) to *Mesohippus* (Early Oligocene) and the sudden transformation of toes II and IV, short but complete, into splints made solely of the metatarsals in the move from three-toed to one-toed Equidae, sometime around the Miocene-Pliocene boundary. In these cases, the new form appeared as an emergent property of the "embryological mechanism", which is confirmed by the reappearance, now considered teratological, of one, or more rarely two, side-toes on certain present-day horses.

11.7 The Brain as the Last Space of Freedom

In the vertebrates, it has long been known that the brain gradually becomes more complex as we move from *Agnatha* through to *Mammalia*, and that this trend has operated in every class. It is very easy to observe in *Mammalia*, where many orders are characterised by a marked increase in the cerebral hemispheres (telencephalisation). This phenomenon is present but not very marked in for example the *Sirenia* or the *Equidae*; it is more pronounced in the *Proboscidia* or the*Cetacea* and it finds its fullest expression in the Primates.

Within the vertebrates, and the mammals in particular, there is therefore a convergent trend towards an increase in brain volume, often referred to as encephalisation. This means that the mechanical constraints limiting the increase in cranial volume are weak. And indeed, during ontogeny, the dermal bones that will make up a large part of the skull are not yet joined: if the brain volume increases at an early stage in embryonic development, the bones can easily follow this increase. This phenomenon, which appeared with bony fish (the first to have dermal bones) has always facilitated increases in brain size, which consequently influences the morphology of the neurocranium, especially at the foetal stage. The final structuring of the adult neurocranium, on the contrary, is determined by the combined action of the muscles of mastication, which are attached to the side walls of the skull, and the muscles involved in head movement, which are attached to the back of the skull.

But it is, of course, with humans that the use of this space of freedom has been the most spectacular. From 2 million years ago (about 8 p.m. on the 31 December) up until today, the average cranial volume has increased from 400cm^3 to 1400 cm^3, an astonishing acceleration on a geological timescale, even if it only represents an increase of 1 cm^3 every 2000 years. The history of Hominids (genera *Australopithecus* and *Homo*) shows that it is above all in our genus that the increase in cranial volume is most visible. It is also significant that the first event in Australopithecus was the filling-in of the space behind the orbits, followed at a later date by the elevation of the frontal bone, which tended to become more vertical or even bulge out in women and children. And when the changes in the form of the human neurocranium are modelled, it is often forgotten that this form is influenced by changes in the volume and form of the brain, during embryonic and foetal stages, and by the impact of the muscles that attach to the neurocranium, during postpartum ontogeny. The human skull is modified from within before it is modified from the outside, and this is the case for all vertebrates, without exception.

11.8 Conclusion

If we only take into account the general body plans represented by phyla, palaeontology shows us that the main features of present-day forms have been highly constrained over the last 500 million years or more. When we consider classes, on the other hand, then the body plans within each phylum are of course more numerous and for the most part more recent. And the same is true for each level, through orders and families down to genera.

Palaeontology also shows us that innovations appear in fits and starts and that their impact is more and more limited over the course of linear time. The major changes in form that occur within increasingly constrained body plans mainly concern non-vital organs like the limbs in vertebrates or the appendages in arthropods.

Embryology teaches us that the very early stages are similar, while subsequent stages diverge all the faster as the taxonomic level is high and the origin of the groups is old. It also teaches us that, if the succession of early ontogenetic stages is not respected, then ontogeny most often comes to a stop. To draw a comparison, this means that to build an organism, powerful constraints must be respected: the foundations before the ground floor, the ground floor before the first floor, etc. Lastly, it teaches us that innovations involving important restructuring of morphology are rare because they have to respect an increasing number of constraints, some of which are obviously vital.

The 3D modelling that is available today, and the diverse analyses of form, help us to understand better the interactions that exist between the different parts of an organism, the nature of an inter- or intra-species morphological space, what separates two sexual dimorphs, etc. But we should never forget that morphological variations must respect biological rules, and understanding these constraints on construction remains essential if we are to grasp, in all its subtlety, the true significance of a biological form.

References

1. Chaline J. and Marchand D. (2002) *Les merveilles de l'évolution*, Éditions Universitaire de Dijon (Dijon), in French.
2. Domning D.P. (2001) The earliest known fully quadrupedal sirenian, *Nature* **413**, 625–627.
3. Geoffroy Saint-Hilaire, E. de. (1822) Philosophie anatomique, *Mem. Mus. Hist. Nat. Paris* **9**, 89–119, in French.
4. Gingerich P.D. and Russell D.E. (1981) Pakicetus inachus, a new Archaeocete (Mammalia, Cetacea) from the Early-Middle Eocene Kuldana Formation of Kohat (Pakistan), *The University of Michigan. Contributions from the Museum of Paleontology* **25**, 235–246.
5. Gingerich P.D., Raza M., Arif M., Anwar M., and Zhou X. (1994) New whale from the Eocene of Pakistan and the origin of cetacean swimming, *Nature* **368**, 844–847.
6. Gingerich P.D., Haq M.U., Zalmout I.S, Khan I.H., and Malkani M.S. (2001) Origin of Whales from the Early Artiodactyls: Hand and Feet of Eocene Protocetidae from Pakistan, *Science* **293**, 2239–2242.
7. Goldschmidt R. (1933) Some aspect of evolution, *Science* **78**, 539–547.
8. Gould S.J. (1977) *Ontogeny and Phylogeny*, Belknap Press of Harvard University Press (Cambridge MA).
9. Gould S.J. (1987) *Time's Arrow, Time's Cycle*, Harvard University Press (Cambridge MA).
10. Gould S.J. (1989) *Wonderful Life: The Burgess Shale and the Nature of History*, W. W. Norton & Co. (New York).
11. Rage J.C. and Escuillié F. (2003) The Cenomaniam: stage of hinlimbed snakes, *Notebooks of Geology* (Maintenon), Article 2003/01.
12. Sagne C. (2001) Halitherium taulannense, nouveau sirénien (Sirenia, Mammalia) de l'Éocène supérieur provenant du domaine Nord-Téthysien (Alpes-de-Haute-Provence, France), *C.R. Acad. Sci. Paris, Sciences de la Terre et des planètes* **333**, 471–476, in French.
13. Thewissen J.G.M., Williams E.E., Roe L.J., and Hussain S.T. (2001) Skeletons of terrestrial cetaceans and the relationship of whales to artiodactyls, *Nature* **413**, 277–281.
14. Tintant H. (1986) La Loi et l'événement. Deux aspects complémentaires des Sciences de la Terre, *Bull. Soc. Géol. de France* **8**, 185–190, in French.

Chapter 12
Forms Emerging from Collective Motion

Hugues Chaté and Guillaume Grégoire

12.1 Introduction

From the smallest to the largest scale, and whether they are natural or artificial, groups of organisms move together. Already, Pliny the Elder [10] bequeathed us his fascination with the collective movements of animals. The evening flight of clouds of starlings is one of the most spectacular of these phenomena. Common around the Mediterranean basin in winter, they have inspired countless literary and poetic descriptions [2]. Scientifically, however, these clouds of birds remain as poorly understood as the remarkable stability and sometimes precise internal order of schools of fish, the dynamics of herds of caribou or wildebeest, etc.

In fact, the whole of the animal kingdom is rich in examples of collective motion, ranging from the simplest occasional gatherings to the most complex hierarchical societies. And even if we limit ourselves to the simplest "non-social" links, there is an incredible ubiquity of herding tendencies in every medium: on land, in the air and in the water. We now know that plants are also capable of collective behaviour [4], but it is still the collective *motion* of animals that most fascinates, because the groups very often evolve in an open environment, and therefore possess a *form*, which can itself evolve (or not, as the case may be). A swarm of mosquitoes, for instance, is often immobile, a sphere without internal texture emerging from the chaotic flight of individuals. In the case of our starlings, on the other hand, their rapid trajectories through the three dimensions of space form superb helices, spirals, clouds of complex forms that are constantly moving and changing shape. Form can also be the purpose of collective motion, notably that of cells during various stages of development, or during the healing of wounds.

Scientists are not immune to this fascination with the apparently coordinated movements of organisms. The way they approach these phenomena depends a great deal on their discipline. Biologists, ethologists and ecologists, who are the most

H. Chaté (✉)
Physicist at CEA-Saclay, France
e-mail: Hugues.Chate@cea.fr

P. Bourgine, A. Lesne (eds.), *Morphogenesis*, DOI 10.1007/978-3-642-13174-5_12,

directly concerned, generally focus on particular cases (each species must be studied, or at least described, in its own right). Typically, they seek to "explain" these behaviours in Darwinian terms, demonstrating *a posteriori* the selective advantage conferred on an individual, a species or a gene by grouping together.

Relatively recently, physicists, working in the modern domains of non-linear dynamics and non-equilibrium statistical physics, like ourselves, have also turned their attention to the problem of collective motion [1]. Naturally attracted by the omnipresence of the phenomenon, they immediately suspected the existence of simple, shared, underlying mechanisms, a "typical" physicist's reaction. Very different from traditional biologists, looking beyond all the specific details, aiming to describe *how* rather than *why* "things" can move and stay together, they have set out to determine the minimum ingredients necessary and sufficient to produce the phenomenon of collective motion. Once these have been understood, we can envisage – and this has been one of the major advances of the statistical physics of non-equilibrium phenomena – the existence of *universal* properties, "emergent" as they are now called, qualitative or semi-quantitative, that can be observed and verified at every scale. In the rest of this chapter, we shall describe some of the stages in this search for the minimum ingredients and universal emergent properties of collective motion, with particular attention to the forms taken by groups of organisms in movement.

The fascination mentioned originates in the remarkable organisation of motion in certain groups of animals, an organisation that appears to occur without any centralised programme to coordinate individual movements. If such a programme does not exist (even for the movements of cells during embryogenesis), we can imagine other simple explanations – "trivial", as the physicist would call them – of the origin of group motion. For example, there may be a leader who stands out from the other individuals and guides them. Or the group may be so spatially confined (in a corridor, or in an aquarium) that the geometrical constraints force its motion. Another trivial explanation could be that each individual in the group can see all the others. Here, as we wish to get to the root of the problem, we shall set aside these situations, together with others less trivial but already too complicated for our purpose, in which collective motion is induced by an external flow or gradient (such as the advection of plankton by ocean currents or chemotaxis). We shall therefore be placing ourselves in the most *a priori* unfavourable conditions for observing the emergence of collective motion: identical individuals, interacting strictly locally within a free, homogeneous space, in the presence of strong and statistically constant disturbances (noise, turbulence) without mediation from external substances.

In this way, we are assured of the maximum possible "non-triviality", corresponding to the most spectacular cases observed in nature: how, in a very dense school of fish, containing up to several hundred thousand individuals, could a fish see a leader, or all of its fellows? We shall see how an unpredictable and at the same time organised collective dynamics, and particularly a dynamics of form, can emerge from this disorderly democracy.

12.2 Towards a Minimal Model

12.2.1 The Ingredients

To identify the ingredients that seem to be relevant and necessary to the appearance of collective motion, we might start by using our experience as demonstrators. Rather distracted, and not having a plan of the route of the demonstration, we roughly follow the average direction taken by our neighbours. In other words, we choose a vector that is, on average, colinear to the vector of the average local flow. This *alignment tendency* is the key ingredient. Its precise mechanism may have different origins depending on the circumstances: for free-swimming tubular bacteria, it is the shocks between the bacteria that give rise to an effective alignment; for fish, depending on the species, it is due to visual and/or acoustic perception of their neighbours, etc.

But if that were all, then there would be nothing to maintain the cohesion of the group (in an unconfined space): the individual aligns with its neighbours, but if it should move slightly further away (because of a fluctuation, for example), it may become detached from the group, and there will be no force to bring it back. Our demonstrator will therefore also be careful not to stray too far from his neighbours, for fear of losing contact with them. On the other hand, if he gets too close, then discomfort is the strongest sensation and he will move away somewhat. This interplay of attraction and repulsion defines a preferential distance between individuals, and if it is intense enough, it can suffice to maintain cohesion.

Alignment and an attraction-repulsion mechanism (Fig. 12.1), together with their respective intensities in relation to the "disorder" (ambient noise, errors made by individuals in the calculation of interactions) are the two indispensable ingredients of non-trivial collective motion [9].

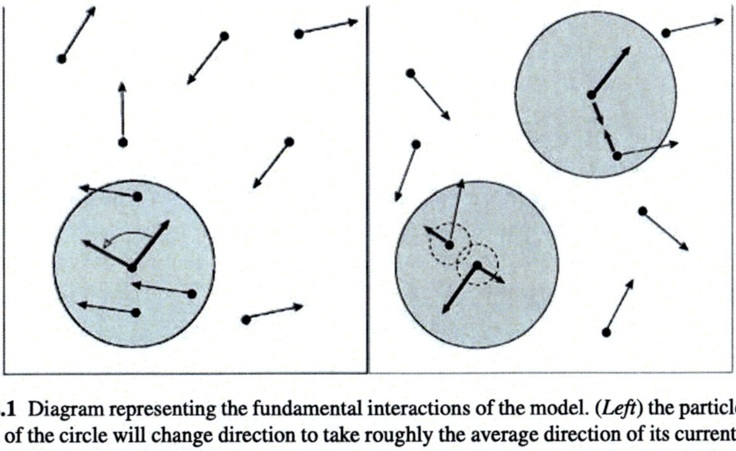

Fig. 12.1 Diagram representing the fundamental interactions of the model. (*Left*) the particle in the middle of the circle will change direction to take roughly the average direction of its current neighbours. (*Right*) two particles too close together repel each other; two particles interacting attract each other

12.2.2 Formalisation

Now let us turn to the mathematical formalisation of the ingredients identified above. This means building models that incorporate these ingredients in a minimal way, with the least possible superfluous details. Our approach does not consist in describing a particular situation, nor in *simulating* it in a realistic way. *A priori*, we should therefore have a wide range of modelling options open to us, but this is illusory, because our desire for minimalism, together with the theoretical advances made in demonstrating the qualitative equivalence of sets of equations, actually rather limits our choice.

To simplify our exposition, we shall limit ourselves to the two-dimensional case, which is not devoid of biological realism. Of course, all land-based animals move over what is essentially a surface, but experiments have recently been conducted in which bacteria [15], or even fish, are confined in a layer of fluid of a thickness similar to their size, so that they are in effect operating in an almost two-dimensional environment.

As is often the case in physics, we distinguish between "microscopic" models, in which the individuals are explicitly represented, and mesoscopic or macroscopic models describing the evolution of continuous fields, representing the local average of a relatively large number of individuals.

12.2.2.1 Continuous Variables

The relevant variables in mesoscopic or macroscopic models are the local density of individuals ρ and their average local speed $\mathbf{v}$. These two fields are coupled and governed by equations with partial non-linear derivatives, *a priori* quite similar to Navier-Stokes equations. For example, to describe a bacterial bath, it is possible to write [3]:

$$\partial_t \rho + \nabla.(\rho \mathbf{v}) = 0, \tag{12.1}$$

$$\partial_t \mathbf{v} + (\mathbf{v}.\nabla)\,\mathbf{v} = -g\nabla h - \frac{1}{\rho_0}\nabla p_{ext} + \nu\nabla^2\mathbf{v} + \frac{1}{\tau v}(v_0 - v)\,\mathbf{v} + \eta. \tag{12.2}$$

The first equation expresses the conservation of the number of bacteria. In the second, corresponding to the conservation of the quantity of movement, the successive terms account for gravity, the pressure due to repulsion, the viscosity, the propulsion of bacteria at speed v_0 and viscous friction, and finally the noise.

Equations of this type are often described solely on the basis of arguments of symmetry and/or under hypotheses of slow and large-scale variation of the fields (hydrodynamic limit) [12]. Given our lack of knowledge about microscopic (i.e. individual) dynamics, computing the continuous limit is both difficult and poorly controlled – whence the presence, even today, of rival equations to describe the same situations – and nobody has succeeded in proving the true relevance of continuous descriptions. In this context, it seems preferable to concentrate on microscopic models, at least to begin with.

12.2.2.2 Discrete Variables

Considering each animal as a punctual particle i, of mass m, speed $\mathbf{v_i}$, subject to the laws of Newtonian dynamics, we can write, for example, following [11]:

$$m \frac{d\mathbf{v_i}}{dt} = \alpha \, (\mathbf{v_i} \wedge \vartheta[\varphi_\mathbf{i}] \wedge \mathbf{v_i}) + \beta \sum_{j \in \mathcal{V}_i} \mathbf{f_{ij}} - \gamma \, \mathbf{v_i} \, . \tag{12.3}$$

The last two terms, representing the interactions of cohesion with neighbours j in the neighbourhood $\mathcal{V}_i$ of i and the usual viscous friction respectively, are quite classic. The first term on the right-hand side, on the contrary, corresponds to an "alignment force". As the operator ϑ represents the unit vector, this term depends on the vector φ_i, the average of the speeds of the $\mathcal{N}_i$ individuals situated in the neighbourhood $\mathcal{V}_i$ of i, defined by:

$$\varphi_i = \frac{1}{\mathcal{N}_i} \sum_{j \in \mathcal{V}_i} \mathbf{v_j},$$

The choice of the form of the alignment interaction is largely arbitrary. We could write it as being proportional to φ_i, but in this case, we would have to re-introduce a term of disorder. As the (12.3) is non-linear, it has a chaotic regime and therefore has no need of an explicit stochastic term: the intrinsic stochasticity is sufficient.

This type of model can be simplified even further without any prejudice to our main objective, that of extracting the minimum ingredients necessary to reproduce the universal properties of collective motion. So let us pare it down further.

If the term of inertia is negligible compared to that of viscosity (an overdamped regime, which may be relevant in the case of bacteria, for example), we can calculate the speed vector directly, and the dynamics is only first order in time. We can simplify still further by ignoring variations in the module of speed, which we fix at a value v_0. The animal modelled like this never gets tired, and always swims, flies or runs at the same speed. The energy thus injected into the system drives it strongly out of equilibrium. The dynamics then consists solely in the determination of the *direction* of animal i's movement, based on the current directions and positions of its neighbours. By reducing the alignment term to the average of the neighbours' speeds and adding an explicit noise term, we obtain, once we have divided time t into discrete intervals Δt for the digital application:

$$\mathbf{v_i}(t + \Delta t) = v_0 \, (\mathcal{R}_\eta \circ \vartheta) \left[\alpha \sum_{j \in \mathcal{V}_i} \mathbf{v_j}(t) + \beta \sum_{j \in \mathcal{V}_i} \mathbf{f_{ij}}(t) \right] \tag{12.4}$$

or equally:

$$\mathbf{v_i}(t + \Delta t) = v_0 \, \vartheta \left[\alpha \sum_{j \in \mathcal{V}_i} \mathbf{v_j}(t) + \beta \sum_{j \in \mathcal{V}_i} \mathbf{f_{ij}}(t) + \mathcal{N}_i \, \xi_\eta \right] \tag{12.5}$$

This equation is then completed by a simple updating of positions $\mathbf{r_i}$:

$$\mathbf{r_i}(t + \Delta t) = \mathbf{r_i}(t) + \Delta t \; \mathbf{v_i}(t + \Delta t) \, . \tag{12.6}$$

The difference between (12.4) and (12.5) resides uniquely in the effective noise term: for (12.4), the noise is added after "calculation" of the new direction, in the form of a random rotation $\mathcal{R}_\eta$ of an angle of a value between $-\eta$ and $+\eta$. We could, for example, imagine the animal only imperfectly implementing the direction calculated. In the case of (12.5), a random vector ξ_η, of module η, is added to the forces of interaction, representing the error committed in calculating each interaction, because of ambient turbulence for example. In fact, as our studies have shown, despite their very different origins and structures, these two noise terms – which can be combined – do not produce different behaviours within the asymptotic limit of very large populations.

12.2.3 The Results of Vicsek et al.

The model defined above is actually derived from that developed by Vicsek and colleagues in the mid-1990s. In [14], the case with "angular" noise (12.4) and without the interaction of cohesion ($\beta = 0$) was studied, neighbours being defined by a maximum distance of interaction r_0. It can be summed up in the equation

$$\mathbf{v_i}(t + \Delta t) = v_0 \; (\mathcal{R}_\eta \circ \vartheta) \left[\sum_{j \in \mathcal{V}_i} \mathbf{v_j}(t) \right] \tag{12.7}$$

completed by (12.6). This "absolutely minimal" version is of great interest, despite its extreme simplicity. There can be no question of maintaining the cohesion of a group placed in an infinite space, with a finite density of animals ρ, but a collective motion can nevertheless emerge: with low or zero noise η, alignment with neighbours is almost-perfect and is propagated from neighbour to neighbour. Within a short time, all the animals are moving in the same direction. On the other hand, for $\eta = \pi$ and more generally with loud noise, the animals are simply random movers, and no coordination of directions is possible.

Only the global parameters ρ and η are important; the others play no role (provided that the condition of locality and continuity of interactions $v_0 \Delta t < r_0$ is respected). What interested Vicsek et al. was the existence, between the two extreme situations described above, of a threshold for the appearance of collective motion, which they described as a *phase transition*. The system spontaneously chooses a direction of collective motion if the noise is low enough and/or the density of animals is high enough. The instantaneous natural order parameter that accounts for the change in symmetry is the average global speed, normalised by v_0:

$$\varphi(t) = \frac{1}{v_0} |\langle \mathbf{v_i}(t) \rangle| \tag{12.8}$$

where $\langle \ldots \rangle$ represents the average over all individuals. In what follows, we shall simply denote by φ the time average of $\varphi(t)$. Thus, $\varphi = 0$ in the completely disordered phase, and $\varphi = 1$ for a perfect alignment of direction. From their computer studies, Vicsek et al. concluded that the emergence of collective motion can be seen as a new type of continuous phase transition, accompanied by new universal critical properties. In particular, φ varies continuously from zero during the emergence of collective motion, following a power law with universal exponent.

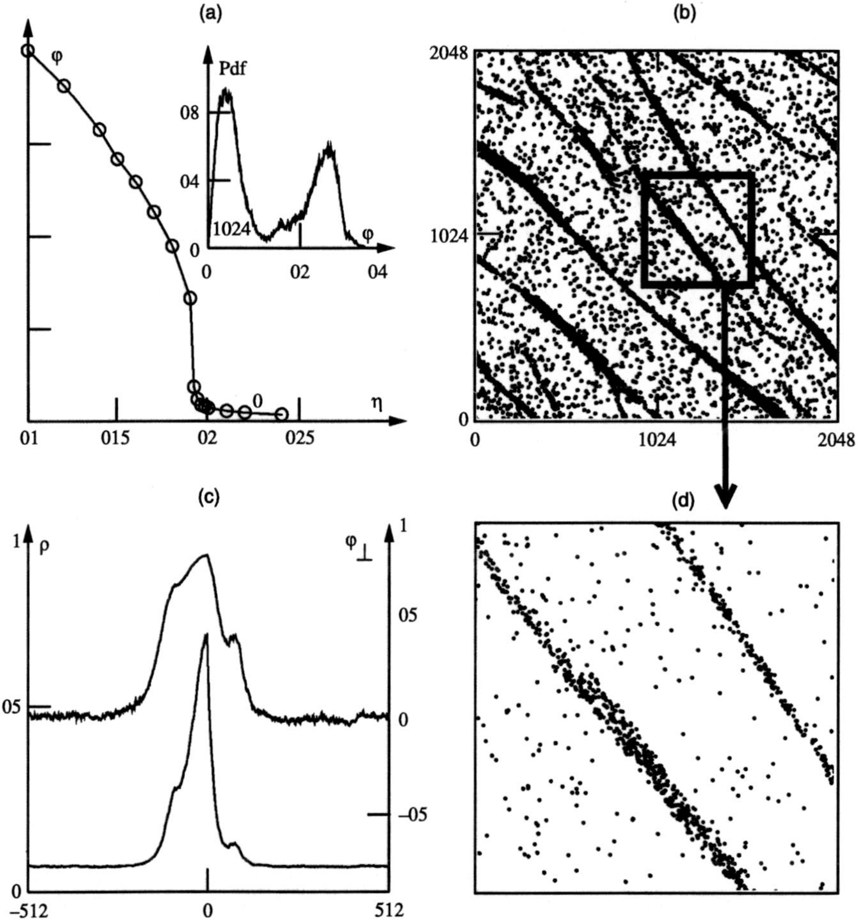

Fig. 12.2 Results of simulations of the model without cohesion and with low density $\rho \sim 0.1$. (**a**) Variation in the order parameter φ with noise intensity and, inset, the bimodal distribution of values (Pdf – probability distribution function) taken by $\varphi(t)$ near the transition point. (**b**) Snapshot of the population of animals in ordered phase. (**c**) Transverse structure of a group of individuals: outside the group, the density (left-hand scale, lower curve) is very low and the average local speed (right-hand scale, upper curve) is zero. (**d**) Close-up of picture (b)

We have since shown that these conclusions are incorrect [6]. If small-sized systems do indeed appear to display a continuous transition, the transition appears to be discontinuous in larger systems (of a greater size than those tested by Vicsek) (Fig. 12.2a). At the transition point, the two possible phases are two states of equivalent stability. A system of finite size therefore has the possibility of changing from one state to the other: the bar chart of the order parameter is bimodal (see the inset in Fig. 12.2a). The discontinuous character of the transition is very robust: it can be observed all the more easily in Vicsek's model when the density ρ is weak, and it is very pronounced when we use the "vector" noise of (12.5).

12.3 Forms in the Absence of Cohesion

12.3.1 Moving in Self-Organised Groups

Another incorrect conclusion drawn from Vicsek's initial studies concerns the nature of the ordered phase, implicitly described as having homogeneous density (although with very great fluctuations). And yet, for the large systems mentioned above, we observe that the animals are organised into quite dense groups, quite the opposite to the image of a homogeneous phase. These groups move, join up, divide, appear and disappear (Fig. 12.2b–12.2d). Within a group, speeds are more or less aligned, but individuals are constantly leaving and joining, so that in between groups we can observe a sort of "residual gas" of disoriented individuals. Over long time scales, the ordered regime is composed of one or more parallel groups. It can therefore be seen as the coexistence of two phases, comparable to the equilibrium of a liquid with its vapour.

Without there being anything, in this case, to keep the individuals together, they are effectively and spontaneously organised into groups: the form appears almost against all expectations. We do not yet understand very much about these groups, particular with regard to their width and their transverse structure. Where does this characteristic size come from? We believe that it may be related to the very particular acoustic properties of this type of environment. It has been shown that the state of homogeneous density is traversed by acoustic waves (in other words, we can observe a propagation of inhomogeneities in density, in the form of longitudinal waves, formally analogous to the sound waves associated with the propagation of zones of air compression), which are anisotropic and globally oriented in the direction of the overall movement [13]. These waves could be the source of the spatial modulation observed.

The emergence of these groups out of a statistically homogeneous and completely disordered initial condition is instructive. An infinite system would take an infinite time to become organised (the speed of information, equal to $r_0/\Delta t$, is finite). In practice, on a computer, we can study systems that are sufficiently large to allow us to observe the dynamics of this progression towards order over quite a long time: fairly small groups appear quite quickly, then grow larger by aggregating while at the same time moving more and more regularly (Fig. 12.3, upper

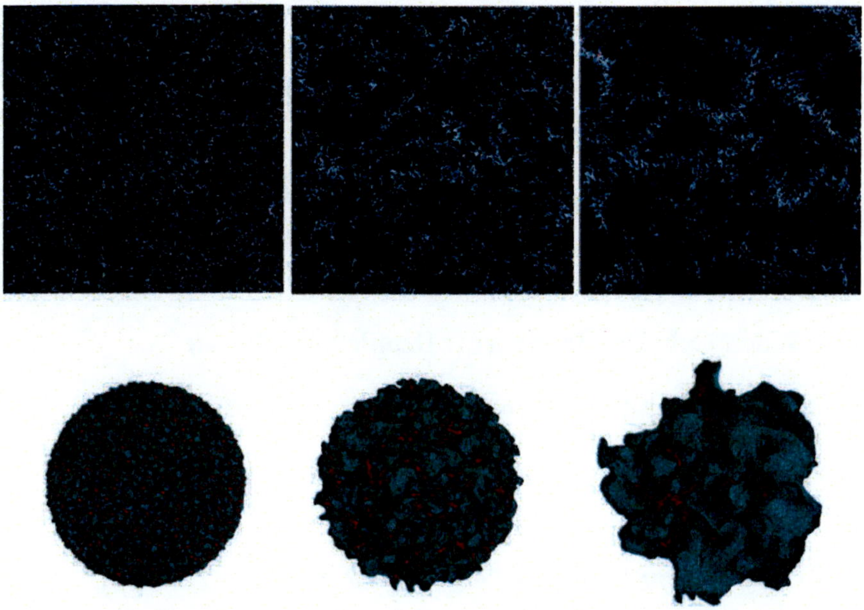

Fig. 12.3 Domain growth in an initially immobile population, but with parameters corresponding to a mobile state. In the top three pictures, there is no cohesive force, whereas there is in the bottom three. From left to right, the three snapshots at taken at times $t = 160$, 320, and 640 for the top row, and $t = 64$, 256 and 1,024 for the bottom row

row). The connected part of the spatial correlation function of the field of locally averaged density decreases exponentially, and the scale of length associated with this exponential decrease (which we take here as the typical size of the groups at the time in question) increases linearly over time. This fast domain growth is typical of the viscous limit of the H-model in the classification of Hohenberg and Halperin [7], which describes the phase separation in a binary fluid, recalling our idea of the coexistence of two effective phases.

12.3.2 Microscopic Trajectories and Forms

The groups emerging in the absence of any cohesive interaction are robust: we have observed them in all the variants of the model studied, in particular independently of the chosen noise term. A parallel can be drawn between this qualitative universality and the characteristics of individual trajectories. In the disordered phase, these trajectories are essentially diffusive; the average squared motion is proportional to time:

$$\langle r^2(\tau)\rangle \equiv \langle |\mathbf{r_i}(t+\tau) - \mathbf{r_i}(t)|^2\rangle \propto \tau \, .$$

No form emerges out of these "Brownian walkers". In the ordered phase, on the contrary, individual trajectories are composed of an alternation between periods spent inside and outside the groups. We can find evidence of the anisotropy of this phase in the properties of diffusion: a super-diffusion of individuals can be observed in the transverse direction: $\langle r_\perp^2 \rangle \propto \tau^{4/3}$. This exponent 4/3, predicted by the analytical calculations of Tu and Toner [13], must be universal, and could be experimentally measured. At the transition point, diffusion abnormal: $\langle r^2 \rangle \propto \tau^{\alpha}$ with $\alpha \simeq 5/3$. In this regime, groups of all sizes come and go, collide with each other, and split up. We have demonstrated that the "trapping" times in these ordered packets are distributed algebraically, with an exponent compatible with the value of α.

12.4 When Cohesion Is Present: Droplets in Motion

We shall now consider the full model, that is to say including the cohesion term ($\beta \neq 0$). In this case, if the cohesion can be maintained, the form of the group will simply be its boundary in space, although internal structures may also emerge.

12.4.1 Phase Diagrams and Form of Droplets

The three ingredients, alignment, cohesion and noise, are not independent. In the case of angular noise (12.4), α/β (or its inverse) and η are independent, whereas for vector noise (12.5), we can limit ourselves to α/η and β/η. Our studies have shown that the phase diagrams obtained in the two cases are topologically equivalent. Here, we shall only present the case of vector noise, with the noise intensity η being kept constant. We have shown that in the limit of zero density (an arbitrarily

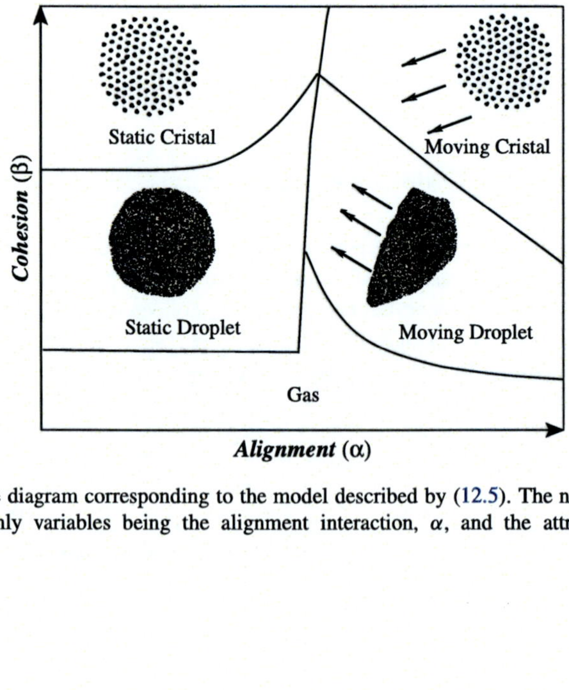

Fig. 12.4 Phase diagram corresponding to the model described by (12.5). The noise intensity, η, is fixed, the only variables being the alignment interaction, α, and the attraction-repulsion interaction, β

large group in an infinite space), the plane of the alignment/cohesion parameters (α, β) is divided up as illustrated in Fig. 12.4. These results were obtained from the interaction f_{ij} as defined in [5], but the general appearance of the diagram does not depend on this choice.

When β is too weak to maintain the cohesion of the group, we observe a gas composed of isolated individuals and small, unstable groups. When β is stronger, we can maintain the cohesion of arbitrarily large groups, made up of individuals placed approximately at the equilibrium distance from each other and often exchanging neighbours. For large values of β, a quasi-crystalline order becomes established in which each individual is maintained at the bottom of the potential well formed by its six neighbours.

Gas-liquid-solid: these transitions are fairly similar to the changes of state encountered in equilibrium thermodynamics. When α is small, nothing spectacular happens: give or take a few fluctuations, the droplet is immobile and spherical, and the individuals of which it is composed diffuse normally: $\langle r^2 \rangle \propto \tau$. Classically, the immobile crystal shows a tendency to form facets. Things change when the alignment interaction is strong enough (large α): the crystal flies and the droplet moves, adopting specific forms. The flying droplet is asymmetrical, and its form bears witness to its global movement: it is usually triangular, with a flattish edge in front and a more or less stretched-out "tail" behind. The activity of individuals is more intense at the front with more agitated individual trajectories, while the "good little soldiers" at the back move more regularly. We can provide a simple argument to explain this triangular form. Because the group moves at a global speed slower than the microscopic speed v_0, an individual arriving at the front edge and not being able to leave the group (due to the cohesive force), changes course to adopt a direction transverse to the global movement. This results in the herd spreading out along a line perpendicular to the overall direction of movement. This phenomenon can be compared with that of the groups observed without cohesion. Indeed, the same transverse superdiffusion $\langle r_\perp^2 \rangle \propto \tau^{4/3}$ can be observed in both cases.

For a very large group, this cohesion and collective motion can take quite a long time to emerge. As in the case without cohesion, we can observe the growth of internal structures with a characteristic size that increases linearly over time, and this characteristic size can also be observed in the increasing roughness of the edges of the herd (Fig. 12.3, lower row).

12.4.2 Cohesion Broken During the Onset of Motion

In the deliberately minimalist framework of our study, the most spectacular forms arise when all the ingredients are evenly matched, during the onset of motion in cohesive droplets. The transition from immobile spherical droplet to moving triangle cannot take place without breaking the cohesion of the group. Under the effect of the intense fluctuations in density and local order then at work, if the group is large enough, the droplet stretches out and takes the form of several subgroups connected by filamentous structures (Fig. 12.5). At least, that is what we observe

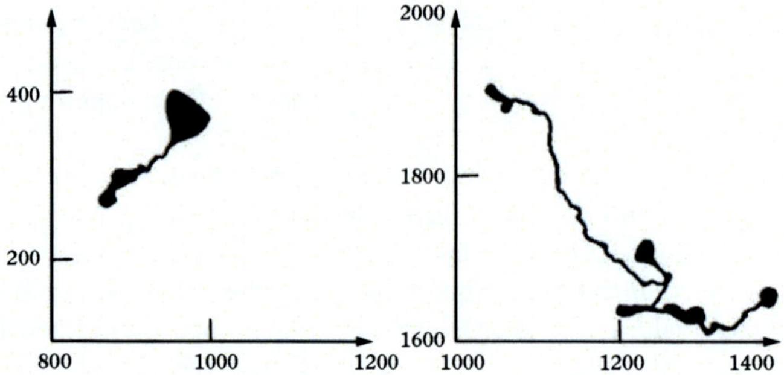

Fig. 12.5 Two filamentous herd configurations

in the cohesive versions of the model. These filaments eventually end up breaking, leaving disconnected subgroups subject to movements of internal rotation.

12.5 Back to Nature

Can our results and predictions be confirmed by observations of real systems? We think they can, but the experimental data or corresponding observations do not yet exist. We hope they will become available in the near future, given the growing interest of researchers in this subject and the advances being made in techniques of image processing and tracking.

Even if, in a framework as minimalist as ours, correspondence with the real world can only at best be semi-quantitative, a good number of the emergent properties described here evoked known phenomena. For example, although most schools of fish do not display positional order, some species form quasi-crystalline patterns. This is the case with old mullet. Young mullet, on the other hand, form schools where the density is very high at the front and decreases towards the back of the school [8], in line with our basic observations of cohesive groups in motion. In *Dictyostellium discoideum*, the colonies of amoebae organise themselves into a sort of slug, within which the cells appear to be more agitated at the front than they are further back.

Minimal models bring to light differences in effective behaviour between initially identical individuals. These differences are reminiscent of observations reported in the literature, but without the need for any additional ingredient to come into play: neither oxygen consumption to explain the decreasing density towards the back of schools of mullet, nor the differentiation due to genetic processes in amoebae. The differentiation is emergent and dynamic. Let us be quite clear: we can say nothing about the reality of any additional mechanism invoked. We are simply pointing out that such ingredients are not necessary.

Beyond these simple qualitative observations, a real semi-quantitative test will become possible with the advent of abundant and systematic data from numerous groups. Again, these data are currently being collected, and we hope soon to obtain experimental measurements of the properties of internal diffusion of individuals in swarms of locusts, schools of fish and even the famous clouds of starlings described in the introduction.

References

1. Albano E. V. (1996) Self-organized Collective Displacements of Self-Driven Individuals, *Phys. Rev. Lett.* **77**, 2119 ; Ben-Jacob E. *et al.* (2000) Cooperative self-organization in microorganisms, *Adv. in Phys.* **49**, 395 ; Levine H. *et al.* (2000) Self-organization in systems of self-propelled particles, *Phys. Rev. Lett.* **63**, 017101 ; Couzin I. *et al.* (2002) Collective Memory and Spatial Sorting in Animal Groups, *J. Theor. Biol.* **218**, 1.
2. Calvino I. (1983) *Palomar*, Einaudi (Turin).
3. Csahòk Z. and Cziròk A. (1997) Hydrodynamics of bacterial motion, *Physica A* **243**, 304.
4. Drouet J.-L. and Moulia B. (1997) Spatial re-orientation of maize leaves affected by initial plant orientation and density, *Agric. For. Meteorol.* **88**, 85.
5. Grégoire G., Chaté H. and Tu Y. (2003) Moving and staying together without a leader, *Physica D* **181**, 157.
6. Grégoire G. and Chaté H. (2004) Onset of Collective and Cohesive Motion, *Phys. Rev. Lett.* **92**, 025702.
7. Hohenberg P. C. and Halperin B. I. (1977) Theory of dynamic critical phenomena, *Rev. Mod. Phys.* **49**, 435.
8. Mc Farland W. and Okubo A. (1997) Metabolic models of fish behavior– the need for quantitative observations, in *Animal Groups in Three Dimensions*, edited by J.K. Parrish and W.M. Hamner, Cambridge University Press (Cambridge).
9. Notice that both ingredients were already used to compute animations for movies, see Reynolds C. W., Flocks (1987) Herds and schools : a distributed behavioural model, *Comput. Graph.* **21**, 25.
10. Pliny the Elder (-79) *Naturalis historia*, English version: *Natural history* translated by John Bostock and H. T. Riley (1855), Taylor & Francis (London).
11. Shimoyama N. *et al.* (1996) Collective Motion in a System of Motile Elements, *Phys. Rev. Lett.* **76**, 3870.
12. Toner J. and Tu Y. (1995) Long-Range Order in a Two-Dimensional Dynamical XY Model : How Birds Fly Together, *Phys. Rev. Lett.* **75**, 4326 ; Simha R. A. and Ramaswamy S. (2002) Hydrodynamic Fluctuations and Instabilities in Ordered Suspensions of Self-Propelled Particles, *Phys. Rev. Lett.* **89**, 058101.
13. Toner J., Tu Y. and Ulm M. (1998) Sound Wave and the Absence of Galilean Invariance in Flocks, *Phys. Rev. Lett.* **80**, 4819.
14. Vicsek T. *et al.* (1995) Novel Type of Phase Transition in a System of Self–Driven Particles, *Phys. Rev. Lett.* **75**, 1226.
15. Wu X.-L. and Libchaber A. (2000) Particle Diffusion in a Quasi-Two-dimensional Bacterial Bath, *Phys. Rev. Lett.* **84**, 3017.

Chapter 13
Systems of Cities and Levels of Organisation

Denise Pumain

It is relatively easy to think of a city as the product of multiple interactions, between the agents involved (local authorities, companies, social groups, inhabitants), their material or symbolic artefacts (housing, infrastructure, institutions, representations), and events or episodes marked by political interventions, economic circumstances, technological innovations, etc. The recent history of writing about the city is punctuated by conceptual formalisations of the production of the urban space [41], models of urban dynamics *à la* Forrester [28], or models of self-organisation inspired by synergetics [51, 77] and dissipative structures [1]. Game theory has inspired a city-building simulation game, SimCity, and its many sequels.

But on another level of organisation, emergent properties appear, stemming from the different forms of exchange and interaction that take place *between* cities, more or less distant from each other. These properties are characteristic of the organisation of cities into networks or systems, on a macro-geographical level, at the scale of a large region, a state or continent, or even the entire world [55]. In turn, they impose constraints on the future of each city, insofar as they constitute spontaneous rules governing the co-evolution of cities [48]. Some of these properties invite comparison between the organisational forms of systems of cities and those of physical systems, because the regularities observed are so general and universal. Can we therefore imagine explaining them with dynamic models, the functioning of which would be independent of time and of the rules specific to each society? Can the structuring of systems of cities thus be described using models of stochastic growth, geometric models (e.g. fractals), or by a-temporal economic models? To what degree are these very particular systems, on the contrary, *social objects* that must be considered within their evolutionary dimension? This might apply, for example, when it comes to interpreting the long-term emergence of structure and its transformations, notably under the effect of technological modifications of space and time. To what extent should we take into account the influential accidents of their history in order to understand the particularities of their form? The greater

D. Pumain (✉)
University of Paris I, Institut Universitaire de France, Paris 75006, France
e-mail: pumain@parisgeo.cnrs.fr

P. Bourgine, A. Lesne (eds.), *Morphogenesis*, DOI 10.1007/978-3-642-13174-5_13,

complexity of the interpretations and models that follow this second path appears to be necessary when it comes to testing hypotheses about the future evolution of cities.

13.1 Three Levels of Observation of the Urban Fact

Forty years ago, the American geographer Brian Berry summed up the conception of a two-tier organisation of geographical space by cities in systemic terms [10]:

> cities as systems within systems of cities.

But this type of representation of urbanisation goes back a lot further in time. In the article written by Jean Reynaud for the section on "Villes" in the Encyclopédie Nouvelle in 1841 [66], for example, we can find the expression *"système général des villes"* (general system of cities), cited by Robic [67]. In addition to these two levels, we must of course add that of the urban agents (users, households, companies, town councils, property developers, chambers of commerce, etc.) who participate in the establishment and functioning of urban systems [57]. These two levels are characterised by particular descriptors, specific to each scale. Generally, the regularities observed, the forms and interpretations that are proposed also differ according to the level under consideration: that of the city or that of the network of cities. Geographers have long since identified, without actually using the term, the emergent properties specific to the level of the objects they study. These properties consist either in new variables, attached to a precise level and without any real significance for other levels of observation, or in forms, structures, configurations, or even questions and types of analysis, that characterise one level of observation and are only formulated with regard to that level (Fig. 13.1).

13.1.1 Emergent Properties at the City Level

At the level of the city, the concepts of *site* and *situation* were developed with the first analyses of cities (e.g. Raoul Blanchard, in his monographs on Annecy and Grenoble *circa* 1910). From the outset, these concepts define the main attributes of the city in relational terms. In the case of the site, these are relations with the natural or man-made elements that make up the immediate environment of the city (relations at a local scale). In the case of the situation, they comprise relations with other cities, with elements of relief like mountains or circulation corridors, or with communication networks and the other cities that form their nodes (relations defined at the regional, national or even worldwide scale, depending on the city under consideration).

Equally characteristic of the city as collective entity are the concepts used to analyse the urban morphology. The form of city layouts, radial-concentric or gridiron, the mono- or polycentric character, its more or less symmetrical form, whether it is compact or on the contrary dendritic, the architectural particularities that com-

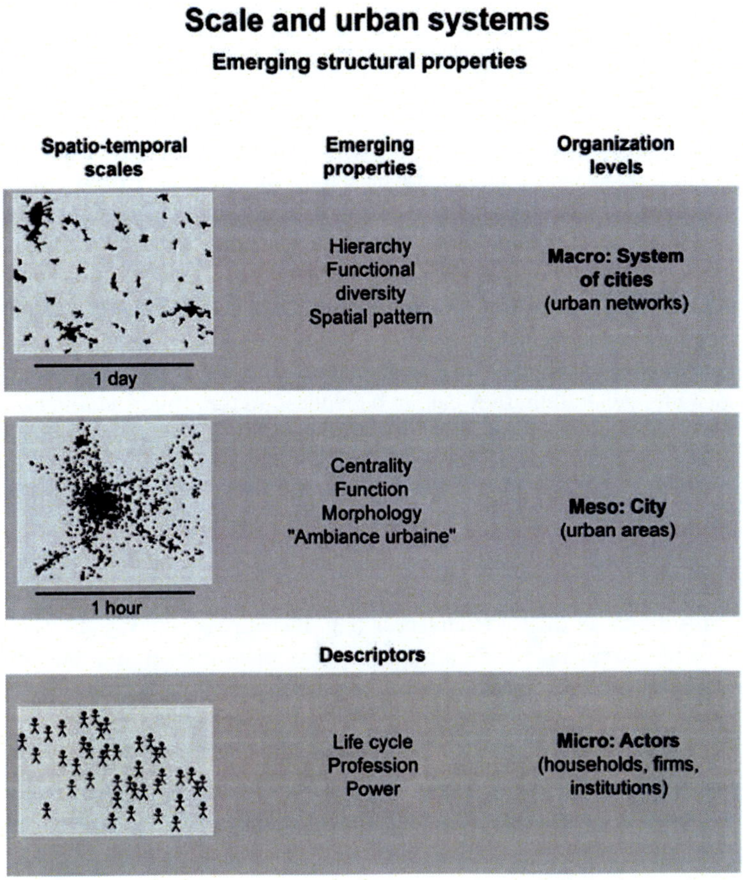

Fig. 13.1 Levels of scale and urban systems: emergent structural properties (after [58])

pose the urban landscape: these are all attributes that refer to the city as a whole entity. The richer description that is given by the concept of urban ambience, which combines objective elements of the urban landscape (volume of buildings and free spaces, light and shade, temperature, noise, wind, odours, etc.) with qualities of sensorial appreciation (open or enclosed nature, calm, movement, etc.), also belongs to this set of global descriptors that are defined for the level of the city. Likewise, the different gradients from the centre to the outskirts (resident population density, land and real estate prices), reflecting the regular decreases in these indicators as exponential functions of the distance from the centre, express the concentric structure of the urban field. Although density can be interpreted on an individual level as an indicator of the space available per inhabitant, and although the sale or rental price of land and real estate are indicators that can be brought down to the level of each building or unit, the distribution of these quantities over the space of the city according to precise configurations, repeated from one town to another, stable over time and summed up in the term *urban field*, is indeed an emergent property, which appears almost systematically during the growth of a city, characterising and shaping it on this individual scale. It is an essential morphological characteristic that persists over time [8, 9].

All these concepts are specific to the level of the city; they refer to this aggregate entity, considered as a whole, and cannot be applied individually to the urban agents. In the case of the social composition of cities, the character of collective property is obviously harder to distinguish from the intentional actions of the agents, but is nevertheless interpreted in the same way. The different spatial configurations observed in relation to social segregation, which may form concentric rings for the distribution of age or family size, more or less wealthy sectors, or complicated mosaics according to income or social status, are structures whose stability often far exceeds the lifespan of a generation; they cannot be reduced to the simple addition of individual decisions to choose any particular district to live in (residential strategies). Likewise, the qualities attributed to different districts are more than just the sum of their objective advantages: they are amplified or diminished by collective effects of social labelling and representations of "good" and "bad" neighbourhoods.

Lastly, other collective qualities of cities, which may be objectively measurable (e.g. economic and social specialisation), or more difficult to objectivise but with a perceptible effect on their attractiveness (e.g. their image, conveyed by more or less precise representations and partly shaped by urban marketing), must also be considered as emergent properties. We shall see that they are produced, to a far greater extent than those mentioned earlier, not only by the interactions between the urban agents present in the city itself, but also by the constraints deriving from the fact that it belongs to networks of cities, at a larger scale. The fact that geographers call specialisation the *urban function* shows how this quality, specific to each city, is an expression of the role it plays in the system of cities.

13.1.2 The Structure of the System of Cities

At the level of the system of cities, emergent properties have fascinated observers for a long time. The fact that in every country in the world and during every epoch, there is a hierarchy whereby the number of cities is inversely proportional to the size of their population, is a remarkable characteristic, invariant with regard to political, economic and cultural systems. The regularity of the distribution of the number of cities as a function of their size was noted as long ago as the nineteenth century (prompting the historian Emile Levasseur to make an analogy with galaxies). It was formalised in 1913 by the geographer Auerbach, who observed that the product of a city's population P multiplied by its rank in the hierarchy r is a constant, $P.r = K$ and used this constant value K as an index of concentration. To put it another way ([6], p.76):

> the number of [inhabited] places in inversely related to their minimum number of inhabitants.

The statistician Lotka applied this regularity to American cities and introduced a graph of city populations as a function of their ranking on two logarithmic scales [42]. The sociologist Goodrich, of the Chicago School, also discussed this statistical regularity [33]. In 1936, the economist Singer noted the resemblance between city-size distribution and the Pareto law developed to describe the distribution of incomes [71], while in France, the statistician R. Gibrat, in a thesis on economic inequalities published in 1931, proposed a different model, with log-normal distribution [32].

However, it is the name of Zipf that remains most closely associated with the "rank-size rule" that he systematised in 1941 [81]. In his first work, *National unity and disunity* [81], Zipf presented himself as the discoverer and totally omitted to mention the contributions of the pioneers, provoking an immediate reaction from Lotka ([43], p.164):

> This discovery is neither new nor perhaps so striking [...] This type of frequency distribution is, in fact, a Pearson type XI distribution, a particular case of the type VI.

In his 1949 work, *Human behaviour and the law of least effort* [82], Zipf only very approximately signalled the existence of predecessors to the expression of his law. The theoretical explanations that Zipf gave of his model (a supposed "equilibrium" between a force of concentration and a force of dispersion) lacks rigour and fails to convince. However, unlike the statisticians of the time, who were looking for a measure of concentration and only used graphs incidentally, Zipf proposed qualitative analyses of the shape of curves, which he related to the political vicissitudes of the construction of the States, thus opening the model up for wider interpretation. Sadly, his graph representation, although certainly useful, was far removed from the canonical forms of representation of statistical distributions, subsequently giving rise to misunderstanding by some of its users (who interpreted it as a linear adjustment between the logarithms of the population of cities and their rank, instead of reading it as a cumulative frequency distribution as a function of size). In its most general form, this model, known as the *rank-size rule*, or *Zipf's law*, is written:

$$P_i = \frac{K}{r_i^a} \tag{13.1}$$

with K=constant (close to the population of the biggest city) and a close to 1. But this statistical regularity in city-size distribution is also accompanied by a certain regularity in their distribution over geographical space, their spatial framework. The size of cities, their importance as a hub and the number and diversity of the activities they host were considered jointly in the writings of engineers dealing with territorial issues of political economics, notably in relation to the planning of rail routes. The principle of dividing the land into hexagonal areas of influence covered by a hierarchy of centres of trades, commerce and administrative functions was first proposed in 1841 by the Saint-Simonianist engineer J. Reynaud [66], who thereby unveiled all the principles of what was to become, in the following century, the "theory of central places" [67], but his theory of the "general system of cities" remains largely unknown. An academic and specialist in rail networks, L. Lalanne, noted in 1863 and 1875 the optimality of a layout of cities at the apices of equilateral triangles, spaced out by multiples of a basic distance, while the number of centres decreases as one moves up the administrative hierarchy (cantons, sub-prefectures, prefectures) [39, 40]. He thereby developed an inductive theory of the localisation of cities, governed by the laws of "equilateralism and multiple distances", which he presented before the Académie des Sciences, while at the same time rediscovering some of the geometric regularities recommended by the German geographer J.G. Kohl in his geography of the circulation [38]. But Lalanne only considered the "order of population centres" according to their position in the administrative and political organisation of the land.

It was the German geographer Walter Christaller who emerged around 1930 as the indisputable inventor of a formalised explanation "of the number, size and spacing of cities", based on economic and geographical considerations [21]. The concept underlying the theory is that of centrality, which sums up the interaction between a *centre* in which the supply of goods and services is concentrated and a *complementary region* in which the demand for those goods and services is localised. The city is therefore a market, the place where the supply and demand of services meet. Several postulates complete the foundations of this theory: the first concerns the spatial behaviour of the consumers, who are assumed to obtain what they want from the nearest supply place. The second concerns the diversity of goods and services, which are ranked into a hierarchy of levels, depending on the frequency of their demand, their spatial range (the maximum distance consumers are willing to travel to obtain them, given that the consumers incur the extra transport costs) and their threshold (minimum size of the market, in terms of population or income, and therefore of the centre, for the supply to be profitable). The third postulate is that goods and services of the same range are supplied in the same centres, and the higher-level centres supply all the goods and services with a lower range than that of their own level. From these premises, Christaller deduced a hierarchical configuration of nested hexagons of centres and market areas. He modelled several forms of this hierarchy of urban functional levels (which comprises six or seven levels), depending on whether the principle defining the number of centres at each level was based on maximising the points of supply, reducing the cost of transport infrastructure or facilitating the administration. Christaller illustrated his theory with geometric constructions, built up from combinations of circles and triangles and taking the form of regular patterns of urban centres surrounded by interlocking, hexagonal market areas. Already predicted, but not drawn by Jean Reynaud a century earlier, these figures were to be consulted around the world for their representation of the organisation of cities in a region. Their capacity to simulate urban configurations observed at a given moment in diverse regions of the world has been quite well verified on many occasions, at least to a first approximation.

However, three key criticisms have been levelled at this theory. Firstly, by stipulating that the consumer chooses the nearest place of supply, it overlooks the importance, increasing with the use of the car and the spread of concentrated forms of distribution, of "multi-purpose journeys" that are believed to account for about 40% of the volume of purchases, and the essential consequence of which is to bypass the smaller centres and so to weaken the lower levels of the urban hierarchy predicted by the theory (which predicts a roughly constant ratio between the consumer volumes of the centres of one level and the next, for all levels of the hierarchy). The second major criticism is connected to the static nature of the explanation proposed, which accounts for the existence of different-sized cities at a given moment in time, but fails to explain how they were founded or how they evolve. Finally, the third shortcoming concerns the reduction of urban activities to "central" functions, providing goods and services to the local population. In doing so, the theory fails to explain the localisation of so-called specialised cities, which have a function of industrial or tourism-based production, for example, and which are governed by other factors of localisation than proximity to customers. The ambition of this theory to explain the size, number and spacing of cities is thereby diminished.

Indeed, another emergent property that is characteristic of systems of cities is their functional diversity. Envisaged very early on in descriptive typologies, such as the one developed by the geographer Aurousseau in 1921 for American cities, or G. Chabot in France, this variety in the economic specialisations and social profiles of cities has since been specified by multivariate classifications (B. Berry & J. Kasarda in the United States, Moser & Scott in Great Britain, P. Pinchemel & F. Carrière in 1964, Pumain & Saint-Julien in 1978 in France) [10, 46, 49, 59]. Although local or regional concentrations of cities of the same type can sometimes be observed (for example near mining deposits or in tourist areas), as a general rule cities with different specialities are interdispersed within each region, testifying to the interdependencies that become established between them, and therefore to the fact that they function as a system for the production and exchange of goods and services. Functional diversity is constituted firstly by differences in the localisation and concentration of economic activities, enabling cities to grow by supplying them with an export basis (theory of the economic basis of cities). It is accompanied by high social inequalities between cities (income inequalities between "working-class" and "middle-class" cities, inequalities in skills or "human capital"), which have strong repercussions on collective representations at a given moment (today, the "image" of cities is the object of deliberately mediatised, politicised and commercialised manipulations). We shall see that the processes affecting the distribution of economic activities between cities is very dependent on this image, on the relative overall situation of each city in the system of cities during a given period, although it is also interpreted partly on the basis of local initiatives undertaken as a response to that situation.

13.2 A Functional Interpretation of the Hierarchical Ordering

From a static point of view, at a given moment, we can observe that each level of the urban system lends itself to a different usage by the inhabitants of a territory, the cities being considered as the place where most of the usual daily activities are pursued, while the system of cities functions for less frequent, but longer-range exchanges, which have a strong impact, a non-negligible probability of modifying the future of each city. *We can thus give a first, naively functional interpretation of the urban organisation into two levels, as if it had been designed for the articulation of territorial entities at different geographic scales, to connect up levels of territorial organisation with different orders of magnitude in space and time.* The settlement system thus allows to manage the co-existence of at least two main levels of space and time in the uses that a society makes of its territory.

13.2.1 Daily Life in the City

The first of these levels is that of the territory frequented on a daily basis, represented by a city and its immediate surroundings (including peri-urban commuter housing and the trading area of urban businesses, which we can group together under the term "*basin of life*", and which often also includes a large proportion of the sub-contractors to city businesses). Here, the term territory loses some of its political and administrative connotations, taking on more of a psychological or sociological significance, as most cities have gradually lost their capacity to enact laws or exercise political power, now the prerogative of states. We still speak of territory, because the community that lives within a same settlement node generally retain physical ownership of the land and also define themselves in terms of various sentiments of belonging to the place in question. The recent SRU law (*Solidarité et Renouvellement Urbain* – Solidarity and Urban Renewal) in France (2002), for example, encourages the setting-up of *communautés d'agglomération*, metropolitan authorities grouping together several districts and exercising multiple competences for the good management of such territories. The strong development and compactness of the central zone, which enjoys the best accessibility and contains a high proportion of the production function, particularly in its more sophisticated and recent forms, pushing part of the residential function out to the periphery, is the chief characteristic of the organisation of these daily territories, especially since they are of a large size. The presence of an urban field, a field polarised around the centre, is also expressed in the very widespread existence of a density gradient running from the centre to the periphery. The dilution in the intensity of buildings and activities on the outskirts, the fractality of this urban structure, which derives from the compromise between the strength of the constraint related to the advantages of centrality (accessibility to numerous and diverse activities) and the necessary "obligation of spacing" ("two buildings cannot occupy the same place" [65]), reinforced by the secondary development of the unbuilt, "natural" environment on the edges of the agglomeration and its advantages [29].

The spatial extension of this daily territory remains highly constrained by internal travelling times, of which the speed is itself reduced by the high density. The time devoted to commuting in France, for example, hardly changed between 1970 and 1990, while the average distance travelled doubled over the same period. The average daily travelling time in a city is about one hour, sometimes between one and two hours, for a sequence of activities situated in an average of slightly more than three different places in the day. Although this time may vary considerably from one individual to another and from one town to another (90 min per day for the Paris agglomeration, compared to 50–60 min in smaller cities like Rennes or Lille), it nevertheless constitutes, in this order of magnitude, an invariant for the organisation of the territories of daily life (10–15% of working time), and represents a strong constraint on their maximum spatial extension [61]. As we shall see, the spread of the car and fast transport techniques have allowed agglomerations to extend somewhat, but the speed of circulation within urban zones remains much lower than that of intercity transports (as an example, it is 16 km/h in Paris, rising to 50 km/h in the wider outskirts), to such a degree that today, the extension of these daily territories rarely exceeds 30 or 40 kilometres around a centre. During historical times, the diameter of one hour's travelling represented the standard extension of the maximum urban field; this time-distance has always covered the major share of activities grouped around an urban centre. Much more extensive agglomerations, such as Los Angeles, still appear as anomalies (although average travelling times are longer in the United States than in Europe).

With the evolution of transport techniques, the maximum radius of cities, which had been about 5 km before mechanisation, has lengthened considerably, allowing the population of the biggest cities to rise into the tens of millions. However, the speed of circulation in urban zones has only increased by a factor of about 5 over the last two centuries. At the same time, the morphology of cities is tending to become slightly less compact than in the past. Since the widespread adoption of the car, a certain morphological duality has even emerged, in cities with a plurisecular history, between the central agglomeration, solidly built-up, which has retained sharp density gradients (a fractal dimension between 1 and 2, and close to 2), and the peripheral urban zones, developed with the use of the car, which are structured by the urban field for distances of up to between 40 and 100 km from the centre, but with much less pronounced gradients (fractal dimension of built-up surfaces less than 1), according to the measurements carried out by M. Guérois for all the cities in Europe [35].

The main functions of this daily territory represented by each urban entity are housing and production (of goods and services). In most cases, however, these two major functions are no longer regulated locally. For the most part, the determinants of their evolution are located elsewhere, in the nodes of the networks of power where decisions are taken. In these networks, certain nodes of power are clearly identified, by a political or administrative capital, or the headquarters of a large multinational. Most often, however, it is through multiple networks of negotiation, competition or cooperation that the modalities of deductions or exchanges, of materials, people and information are governed, ensuring the functioning of the territories on a higher level of organisation, at a regional, national or even supranational scale.

13.2.2 The Functions of the System of Cities

It is the system of cities that operates these networks. On this other level of territorial organisation, the spatial framework of the settlement system is regulated by the modalities of circulation, following a network morphology defined by the size and spacing of cities. As the geographer Élisée Reclus observed as long ago as 1895 [64], it is approximately the distance that can be travelled in one day that determines the spacing between the major nodes of the system, the ones that concentrate a certain degree of decision-making power (see, for example, the *chefs-lieux* of French *départements* at the beginning of the nineteenth century, the regional capitals or even the large European metropolises of today). This time limit of a day (obviously, it is no more than an order of magnitude, difficult to determine more precisely), considered on average for relatively frequent negotiations, confers on the territories of power the possibility of much wider spatial extension than that available to the territories of daily life. Not only is the travelling time governing the spacing of nodes in each territory much longer, but the speed of circulation between those nodes, at any given moment in the history of transport, is generally much faster than anything possible within each node [54].

Nevertheless, it remains very much harder to determine a typical spatial extension for systems of cities than it is for individual cities. Defining the boundaries of a system of cities is a necessary operation, but never entirely satisfactory. If we attempt to apply a criterion often used in systems analysis, consisting in grouping together all the subsystems that have more exchanges with each other than with the system's environment, we come up against the particularity of the territorial functioning of cities, which confers on them a variable spatial range depending on their position in the urban hierarchy: a small city will generally have local exchanges, a large city will have regional exchanges, while metropolises can have important exchanges at far greater distance. In fact, it is important to determine the types of exchanges that we are going to take into consideration for this definition: thus, the map of zones of influence of French cities, recently drawn up by INSEE[1] with the help of IGN[2] and based on the results of the *Inventaire Communal*, does not show much variability in the spatial ranges of exchanges for everyday services (according to customer patronage of the nearest centre), although they tend to increase with city size. Residential migration clearly brings to light two scales of migratory exchanges for French cities: the national scale for Paris (and to a lesser extent for certain cities on the Côte d'Azur), and the regional scale for the other agglomerations of more than 50,000 inhabitants [60]. If we turn our attention to more abstract relations, the system of cities is often considered within the context of a national territory, because this is the level at which the rules are defined that ensure homogeneity in the conditions of social and economic functioning. Moreover, due to the "barriers" constituted by frontiers, which are not only legal but also linguistic and cultural, the presence of an international frontier strongly reduces the flows of exchanges between the cities of two neighbouring states: even today in Europe, these flows are reduced by a factor of five or six, for example, between Belgium and France [18]. Finally, if we turn to even more abstract relations that are decisive for understand-

[1] French National Institute of Statistics and Economic Studies.
[2] Franch National Geographic Institute.

ing and predicting the future of cities, such as investment decisions, or decisions affecting relative urban dynamics, the competition brought into play to identify an interdependent evolution, characteristic of a system, are not defined at the same scale for all cities: for example the decisive interdependencies for regional French metropolises are played out sometimes between themselves, sometimes with other large European cities, whereas Paris measures up against London and other global cities. The geographical situation of smaller cities is generally evaluated in relation to other cities in the same region. (However, even if these average ranges can be evaluated with a certain degree of approximation, it is clear that very long-range decisions, e.g. involving the relocation of a Japanese or American multinational, can have a decisive impact on the future of a small city). It is therefore very difficult to consider a system of cities as an autonomous, delimited entity. We should rather look at it as a network, unequally connected by relations of varying intensity, which taken as a whole give a certain coherence to the co-evolution of its constituent cities.

The dimension of these connected groups has varied over time. Today, they are of the order of magnitude of a large state, or even, for the largest cities (200,000 inhabitants) extended for example over the whole of Europe, so from one to several thousand kilometres. But up until the industrial revolution, this extension was much more limited, often hardly exceeding a few dozen kilometres (at the beginning of the nineteenth century, von Thünen established a diameter of 350 kilometres as the maximum extension of his "isolated state", determined by the maximum range of urban relations at the time). For the mid-18th century, B. Lepetit evoked the "unthinkable network" with regard to the royal highways, seen as radiating lines joining Paris to the frontiers of the kingdom, but not as a system of connections between the various different population centres. It is the rising speed of intercity transport that has allowed this gigantic increase in the diameter of urban networks, reaching across the whole planet in the case of the global cities. E. Reclus had already noted the regularity of the distances between stage towns, adapted to the pace of the horses, and the shortening of those distances with the modernisation of transport [64]. Over the last two centuries, the maximum speed of intercity transport has been multiplied on average by a factor of 40 (up to 100 in the case of plane travel), while the speed of intracity transport has only risen by a factor of 5.

13.3 The Interactions that Construct the Levels

Complex systems theory, which is particularly applied to living systems and social systems, interprets the appearance in these systems of persistent and identifiable levels of organisation, with their emergent properties, on the basis of the interactions between the elements or subsystems of a lower level, the level thus constituted in turn orienting, limiting or constraining these interactions. However, between the two main hierarchical levels of the urban systems, that of the city and that of the system of cities, interactions occur that bring into play more than just each of the two entities. For example, urban agents may intervene simultaneously at every scale, over the whole or only part of each system. Urban hierarchies therefore belong to the models of "heterarchy", rather than the "inclusive hierarchy" of biological organisms [58]. It is in the interests of simplification that we identify the interactions that play a role primarily in the constitution of one of the two levels of organisation, bearing in mind that the overall structure almost certainly has more in common with the structure of networks organised into "small worlds" [76]. In geography, we are more particularly interested in the interactions that construct spatial structures. We have represented what we believe to be the most important of these in Fig. 13.2.

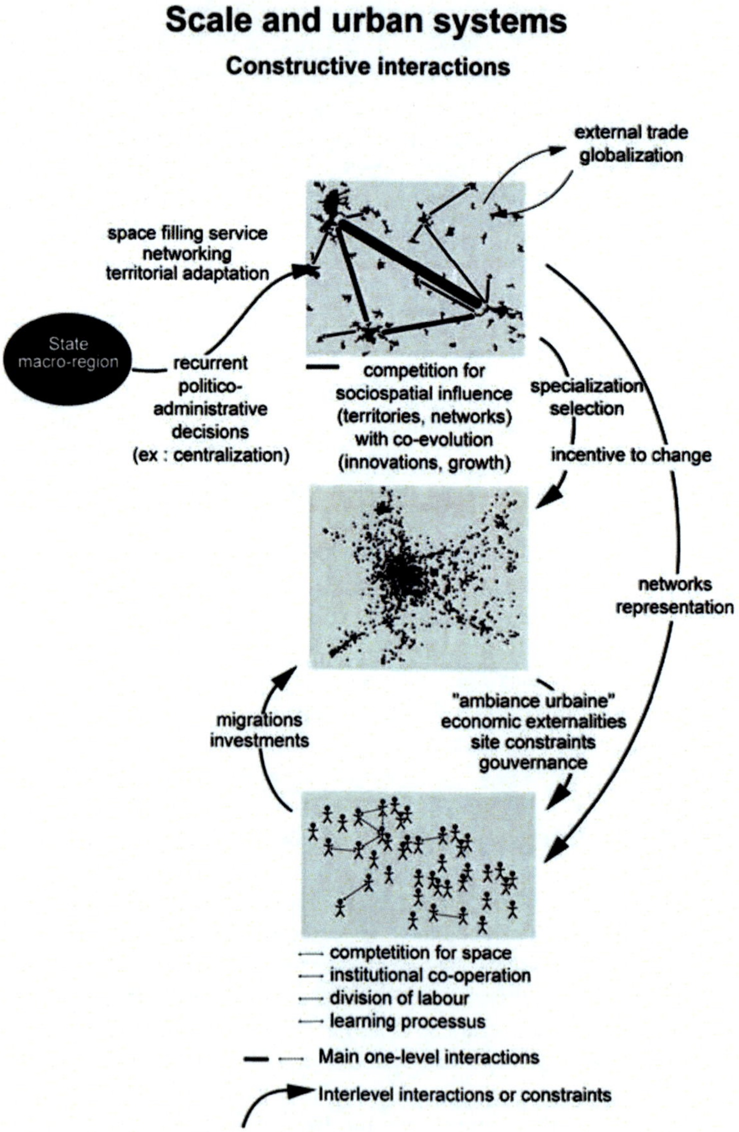

Fig. 13.2 Scales and urban systems: constituent interactions

It is self-evident that for each aspect of the structure analysed, certain interactions are more decisive than others, and it is these explanatory relations that have been formalised in urban models.

13.3.1 The Constituent Interactions of City Forms

Several interpretations have been proposed, for example, to explain the structure of the urban field, in concentric rings of decreasing intensity, with a strong gradient from the centre to the peripheries. The economist W. Alonso [3] showed how this structure could be formed, at the city scale, from the trade-off made by individual households between their housing expenditure and their transport costs. There have also been interpretations similar to the model proposed by von Thünen for agricultural uses of the land. The selection of activities according to their closeness to the centre, which is more accessible, is made through the differential rent (the profit from renting or ownership of the land) that can be obtained from different land uses: high value-added services (banks, financial services, certain services to business, auditing or consulting), that can pay high rents, are located nearest the centre, followed by high-level commercial activities, while housing, and even more so industry, which takes up more space but is today less profitable (and generates pollution) is established on the peripheries. Here, the inter-individual interactions generated by this centre-periphery structure are primarily competition between unequally "profitable" activities or land uses to occupy the available space. This explains the formation, as early as the beginning of the nineteenth century in the largest cities, of the "central crater" in the distribution curve of residential densities as a function of the distance from the centre, even the most well-off populations having lost the competition against economic activities for occupancy of the central districts (see, for example, the banks in the 8th *arrondissement* in Paris).

But this dynamics is that of a complex system: *the same types of interaction are capable of generating different forms when other interactions or constraints intervene*. The quality of a site, for example, may exert constraints that cause the price surface to diverge from its theoretical configuration for the same part of the city, enhancing the value of a hillside or riverside or a site with greater accessibility (crossroads), or the neighbourhood of an historic monument (or a prestigious school). Some of these constraints, generally stemming from another level than that of the city, can profoundly modify the configuration. In North America, for example, legal conditions that facilitate real estate transactions (and explain a rate of residential mobility twice as high, on average, as that observed in Europe or Japan), combined with a cultural appreciation of the suburbs (collective values favourable to rural space, that can also be found in the United Kingdom), modify the price surface and flatten its gradient considerably, although without completely inverting it. These institutional "data" or collective "values" represent an interaction between the level of urban agents and a more encompassing level, corresponding here to a State and a whole society.

Once formed, the city exerts a collective effect, known as a *context effect*, or constraint, at the level of the urban agents. This "constraint" has more than just a negative or limiting regulatory effect: for example, urban economics defines the city as a place that provides companies with a particular type of external economies (in addition to the economies of scale linked to large-sized companies producing at a lower cost per unit than smaller ones). These external economies specific to the urban setting are called economies of agglomeration. They represent profits not directly linked to the activity of the firm, resulting from its localisation, on the one hand close to other firms (which can facilitate subcontracting and synergies between companies in the same sector), and on the other hand close to a large market, representing an immediate outlet for its production and a labour market likely to facilitate the recruitment of skilled workers (human capital). Such profits can also derive from what are called economies of urbanisation, consisting in the undivided use by companies of infrastructure partly financed by the public authorities (airports, convention centres, universities, cultural amenities, railway stations, etc.). The quality of urban life, the urban ambience, even the image of cities, can all enter into this category of "economies of agglomeration", representing for each firm the beneficial effect it obtains from its localisation in a particular city. Increasing interest is being taken in "learning effects" in companies, which reinforce their capacity for innovation through exchanges because of the proximity of other urban agents, and which constitute the decisive advantage of clusters.

In this respect, another process emerges out of the interactions created in cities, and appears to have done so ever since cities first existed. This process is the social division of labour, which makes cities the leading players in social change, according to sociological theories of the city. This is also testified to by the work of historians who have shown the emergence, if not continuous then at least tending to increase, of new professions and social categories or new types of individuals and the invention of new life styles, this movement of creation being characteristic of urban populations, rather than rural ones. Where does this creativity come from, that calls for the invention of new statistical categories to represent the new social diversity (think, for example, of the appearance of technicians in the nomenclature of professions and social categories, or single-parent families)? A first explanation lies in the feedback effect of the city itself: if the size of a city comprises n people, the number of interactions that these people are likely to have with each other varies with n^2. And very often, novelty is born out of encounter, dialogue and exchange. It is in this sense that P. Claval, in *La logique des villes* [19], wrote of cities as places that "maximise social interaction". But a second explanation can also be sought in the formation of networks of cities, in that higher level that is the system of cities, which represents a constraint for each city and for the agents it contains. At this level, the incentive to change derives from the competition between cities, in which they seek to capture the profits from innovation and from the growth (economic and/or demographic) associated with it (Fig. 13.2). Now we must identify the interactions that construct this level.

13.3.2 The Constituent Interactions of Systems of Cities

The co-evolution of cities belonging to the same system comprises a set of processes describing the interdependencies of their quantitative and qualitative evolution. Here, models exist that bring clearly to light the interactions between cities and the structures that result from them at the level of urban networks.

13.3.2.1 A Process of Distributed Growth

The size of cities, as measured by their population, is a synthetic indicator of their importance. When we extend the comparison to cities in countries at different stages of economic development, however, it must be completed by other measurements. Measuring the population of a city is tricky because of the difficulties in delimiting urban entities, which grow in surface area as the number of inhabitants increases. Official statistics are not always comparable, and data harmonisation is a pre-requisite to any analysis of city evolution in time and space. Comparable data bases have allowed to establish the universality of the statistical form of distribution of the size of human establishments in all regions of the world and over historical times [7, 27, 45, 75]. The slope of the rank-size rule, interpreted as an index of the inequality in city size, varies between fairly narrow limits, tending to be higher in recently settled countries (which have founded cities in territories with little or no rural settlement, and in an age when available means of transport far outstrip pedestrian or animal transport). It also appears that city-size inequality has increased over the last two centuries in several regions of the world.

The statistician Gibrat [32] showed that a log-normal distribution of city size could be explained by a process of exponential population growth in cities, in which growth rates (relative variations in city populations) are random variables independent of city size and from one period to the next. Another growth model, which could be interpreted as a deterministic version of the process described by Gibrat, but leading to a power law distribution, was proposed , after others [78], by H. Simon [70]. Gibrat's model has been tested successful for numerous countries and different epochs [34, 45, 53, 68]. Another example is that of European cities, which have increased considerably in number (less than 1,600 cities of more than 10,000 inhabitants in 1,800, more than 5,000 in 2,000) and the population of which has also greatly increased over the last two centuries (from 22–380 million inhabitants), while the system they form has kept essentially the same spatial configuration. This morphological stability can be explained by the fact that the different parts of the system of cities have all developed [11, 14, 62]. This is a process of distributed growth, which can be interpreted as the result of competition between cities to attract inhabitants, and which therefore depends on their interactions (especially in terms of the adoption of innovations or adjustment to change). In this sense, the property of *hierarchical differentiation*, a macro-geographic characteristic of systems of cities, can be interpreted as an emergent property resulting from the interactions between cities at a meso-geographic scale.

In Gibrat's statistical model, it is true that the interactions between cities remain implicit: it is assumed that they grow with the same long-term average speeds and rates because they are interdependent and linked to each other by all sorts of networks [53]. Efforts are being made today to produce more complete versions of this distributed growth model, for example by simulating the construction of networks of relations between cities according to simple rules which have a similar effect to those of Gibrat, but which allow for the spatialisation of growth. However, although the models proposed in economics by Cordoba [22] or Gabaix & Ioannides [31] to explain the interactions that generate Gibrat's model are certainly capable of reproducing a city-size distribution consistent with the observations (as indeed can many other models), their hypotheses concerning the interactions have not been tested empirically. The economic literature is stymied by the problem of the structuring of systems of cities, to such an extent that Fujita, Krugman & Venables [30] devote a whole chapter of their book on regional and urban economy to "an empirical digression: the size of cities"! They take up a proposal by Krugman that an analogy can be drawn between urban networks and hydrographic networks, forgetting to cite the geographers Chorley & Haggett who had already undertaken such an exercise in 1967 [20]. Above all, Krugman suggested applying the physical theory of percolation to model the emergence of urban hierarchies as the result of a process of diffusion, overlooking the fact that the processes of urban diffusion are often "hierarchical", in the sense that the adoption of innovations often "jumps" from one large city to another before reaching closer, smaller cities [52], which in no way corresponds to a passive process of diffusion from neighbour to neighbour. The models of "scale free networks" are also used by sociologists or analysts of large networks like the Internet, in which they discover "small worlds" (which are in fact hierarchical structures) [5].

13.3.2.2 Adoption of Innovations and Structural Change

For it has also been demonstrated that more qualitative transformations of cities, for example in their profile of activities (economic specialisations) or in their social composition, follow modalities of co-evolution similar to those of their quantitative growth. Thus, Fig. 13.3, produced by F. Paulus, illustrates how the main urban areas in France changed position, between 1962 and 1990, with regard to their structure of economic activity (represented by the first two axes of an analysis in terms of the main components of the profiles of urban economic activity observed at the dates of five surveys) [47]. Each city is represented on the graph by five points, connected by an arrow to illustrate the trajectory defined by their structure of activity. What is remarkable about this figure is the high degree of parallelism between all the trajectories. This means that the structure of activity of all the cities has evolved in the same way, with the loss of industrial employment and the rise of the service sector, in which new activities, like services to businesses, have taken over from older ones, like retail trade. Paris, which was initially "in advance" of the other cities in the general direction of the transformation, was in the same relative position at the end of the period studied. So these transformations, affecting all the cities at pretty

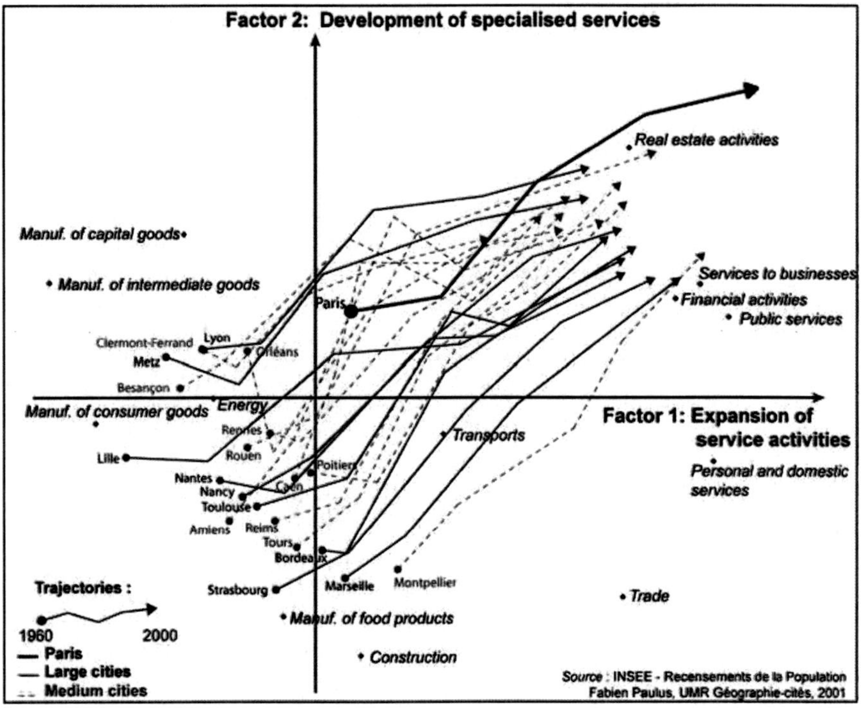

Fig. 13.3 Trajectories of the main French cities in the space of economic activities (1962–1990); after [47]

much the same time and reflecting the spatial diffusion of innovations, which is now very rapid, explain the persistence of what we call the economic structure of the system of cities, in other words the main factors that differentiate the activity profiles of the cities: the initial inequalities remain unchanged, if all the cities experience the same changes. These modalities of change, at the scale of the system of cities, can be explained by the interactions between cities, which lead to this co-evolution.

These interactions take place in a context of socio-spatial competition between cities. To continue to subsist, and therefore to maintain their market share (territorial influence, key positions in networks), the cities (or more precisely, the agents present in cities who wish to ensure a return on their investments) are obliged to adapt to change, or even to anticipate it. Thus, through imitation of the innovations appearing in other cities, or through their own participation in innovation, cities all change in more or less the same way. The thirty-year post-war period of steady growth in France, for example, saw innovations such as the development of industrial zones in the 1960s, the creation of pedestrian and shopping areas in city centres in the 1970s alongside the development of commercial centres on the outskirts, the construction of convention centres and administrative zones, before the mushrooming of residential districts into the rural villages on their periphery. But everywhere, at the same time, there was a fall in unskilled jobs and a rise in more highly qualified populations and advances in health and education services. Detailed examination of the change shows that it did not occur all in one go, but incrementally, through successive adjustments: in each survey one or another city might be in advance or lagging behind the overall process. But most of these fluctuations disappear by the time of the next survey and do not, therefore, have any lasting effect on the relative position of each city within the system.

It is only over long periods, and at particular moments, that the structure of the system of cities, maintained by these interactions, may be transformed by them. Thus, certain cycles of innovation bring new specialisations, because the activities they comport do not lend themselves to universal diffusion. This was the case during the first industrial revolution, when activities like mining and metallurgy prospered, causing certain cities in the north of France to emerge or grow in a quite exceptional way. Many of these cities had been nothing more than simple villages, and found themselves with populations of several tens of thousands within the space of a few decades (Lens-Liévin, Denain, Béthune, Valenciennes). During the same century, tourist specialisations developed in cities like Nice or Cannes, likewise contributing considerably to their growth. Such pronounced specialisations take a long time to be reabsorbed. Thus, in the 1950s, a hundred years after the industrial revolution, the mining cities, along with textile and metal-working cities, were still the most specialised cities in France, with very high proportions of working-class populations, low incomes (although wages could be high, for example in steel-working cities like Dunkirk, capital income was low), and an under-provision of services that was only very gradually rectified, despite a certain "catching up" over the next forty years. Conversely, certain cities in the south or the west, like Montpellier and Rennes, as well as the tourist cities, had received very little investment during the industrial revolution, but as middle-class, university cities, they proved to be attractive to the high-tech capital of the 1970s and following decades. In both cases, this first dimension of the structure of the system of French cities persisted, but with the cities' relation to growth and their brand image being inverted between the nineteenth and twentieth centuries.

So if we have decided to draw on the theories of self-organisation to formulate a new interpretation of the future of urban systems, it is not simply because of the intellectual seductiveness of these theories, but because of the similarities between the dynamic processes they describe and those that we have observed in the context of urban change [53, 60]. Our aim is to build the necessary models to conduct simulation experiments capable of testing what we think we have understood about the evolution of these systems, and possibly to allow us to make predictions [56, 58, 69].

13.4 Complex Systems Models for Urban Morphogenesis

The conception of a "network of cities" has allowed great progress to be made in the analysis of the determinants of the evolution of cities, bringing to light their strong interdependencies, and an almost complete autonomy of the system in the self-perpetuation of its hierarchical and spatial organisation, over and above the multiple transformations of economies, modes and standards of living, and conditions of transport and housing. However, this conception has the disadvantage of considering cities as so many islands emerging out of an undifferentiated ocean, and the "system of cities" as an archipelago isolated from the rest of the territory [74].

13.4.1 Cities as Spatial Objects

Today, complex systems theory invites us to rethink our whole approach. Another way to consider space consists in making it an intrinsic property of objects, constructed from the interactions between each object, its environment and other objects. The concept of the production of space proposed by H. Lefebvre [41] already followed this line of thinking. P. and G. Pinchemel [50] distinguished two types of interactions, depending on whether they involve *artificialisation*, the transformation of the natural environment by societies, or *spatialisation*, the development of territories according to the rules of spatial organisation of societies. Studying the former process with the help of archaeological evidence, S. Van der Leeuw [72] saw an important positive feedback in the interaction between society and environment, and analysed this feedback in terms of four nesting levels of spatial scale, from the house to the habitat. He shows how cognitive reiteration, by specifying the categories, leads to material substantialisation of the conception of space.

Little by little, a conception of space emerges which is not a support, nor a framework of objects, nor even a separation of which importance varies with the range of interactions, but which integrates time (in the sense of the mathematical integral) by allowing its materialisation by accumulation in an object. This is the type of conception for which H. Reymond [65] argued, when he evoked the constitution of geographical objects through the "obligation of spacing", the necessary deployment over space of human societies. According to Reymond (p. 22, our translation):

> The geographical act distributes or redistributes spacing by deliberately or subconsciously using the freedom of disposition contained in the power of surfaces. In so doing, it constructs geographical objects.

In the case of cities ([65], p. 203):

> The need for hierarchy is inherent to an organisation of spacing based on central points.

A number of models, still very theoretical, have explored various paths for simulating the construction of geographical objects as spatial entities from elementary interactions. Amson [4] used a plasma model to represent urban growth, Couclelis [23] based his model on cellular automata, and Zhang [80] on the theory of diffusion equations ... All these models call for us to reflect on the form of the interactions underlying the genesis and evolution of urban objects. They need to be transcribed, if possible, into "practical models", and tested experimentally, to determine whether they can provide a better description of the urban dynamics. It is pertinent, for example, to observe, like Dendrinos and Mullaly [25], that chaotic dynamics requires parameter values that are perfectly unrealistic on the scales of space and time considered in a model of interurban dynamics. This consistency with observations is not sufficient to validate the model, but at least it gives some assurance of its aptitude to keep in line with experience rather than a "fashionable" interpretation.

13.4.2 Cities and Fractal Objects

Can the reunification of the conception of city-objects, understood as a particular form of land use and at the same time as differentiated collections of constructions, activities, and population on a territory, be achieved through a reflection on the fractal structure of cities, taken in their materiality [29]? The internal homothety of the structures of geographical systems, interpreted by the geographer Philbrick as an alternation between "homogeneous" spaces and hierarchical "polarised" spaces, has been described in the theory of central places as a relatively rigid, nested configuration of nodes, ordered into a hierarchy according to their dimensions, their spacing and the range of their interactions. Rather than considering "deformations" of this cellular structure as a function of variations in density or diverse accidents of the environment, fractals allow to posit from the start a structure endowed with a hierarchy, but not systematically nested, admitting stochastic variations. The fractal description is, however, insufficient, because the fractal object, which has trivial homomorphism between levels, is no more than a

> pale imitation of the paradoxical form that natural systems reveal, with the existence of loops between the different levels, which confuse and entangle the hierarchy

as Varela puts it ([73], p. 91, our translation). Here again, a more complete investigation of the interactions could enrich the conception, still being formed, of the spatio-temporal construction of geographical objects, and of the necessary intervention of interlinking hierarchical structures in the genesis of these systems [58]. The production of forms based on spatial interactions, the identification of more or less stable – and therefore identifiable – spatio-temporal constructions, rejoins the reflection of R. Brunet [16] on choremes. Here, the question is not whether choremes are the result of "deliberate strategies" or "enshrined random choices" ([16], p. 264), because real strategies always include an element of uncertainty, but rather to understand how the observed result of strategies, with or without a deliberate finality, are maintained in a particular form, develop along a certain trajectory, and under what conditions this dynamics endures. This entails going beyond observation of the form, to measure the scales of time and space incorporated by those forms, and the parameters of order underlying their appearance.

13.4.3 From Support Space to Relational and Conforming Space

The majority of dynamic urban models refer, most often implicitly, to the conception of space as support. In the "predator-prey" urban model of Dendrinos and Mullaly [25], any consideration of space in the sense of distance is excluded, as this model of inter-city competition only considers the relative dynamics of the demographic weights and incomes of cities. Most often, in models of intra-urban dynamics, the division of space is assumed to be pre-determined and invariant over time: the dynamics concerns the relative growth of different zones, which may possess unequal resources, with or without the explicit consideration of exchanges in population and activities between the zones. In these models, quantities are linked to localities by a simple zone index, whether the dynamics is of the predator-prey type between the centre and the outskirts [24], or a result of the relative attractiveness of localities as a function of interactions that are assumed to decrease exponentially with distance. The inter-zone distance may then be a physical distance [2], a time-distance calculated over a network [26], or a cost-distance [79].

Dynamic urban models using differential equations thus face two difficulties: firstly, they refer to a simple support space to measure the interactions between cities, and secondly, they only represent one type of interaction, a power or exponential function of a distance, sometimes modulated by a parameter for variable weighting of the dissuasive effect of the distance according to the type of population concerned [56].

However, models must also take into account a relative conception of the urban space, correctly describing the process that has been called the "convergence", "contraction", or "compression" of space-time. This idea has been around for a long time, either in a descriptive form as the "modification of relations between different parts of the Earth by acceleration of the speeds necessary to cover the same distances" proposed by Jean Bruhnes, or more formally as the process of "spatial reorganisation" described by Janelle [37]. The space-time dialectic of cities is inseparable from the state of communication techniques. Technological advances lead to the reduction of space-time, thanks to the speed of means of communication. The historian Henri Pirenne pointed out the importance of improvements in transport in the revival of development in European cities in the Middle Ages. A. Anderson took up this idea in describing the four "logistical" revolutions (involving techniques for transporting goods, capital, people and information) that have had an important influence on urban development and selection among cities between the years 1,000 and 2,000.

The consequences of this law of accelerating interactions for the form of cities and for the structure of urban systems are well-known. Although the historical increase in the speed of communications has been neither linear nor homogeneous, it has been monotone, in other words it has always moved in the same direction. Its highly concentrating effects on settlement structure have been demonstrated many times. Starting with a pattern of villages separated by small distances, defined by the radius of action of the most basic techniques, the selection of cities, points of settlement concentration, is governed by travelling speeds on foot, by horse, by stagecoach, by train, by car, by plane or by high-speed train. The small centres are bypassed, the relative concentration of the population in the larger centres increases, and the hierarchy of settlement is simplified from the bottom: the smallest villages are the first to disappear, the larger ones resisting longer, and then the small towns, which have lost relative importance, experiencing small increases in population, but growing much more slowly than the big cities [11, 34, 53]. Their decline may be momentarily concealed if they are within the circle of influence of a larger city that spreads out to incorporate them into its system of habitat. The steepening of the Pareto distribution of city size, which has been particularly noticeable in many countries for nearly two centuries [34], could be explained not so much by the disproportionate growth of the large cities as by the contraction of the space of the settlement system, which causes the smallest centres to disappear [11, 12, 15].

Geographical space is therefore in no way an immovable container. Its gradual shrinkage, measured by the yardstick of travelling speeds, could be taken into account in models that simulate the future of a system of central places, other than by the filling-in of places and the heightening of their hierarchical order as in the model proposed by Allen and Sanglier [2], or by a simple increase in the mobility of the population as, for example, in the migration model of Weidlich & Haag [36, 69]. Current research, notably with the SIMPOP model (Géographie-cités laboratory), uses multi-agent systems to simulate the evolution of social interactions that are constructing a geographical space ever more intensely connected by relations of increasingly long range [13, 17].

This type of computer modelling allows to represent interactions of different types, depending on whether they operate as a function of proximity, or are constrained by political or administrative limits, or whether they use functional networks of economic interdependence, defying the rules of proximity and frontiers. These differentiated spatial interactions all have ranges that vary over time, following variable trajectories, accentuated by innovation cycles that transform means of communication.

Dynamic models built in this way allow to test hypotheses about the conforming power of the forms created by interactions in space-time. What is the role of the "historical chain of events" [63] in differentiated urban forms? To what extent are the different forms generated by similar processes? What degree of freedom does an urban form – either local (city) or continental (system of cities) – have in relation to the standardising processes of globalisation? Such are the essential questions that give these new models all their relevance.

References

1. Allen P. (1997) *Cities and regions as self-organizing systems: models of complexity*, Gordon & Breach (Amsterdam).
2. Allen P. and Sanglier M. (1979) A dynamic model of growth in a central place system, *Geog. Anal.* **11**, 256–272.
3. Alonso W. (1964) *Location and land use*, Harvard University Press (Cambridge MA).
4. Amson, J.C. (1975) Catastrophe theory: a contribution to the study of urban systems, *Environ. Plann.* **2**, 175–221.

5. Anderson C., Hellervik A., Hagson A., and Tornberg J. (2003) The urban economy as a scale-free network, *Phys. Rev. E* **68**, 036124.
6. Auerbach F. (1913) Das Gesetz der Bevölkerungskonzentration, *Petermans Mitteilungen* **1**, 74–76, in German.
7. Bairoch P., Batou J., and Chèvre P. (1988) *La population des villes européennes. Banque de données et analyse sommaire des résultats: 800–1850*, Centre d'Histoire Économique International de l'Université de Genève, in French.
8. Batty M. and Longley P. (1994) *Fractal cities*, Academic Press (London).
9. Batty M. and Xie Y. (1996) Preliminary evidence for a theory of the fractal city, *Environ. Plann. A* **28**, 1745–1762.
10. Berry B.J.L. (1964) Cities as systems within systems of cities, *Papers of the Regional Science Association* **13**, 147–163.
11. Bretagnolle A. (1999) *Espace-temps et système de villes: effets de l'augmentation de la vitesse de circulation sur l'espacement et l'étalement des villes*, Université Paris I, thèse de doctorat, in French.
12. Bretagnolle A. (2003) Vitesse des transports et sélection hiérarchique entre les villes françaises, in *Données Urbaines*, edited by D. Pumain and M.-F. Mattei, vol. 4, pp. 309–322, in French.
13. Bretagnolle A. and Pumain D. (2005) Artificial intelligence and collective agents: a generic multi-agent model for simulating the evolution of urban systems (Simpop2), poster presented at the *European Conference in Complex System (ECCS'05)*, Paris.
14. Bretagnolle A., Mathian H., Pumain D. and Rozenblat C. (2000) Long-term dynamics of European towns and cities: towards a spatial model of urban growth, *Cybergeo* **131** (http://www.cybergeo.presse.fr)
15. Bretagnolle A., Paulus F., and Pumain D. (2002) Time and space scales for measuring urban growth, *Cybergeo* **219** (http://www.cybergeo.presse.fr)
16. Brunet R. (1980) La composition des modèles dans l'analyse spatiale, *L'Espace Géographique* **4**, 253–264, in French.
17. Bura S., Guérin-Pace F., Mathian H., Pumain D., and Sanders L. (1996) Multi-agent systems and the dynamics of a settlement system, *Geographical Analysis* **2**, 161–178.
18. Cattan N., Pumain D., Rozenblat C., and Saint-Julien T. (1999) *Le système des villes européennes*, 2nd edition, Anthropos (Paris), in French.
19. Claval P. (1982) *La logique des villes*, LITEC (Paris), in French.
20. Chorley R. and Haggett P. (1967) *Models in geography*, Methuen (London).
21. Christaller W. (1933) *Die Zentralen Orte in Süddeutschland*, Fischer (Iena), in German.
22. Cordoba J.C. (2003) On the Distribution of City Sizes, *J. of Urban Econ.* **63**, 177–197.
23. Couclelis, H. (1985) Cellular worlds: a framework for modeling micro-macro dynamics, *Environ. Plann.* **A17**, 585–596.
24. Dendrinos, D.S. and Haag G. (1984) Toward a stochastic dynamical theory of location: empirical evidence, *Geog. Anal.* **16**, 287–300.
25. Dendrinos D. and Mullaly H. (1985) *Urban evolution*, Oxford University Press (Oxford).
26. Engelen G. (1988) The theory of self-organisation and modelling complex urban system, *Eur. J. Oper. Res.* **37**, 42–57.
27. Fletscher R. (1986) Settlement in Archaeology: world-wide comparison, *World Archaeology* **18**, 59–83.
28. Forrester J. (1964) *Urban dynamics*, MIT Press (Cambridge MA).
29. Frankhauser P. (1993) *La fractalité des structures urbaines*, Anthropos (Paris), in French.
30. Fujita M., Krugman P., and Venables A. (1999) *The spatial economy: cities, regions and the international trade*, MIT Press (Cambridge MA).
31. Gabaix X. and Ioannides Y.M. (2004) The evolution of city size distributions, in *Handbook of Regional and Urban Economics*, edited by V. Henderson and J-F. Thisse, vol. 4, chap 53, North-Holland (Amsterdam), pp. 2341–2378.
32. Gibrat R. (1931) *Les inégalités économiques*, Sirey (Paris), in French.

33. Goodrich E.P. (1926) The statistical relationship between population and the city plan, in *The urban community*, edited by E.N. Burgess, University of Chicago Press (Chicago), pp. 144–150.
34. Guérin-Pace F. (1993) *Deux siècles de croissance urbaine* **4**, Anthropos (Paris), pp. 411–425.
35. Guérois M. (2003) Les villes d'Europe vues du ciel, in *Données urbaines*, edited by D. Pumain and M.-F. Mattei, vol. 4, Anthropos (Paris), pp. 411–425, in French.
36. Haag G., M'unz M., Pumain D., Sanders L., and Saint-Julien T. (1992) Inter-urban Migration and the Dynamics of a system of cities. Part I: the Stochastic framework with an application to the French urban system, *Environ. Plann.* **24**, 181–188.
37. Janelle D.G. (1969) Spatial reorganisation, a model and concept, *Annals of the Association of American Geographers* **59**, 348–364.
38. Kohl J.G. (1841) *Der Verkehr und die Ansiedlungen der Menschen in ihrer Abhängigkeit des Gestaltung der Erdoberfläsche*, Arnold (Leipzig), in German.
39. Lalanne L. (1863) Essai d'une théorie des réseaux de chemins de fer, fondée sur l'observation des faits et sur les lois primordiales qui président au groupement des populations, *Comptes rendus des séances de l'Académie des Sciences* **57**, 2e semestre, 206–210, in French.
40. Lalanne L. (1875) Note sur les faits d'alignements naturels dans leurs relations avec les lois qui président à la répartition des centres de population à la surface du globe, *Comptes rendus du 2e Congrès International de Géographie*, Martinet (Paris), vol. 2, pp. 45–55. Reprinted in *Deux siècles de géographie française*, edited by P. Pinchemel, M.-C. Robic, and J.-L. Tissier CTHS (Paris), *Mémoire de la section de géographie*, vol. 13, pp. 57–64 (1984), in French.
41. Lefebvre H. (1974) *La production de l'espace*, Anthropos (Paris), in French.
42. Lotka A.J. (1924) *Elements of physical biology*, Williams & Wilkins (Baltimore).
43. Lotka A.J. (1941) The law of urban concentration, *Science* **94**, 164.
44. Marchetti C. (1991) Voyager dans le temps, *Futuribles* **156**, 19–29, in French.
45. Moriconi-Ebrard F. (1993) *L'urbanisation du Monde depuis 1950*, Anthropos (Paris), in French.
46. Moser C.A. and Scott W. (1961) *British towns. A statistical study of their social and economic differences*, Oliver and Boyd (Edinburgh).
47. Paulus F. (2004) *Coévolution dans les systèmes de villes: croissance et spécialisation des aires urbaines françaises de 1950 à 2000*, Université Paris 1, thèse de doctorat, in French.
48. Paulus F. and Pumain D. (2000) Trajectoires de villes dans le système urbain, in *Données urbaines*, edited by M.-F. Mattei and D. Pumain, vol. 3, pp. 363–372.
49. Pinchemel P. and Carrière F. (1964) *Le fait urbain en France*, A. Colin (Paris), in French.
50. Pinchemel P. and Pinchemel G. (1988) *La face de la terre*, A. Colin (Paris), in French.
51. Portugali J. (2000) *Self-organisation and the city*, Springer (Berlin).
52. Pred A. (1977) *City systems in advanced economies*, Hutchinson (London).
53. Pumain D. (1982) *La dynamique des villes*, Économica (Paris), in French.
54. Pumain D. (1993) L'espace, le temps et la matérialité des villes, in *Temporalités urbaines*, edited by B. Lepetit and D. Pumain, Anthropos (Paris), 133–157, in French.
55. Pumain D. (1997) Vers une théorie évolutive des villes, *L'Espace Géographique* **2**, 119–134, in French.
56. Pumain D. (1998) Urban research and complexity, in *The City and Its Sciences*, edited by C.S. Bertuglia, G. Bianchi, and A. Mela, Physica Verlag (Heidelberg), pp. 323–361.
57. Pumain D. (2001) Villes, agents et acteurs en géographie, *Revue européenne des sciences sociales* **121**, 81–93, in French.
58. Pumain D. (ed.) (2006) *Hierarchy in natural and social sciences*, Methodos Series vol. 3, Springer (Berlin).
59. Pumain D. and Saint-Julien T. (1978) *Les dimensions du changement urbain*, Mémoires et Documents du CNRS (Paris), in French.
60. Pumain D. and Saint-Julien T. (eds.) (1995) *L'espace des villes*, La Documentation française RECLUS, Atlas de France, tome 12, in French.

61. Pumain D., Bretagnolle A., and Degorge-Lavagne M. (1999) La ville et la croissance urbaine dans l'espace-temps, *Mappemonde* **3**, 38–42, in French.
62. Pumain D., Bretagnolle A., and Rozenblat C. (1999) Croissance et sélection dans le système des villes européennes (1600–2000). *Travaux de l'Institut de Géographie de Reims* **26**, 105–135, in French.
63. Pumain D., Paquot T., and Kleinschmager R. (2006) *Dictionnaire La ville et l'urbain*, Anthropos (Paris), in French.
64. Reclus É. (1895) The evolution of cities, *The Contemporary Review* **67**, 246–264.
65. Reymond H. (1981) Une problématique théorique, in *Problématiques de la géographie*, edited by H. Isnard, J.-B. Racine, and H. Reymond, PUF (Paris), in French.
66. Reynaud J. (1841) "Villes", *Encyclopédie Nouvelle*, tome VIII, Gosselin (Paris), pp. 670–687, in French.
67. Robic M.-C. (1982) Cent ans avant Christaller, une théorie des lieux centraux, *L'Espace Géographique* **1**, 5–12, in French.
68. Robson B.T. (1973) *Urban growth, an approach*, Methuen (London).
69. Sanders L. (1992) *Système de villes et synergétique*, Anthropos (Paris), in French.
70. Simon H. (1955) On a class of skew distributions, *Biometrika* **42**, 425–440.
71. Singer H.W. (1936) The "courbe des populations": a parallel to Pareto's law, *Economical Journal* **46**, 254–263.
72. Van Der Leeuw S. and Fiches J.L. (eds.) (1990) *Archéologie et Espaces*, Éditions APCDA (Antibes), in French.
73. Varela F. (1983) L'auto-organisation, de l'apparence au mécanisme, in *L'auto-organisation, de la physique au politique*, edited by P. Dumouchel and J.P. Dupuy, Seuil (Paris), pp. 147–165, in French.
74. Veltz P. (1996) *Mondialisation, villes et territoires*, PUF (Paris), in French.
75. de Vries J. (1984) *European urbanisation: 1500-1800*, Methuen (London).
76. Watts D.J. and Strogatz S.H. (1998) Collective dynamics of small world networks, *Nature* **393**, 4.
77. Weidlich W. (2000) *Sociodynamics*, Taylor & Francis (London).
78. Willis J. and Yule G. (1922) Some statistics of evolution and geographical distribution in plants and animals, and their significance, *Nature* **109**, 177–179.
79. Wilson A. (1981) *Catastrophe theory and bifurcation*, Croom Helm (London).
80. Zhang W.B. (1990) Stability versus instability in urban pattern formation, *Socio-spatial dynamics* **1**, 41–56.
81. Zipf G.K. (1941) *National Unity and Disunity*, Principia Press (Bloomington Indiana).
82. Zipf G.K. (1949) *Human Behaviour and the Principle of least effort*, Addison-Wesley (Reading MA).

Chapter 14
Levels of Organisation and Morphogenesis from the Perspective of D'Arcy Thompson

Yves Bouligand

An illustrious scientist who was never really sure whether he was the author of his own works! Such was the impression D'Arcy Thompson gave when he presented his results. He compared them to the observations of his predecessors, as one should, but without really highlighting those points that would assure him priority. If an idea came into his head that he found interesting, it was as if, deep inside, a voice whispered to him that others had already had the same idea, maybe long ago. He searched for the source of the idea in more or less recent conversations with his family and friends, or buried himself in his books, including Ancient Greek literature, in which he was an eminent specialist at an early age, while his profession was teaching natural history to the students of Dundee University in Scotland. His biology lectures were well-researched and constituted the point of departure for profound reflections in the domains of mathematics, physics and chemistry, not to mention historical references ranging from Plato to Galileo or other, more recent authors [8, 11, 41]. This all bears witness to a fine eclecticism, the multiple facets of which have been rediscovered over the last forty years [3, 20, 36].

One of his books, *On Growth and Form*, published in 1917, proposes a wide collection of facts of morphological interest, chosen among the most striking examples in both the living and the nonliving worlds [39]. In this way, the author explored all the natural phenomena that give rise to defined forms, with their transformations. He highlighted problems of dimension and the effective presentation of statistical data. The predominant topics include hydrodynamics and visco-elasticity, solubility and diffusion of chemical compounds, semi-permeable precipitation membranes and osmotic forces, interfacial tension, convection phenomena, phase transitions and the main states of matter, including liquid crystals and various colloids. Biomechanics and the organisation of skeletal tissue also occupy an important place. The author compares the geometrical patterns of the subunits that play a part in the construction of living beings. The book suggests that multiple mechanisms are brought into play for one sole morphogenetic process, perhaps because many different stages

Y. Bouligand (✉)
École Pratique des Hautes Études, Paris, France
e-mail: yves.bouligand@univ-angers.fr

P. Bourgine, A. Lesne (eds.), *Morphogenesis*, DOI 10.1007/978-3-642-13174-5_14,

are indispensable to the future of the embryo. If some of these stages fail in their purpose, there must be other mechanisms present that can step in, so to speak, and take their place. This is seen as a way to avoid putting all the eggs in one basket, and of conceiving a synergy of processes. In reading this book, we understand how nature itself is “eclectic”, and how we can draw inspiration from it to find analogies and imagine solutions to our problems. Indeed, there are many engineers and architects among the readers of D’Arcy Thompson [35]. In this chapter, we shall present several of the ideas explored in his book, setting them in a new context, in the light of much more recent discoveries.

14.1 Games of Construction

The mechanisms of morphogenesis are multiple and change with scale. The factors that govern the form of molecules have nothing in common with those that structure intracellular organelles, or with those that determine the architecture of cells. Other processes intervene in the organisation of tissues and their incorporation into organs or individuals. We can thereby distinguish different *levels of organisation*, but numerous mechanisms come into play in the construction of one sole tissue. D’Arcy Thompson’s book proposed a research programme for his epoch that remains highly topical to this day, covering the factors of morphogenesis and their diversity [8, 9]. In addition, he formulated the problem of the coordination of all these morphogeneses, in the context of the development and growth of the individual.

Morphogenesis involves *combinatorial* aspects, bringing to mind the principle of certain games of construction, which generally come into play at the molecular and supramolecular scales, although they may also appear at higher levels. But there are also phenomena of morphogenesis linked to transitory morphologies occurring in fluid mechanics, and many of these *fluid morphogeneses* are *stabilised*, through sol-gel transitions, for example. Beyond these two physico-chemical types of morphogenesis, there are all the mechanisms of reorganisation and control by cells, in which the chief participants are *genes* and *enzymes*.

14.1.1 Chemical Syntheses and Biosyntheses

D’Arcy Thompson compared the biological morphogenesis of molecules with the syntheses carried out by organic chemists, but he found their toolbox, even endowed with certain *catalysts*, to remain truly lacking, as long as it did not also contain biological catalysts, that is to say *enzymes*. He also recalled that the molecules of living matter are mainly asymmetrical, as Pasteur had demonstrated, from the moment that they are slightly complex, in other words they cannot be superimposed on their image in a mirror. Just as we distinguish our right hand from our left, many chemical compounds exist in left- and right-handed versions, called *enantiomers* or optical isomers. The difficulties encountered in asymmetrical synthesis had been pointed out by Pasteur, and this point of view has lost none of its relevance [24]. The problem of the origin of asymmetrical compounds therefore had to be faced, and D'Arcy Thompson suggested above all the selective photochemical destruction of an enantiomer. Indeed, the circular dichroism corresponds to the perceptible difference in the absorption of circularly polarised light, according to whether it is right or left, and our terrestrial landscape is full of sites that favour one or the other of these polarisations. It is therefore a principle of selection applied to chiral molecules that are sensitive to light or to other wavelengths.

At the time of D'Arcy Thompson, it was clearly known how to synthesise a good number of biological molecules, without necessarily reproducing the more complex paths followed by cells, which we call *biosynthesis*. Thus, the synthesis of amino acids in bacteria or other organisms is controlled by enzymes and is different from the much more random synthesis that can be obtained under the effect of a simple spark in an atmosphere of N_2, H_2O, CH_4 and NH_3 – an experiment conducted to simulate what might have been the first syntheses in prebiotic conditions, in the remote past and without the help of enzymes [32]. The amino acids obtained were mainly symmetrical glycine and asymmetrical alanine, but the two enantiomers had to be present in equal proportions. These same amino acids and others can unite through peptide bonding and the elimination of water, by moderate dry heating. In the very early 1900s, Fischer had already developed methods for synthesising polypeptides by purely chemical means and without enzymes.

Clearly, although the spark method produces a whole range of interesting chemistry, it is not necessarily suitable for other constructions expected of chemistry or biochemistry. The synthesis of proteins is achieved by very different paths in organisms, with the intervention of ribosomes, transfer RNA etc., through mechanisms that have been well-explored [15] and continue to be studied. Moreover, many of these long chains of amino acids do not spontaneously assume the three-dimensional forms necessary to their functions, and they often need the assistance of "molecular chaperones" or other, more complex molecular systems [12].

The biosynthesis of nucleic acids has also been the subject of in-depth research, that is far from complete [21]. It consists in producing strands that are complementary to other, existing strands, adopting multiple helical morphologies. This involves mechanisms very different from those that can be obtained with the sole use of the tools of organic chemistry. The cellular machinery provides a demonstration of its combinatorial talents, with the diversity of proteins produced, notably antibodies, and at the same time of its capacity to control the doses produced.

The study of the biosynthesis of *steroids*, in other words molecules related to cholesterol, was carried out much earlier than that of proteins or nucleic acids. It was the example chosen by Needham in 1936 to highlight the functional differences between related molecules. Needham was also one of the great continuators of the work of D'Arcy Thompson and one of the pioneers of biochemistry at different levels of organisation in the embryo [33].

The first stage in the production of a steroid by a cell involves the synthesis of a *carotenoid* called *squalene*, a long, non-chiral molecule with a series of conjugated double bonds [2]. The squalene then folds up, shaping the surrounds of the *sterol* nucleus, and stabilises by forming four adjacent rings with new covalent bonds, created in the presence of enzymes and other compounds. The addition of different lateral groups produces molecules with highly differentiated functions: bile salts, certain vitamins, male and female sex hormones and many other anti-inflammatory endocrine factors like cortisol or corticosterone, or those involved in the hydrolysis of glycogen, or in the reabsorption of Na^+ ions in the kidneys. In arthropods too, the moult hormone is a steroid. The usual syntheses of chemistry give form to molecules, but without being able to provide them with all the characteristics they need to perform their functions in the organism. For example, the enzymes involved in biosynthesis are asymmetrical, including with regard to their stereospecific sites, which allows the synthesis of one sole type of enantiomer. This would not be possible in enzyme-free reactions. Many living beings also use, in their nutrition, molecules obtained from the breaking-down of biological tissues in the food chain, such as amino acids and many other molecules that are most often asymmetrical with the appropriate orientation, and this is an aspect of the maintenance of molecular chirality in biosynthesis.

Molecular morphogenesis is therefore achieved through an immense variety of reaction mechanisms, in the presence of enzymes and other chiral systems like ribosomes and chaperones, which offer a vast range of combinatorial possibilities, and all within a context of rigorous control.

14.1.2 Supramolecular Assemblies and their Lattices

Many identical macromolecules, produced in large numbers by cells, are assembled by weak links, form lattices just as precise as those of real crystals. Many cells contain such *molecular crystals*, like those produced by ferritin, for example. But there are other lattices, often two-dimensional, formed on various surfaces, *cylindrical*, as in the case of bacterial flagella, certain viruses and microtubules, or *polyhedral* in many other viruses. Bacteriophages contain the two types of lattice. We know how to reconstitute cylindrical bacterial flagella from their separate subunits, each of which is a protein called flagellin. These two-dimensional, cylindrical or polyhedral assemblies share with crystallisations the fact that once their subunits are assembled, they are located in a lattice and they remain there, like the molecules in a molecular crystal.

By using solutions with a sufficiently high *ionic force*, we can break viruses down into their constituent macromolecules (proteins and nucleic acids), which can then be separated and purified. It is then possible to reconstitute the viral particles, with all their initial virulence, simply by putting them back together under the appropriate physico-chemical conditions. This experiment has been carried out successfully on the tobacco mosaic virus [17].

Assemblies can occur spontaneously, as they do in crystallisations, although these are not crystals in the proper sense of the term. We then say that there is *self-assembly*, and this was a very active field of research from the 1950s to the 1970s [17]. But some assemblies also involve the rearrangement of some covalent bonds, catalysed by enzymes; they are no longer spontaneous, and we speak rather of *assisted assemblies*. These highly diverse constructions are developed both inside and outside cells.

The weak bonds that come into play between macromolecules during assemblies are the same as those involved in many ordinary crystallisations and in the cohesion of liquids, but there is one essential difference. Many of these assemblies are *endothermic*, particularly by releasing large numbers of bonds between water and the macromolecules involved. But the *reorganisation of covalent bonds* also plays a role in the subunits before their incorporation into the network. This suggests that a source of energy is being drawn on, in which case the assembly is not spontaneous. This reorganisation can intervene in several ways.

For many viruses, proteins synthesised by the host cell must first be amputated by a precise fraction, by hydrolysis of peptide bonds, and it is only when they are in this new state that they can join the assembly in progress [25]. Cycles of viruses thus contain veritable *programmes of construction*, with a precise order of synthesis of the constituents and their possible modification by enzymes. In this way, we can succeed in reconstituting by self-assembly bacteriophages of very complex form, provided that the subunits, ready to be integrated into the construction, are added in the right order.

Ordinary crystals, on the other hand, are the result of self-assemblies, of which the subunits – often ionised – are connected by weak bonds; *biological mineralisation* is one example. This is mainly involved in the building of the skeleton, as in crustacean shells. Calcium carbonate is deposited in the form of calcite between the fibrils of the organic matrix, but phosphate ions are also present and act as a real poison to the growth of the calcite. Alkaline phosphatases are also present, as if to exacerbate this disadvantage, facilitating the separation of phosphates that attach themselves to the growth front of the crystals. The nucleation of crystals, their growth and even their defects, as well as the crystalline mosaic texture are largely influenced by the fibrillar matrix, constructed well before the beginning of mineralisation [10].

Fibrillar actin is one of the essential muscular proteins, made of filaments that can be broken down into their subunits, mainly actin G. When this latter is purified and put back under the appropriate physico-chemical conditions, it reassembles into the original F form, that is to say filaments in a double helix, comparable to those observed in muscles and in the cytoskeleton [17]. But this reconstruction is favoured when ATP is bound to the actin, and the growth of the filament leads to its hydrolysis in the form of ADP, the precise process being even more complex [13].

Likewise, the assembly of *microtubules* from *tubulin dimers* present in the cells involves the bonding of tubulin to GTP molecules, and the partial hydrolysis of the latter into GDP occurs during the insertion of new subunits at one of the ends of the growing microtubule. This assembly is then accompanied by the hydrolysis of certain covalent bonds. The role of GTP has been tested by using non-hydrolysable GTP analogues, and the assembly takes place all the same. This suggests that we are dealing with a case of self-assembly [16]. The use of GTP itself leads to a dynamics of assembly at one of the ends of the microtubule and simultaneous dissociation at the other end, and under certain conditions, this leads to periodic oscillations in assembly, linked to a hydrolysis reaction, as in certain chemical clocks or other dissipative systems [14]. So it is not easy to determine whether certain assemblies are really spontaneous or not.

14.1.3 Molecular and Supramolecular Models

We have permitted ourselves a brief foray into the forms of biological molecules, which show wide diversity around certain favoured patterns. Many of these *macromolecular forms* are essential at higher levels of organisation; a typical example is that of proteins that form part of *complex assemblies. Rod- or disc-shaped molecular forms* play a role in other assemblies, notably those of *liquid crystalline phases*, often with *asymmetrical or helical forms* necessary in *cholesteric phases*, or other equally chiral forms.

Chemistry disposes of all sorts of atoms, which bond together or separate, as if to produce forms. Some of them were probably conducive to the emergence of life, and during evolution functions were gradually assigned to these molecular forms. One essential aspect of morphogenesis is this *game of constructing molecules* out of about a hundred different atoms. Chemists use to make models of these molecules, which can still be seen in their laboratories, but these models have too often been abandoned in favour of software, displaying molecules that rotate on the screen, so that they can be examined from every angle. This is a magnificent tool, but to fully appreciate a form through models, what better than being able to touch them? Life has chosen certain atoms: C, H, O, N, P etc. and shows a preference for certain asymmetries between left and right. At a slightly higher scale, another game of construction comes into play, with macromolecules connected by weaker bonds. Here we are closer to *games of "Lego"*, but quite particular ones, because there are also motors and all sorts of subtleties, for example in muscles, cilia, flagella and the wide diversity of viruses.

14.2 Water Games

Nature can be playful, and we have seen some examples with these games of construction in the shape of crystallisations and network assemblies, whether physical or biological. But there are other ways of playing with form, to which liquids lend themselves marvellously well, by moulding themselves exactly to the inner form of their containers. They also display phenomena of diffusion, and by heating we can easily produce convection. The most beautiful fountains in our cities also present water games, in hydrodynamic forms. But let us start with the most stable forms, without stream-lines.

14.2.1 Hydrostatic Forms

Examples abound, with the fluid interfaces observed between immiscible liquids, or between liquids and gases. There is the flat surface of still water, the spherical form of a droplet of water deposited on a non-wettable surface, or the polyhedral partition of space by a foam of soap bubbles. The films obtained by dipping fixed metal frames into soapy water have attracted the attention of mathematicians, because of the minimal surfaces produced when these films stabilise.

We say that these forms are hydrostatic, because they remain in place sometimes for long periods and the theories or calculations explaining them do not introduce hydrodynamic considerations in the usual sense. We can, however, observe fluid movements taking place within these films, particularly towards triple junction lines, or towards the supporting frame. Likewise, a foam of bubbles is only stationary temporarily: from time to time it is shaken by the bursting of a wall between two bubbles, which unite in the process. The evolution of an isolated foam is therefore irreversible. Conversely, many living tissues, especially in embryos, are made up of cells with adjacent membranes forming partitioned structures quite similar to those found in foam, but instead of uniting with each other, the cells divide by creating new partitions. From this point of view, foams do evolve like certain living tissues, but in reverse time. This comparison should not, however, divert us from the fact that cell systems are living things, and the site of multiple dynamics, unlike foams.

We have also seen that liquid crystals present a remarkable ability to produce forms, defined not only by their bounding surfaces, but also in depth, by their distribution of molecular orientations and patterns of singularities, in the form of lines laid out systematically to form a wide variety of textures. Analogous geometries can be found in biological liquid crystals, often in a stabilised form, but with differences, many of which can be ascribed to cellular activities. The large majority of these morphologies, described in Chap. 4, are hydrostatic, in other words they remain at rest, without moving, for hours, days or even months, because they are at equilibrium. Of course, the operations needed to obtain these preparations have involved fluid movements, but they intervene as they would at the beginning of any hydrostatic experiment. To observe the flatness of the surface of still water, for example, we start by pouring water into a crystallising dish, but the movements die out and we wait for the water to settle to obtain hydrostatic conditions.

Among the static forms of liquid crystals, those of membranes with two hydrophilic layers enclosing a hydrophobic layer are of particular interest. In zones of elliptical curvature, convex or concave, the intermediate layer covers a smaller surface than the average of the areas of the two corresponding hydrophilic zones, whereas in hyperbolic forms, or saddles, it has a larger surface area. Liquid crystalline assemblies play a role in biology on all supramolecular levels, at the scale of intracellular organelles, like those made of membrane material (like the endoplasmic reticulum, the nuclear envelope, Golgi bodies, mitochondria or chloroplasts), or chromosomes with a fibrillar structure. At higher scales, stabilised analogues of liquid crystals play a role in the construction of muscles and skeletal structures. Physics should be able to establish the fact that certain aspects of life, such as specific assimilation and reproduction, require transition by states of matter that are neither solids nor ordinary liquids. After Lehmann [28] and Haeckel [22], it was mainly D'Arcy Thompson [39] and Needham [33] who underlined the predominant role of liquid crystals in biology.

14.2.2 Hydrodynamic Figures

The similarities between arrangements of living cells and aggregates of bubbles drew the attention of D'Arcy Thompson [39], but he was above all interested in the hydrodynamic processes leading to such patterns, such as Bénard convection cells, which appear in paraffin oil subjected to a heat gradient parallel to and in the same direction as the vector of gravity. The movements of convection can be visualised using tiny graphite plates.

The most elaborate morphologies of hydrodynamics occur in high-speed phenomena, and D'Arcy Thompson discussed a wide series of examples. For instance, take some coloured water and inject it under pressure into a tank of clear water, through a narrow tube pointing upwards. The jet of coloured water meets resistance and takes the shape of a mushroom, with a vortex ring around the edge of the cap. The stem displays the same rotational symmetry and develops regularly spaced rings.

There is also the effect of a drop falling onto a liquid surface, which is observed using a high-speed camera. The best pictures have been taken with milk. A circular wrinkle grows upwards, decorated with regularly spaced spurs, each of which produces a small droplet. The formation of this coronet is followed by a long axial spike, which also releases one or more droplets. This complex behaviour involves interactions between surface tension and hydrodynamic forces, giving rise to instabilities and singularities in the distribution of stream-lines. Today, the physics of this process is largely understood, but we are often unaware that these fleeting forms appear to be adopted and stabilised by certain living beings. D'Arcy Thompson presented several examples, drawn chiefly from the protective calices of the polyps of certain coelenterates, but other illustrations also exist, such as the skeleton of certain diatoms.

The effect of a droplet falling into a pool can also be examined from the interior, replacing the milk by pure, transparent water and the falling droplet by a drop of Indian ink. This droplet has a higher density and a different composition from the water. When it falls into the water this creates an annular vortex, with stream-lines like those observed in smoke rings. But this annular vortex is not stable. It becomes sinuous and the deepest parts accelerate their fall, forming vertical columns corresponding stream-lines that produce other annular vortices. Likewise, a drop of fusel oil in water produces medusoid forms like jellyfish, as D'Arcy Thompson observed [39]. Other experiments were conducted by letting drops of melted gelatin fall into water containing coagulants, which gives rise to stabilised medusoid forms. The coagulants partially dehydrate the droplet, adding an ordered pattern of wrinkles to its original form.

14.2.3 Morphological Adaptations to the Hydrodynamics of the Environment

Life creates forms that are often similar to hydrodynamic figures, but it builds them very gradually over the course of embryogenesis or later stages that also take time, while a drop of water falling into a pool is immediately followed by subtle external decorations and a chain of internal vortices. This suggested to René Thom that the detailed analysis of certain well-controlled biological morphogeneses could improve our approach to the study of some of the more complex and high-speed hydrodynamic evolutions [36].

For his part, D'Arcy Thompson sought the reasons for these similarities between the forms of living beings and those produced by purely physical processes. He also considered the visible or hidden forms characteristic of each natural biotope, encountered at any moment, static or dynamic, lasting or transient, and even recommended that they should be listed. He gave examples, but there remains a vast programme of work to do.

Along these lines, and precisely in an aquatic environment, Houssay had explored the positions of vortices generated by fish swimming [23]. One of his models involved filling a cylindrical bladder, closing it and maintaining it by a string in a fairly powerful, regular stream of water. Vortices form in the water on contact with the bladder, which adopts a sinuous, undulating form like that observed in fish.

The similarity often observed between hydrodynamic forms and living forms (particularly but not exclusively in an aquatic environment) can be considered from another perspective. All the vector fields in three-dimensional physical space resemble each other enormously. A few precise morphological differences can be found, especially at the level of the singularities they contain, but the overall appearance is often the same [7]. For example, we have defined a field of vectors linked to the orientations of fibrils in crab shells. We can observe the same field in a cholesteric liquid crystal, and it is also the pattern produced by an electric field at a given moment in a circularly polarised wave-train of light. Generally speaking, in many fibrillar structures, the lines of force that follow the fibrils allow to define vector fields, and many of these tissues owe their genesis to liquid crystalline phases occurring at certain stages in their development, as explained in Chap. 4.

In other words, there is no reason to be surprised that the organisation of living beings involves stabilised morphologies similar to the often ephemeral morphologies of their hydrodynamic environment. These similarities suggest that evolution may have selected the development of morphologies adapted to the most common hydrodynamic situations in the environment and to the scales of the individual. This was one of the speculative themes of evolutionary biology, in plants and animals, but it has since been abandoned due to the experimental difficulties.

14.3 The Fragile Architectures of Diffusion

Phenomena of diffusion can be found in all three states of matter, but here we shall concentrate on liquids. D'Arcy Thompson describes various series of experiments, known above all at that time because of the analogies perceived with certain aspects of biological morphogenesis. We shall group these examples into three series, slightly artificially as these phenomena often act together.

14.3.1 Hydrostatic Diffusion

Figures of diffusion were obtained by Leduc [39] by introducing drops of salty, coloured water into a crystallising dish or a Petri dish containing water that was less salty, leading to cell partitioning. Leduc also performed *artificial mitoses*, similar in appearance to real mitoses, but at much larger scale, by introducing a drop of Indian ink, flanked by two other drops of salty water with a lower concentration of Indian ink, into a dish of moderately salty water. We now know that tubulin dimers diffuse according to relatively precise migrations during the construction of the mitotic spindle, but many other factors are involved in dividing cells.

These days, we also know that the fluidity of membranes allows the diffusion of phospholipids, of proteins and its other constituents within its bilayer structure. This fluid structure enables numerous morphological reorganisations, by means of more or less pronounced extensions of the two hydrophilic layers and the intermediate lipophilic layer, which are controlled by the physical chemistry of the medium (see Chap. 4). For given morphological conditions, in the absence of membrane recombinations, this leads to morphologies in equilibrium, and therefore hydrostatic.

D'Arcy Thompson also mentions Liesegang rings, which we now know to be produced by oscillating chemical reactions, capable of determining waves of chemical concentration, or even certain stationary patterns, obtained following a theoretical principle conceived by Turing. The main references are given in Chap. 5. When a chemical reagent plays the role of an *inhibitor* and diffuses faster than another that acts as an *activator*, zones with a higher concentration of the latter may become differentiated. This occurs far from chemical equilibrium, but under quasi-hydrostatic conditions of the fluid medium, because in general, no stream-line emerges. Chemists have developed beautiful experimental illustrations of systems with stationary patterns or allowing the propagation of chemical waves, in concentric rings or spirals, notably with the reactions of Belousov and Zhabotinski [26]. Biological examples include glycolytic oscillations and the assembly of tubulin dimers mentioned above (see Chap. 5).

We qualify as *dissipative* all these systems maintained by an input of energy or matter (or both) and which generate dynamic structures that are well-defined and often stationary. Bénard convection, which functions with one sole input of thermal energy in a field of gravity, is one example. But the two types of exchange, matter and energy, generally coexist, and this is the case for each living being, which we can consider as a particular dissipative system. Each cell on its own also constitutes a dissipative system, or even each mitochondrion, with its own metabolism, etc.

Among dissipative systems, reaction-diffusion has been the object of many mathematical models, or *in silico* simulations. Scientists have searched for the conditions allowing to obtain just about all types of morphologies encountered in living beings. This has been a large undertaking [31], but its application to biology still poses certain problems. It is not easy to identify, within the highly complex biochemistry of embryonic tissues, the substances that act as activators and inhibitors, also called *morphogens* (not to be confused with *morphogenes*, the term used to denote certain genes involved in morphogenesis).

When applied to liquid crystalline phases, one would expect the reaction-diffusion principle to generate much more elaborate morphologies than those obtained or even simulated in models of ordinary liquids. It would also be worth experimenting in two dimensions, in bilayers or cell membranes, and also in more solid liquid crystalline phases, extended to three dimensions. Another question that deserves more attention is that of the growth and stabilisation of structures generated after their differentiation by reaction-diffusion.

14.3.2 Hydrodynamic Diffusion

Above, we considered examples of diffusion of Indian ink in ordinary liquid, during sudden evolutions, at least locally, when a drop of water strikes a larger body of water. It would be interesting to extend this research to the case of reaction-diffusion, within a slower hydrodynamic context, including in liquid crystalline media, because we believe that such situations probably have their biological counterparts; moreover, such experiments should be accessible, both in ordinary liquid and in liquid crystalline media. On the other hand, the calculations or simulations are likely to be impossible to accomplish, especially in liquid crystals, because the diffusion and viscosity in a simple nematic are represented respectively by tensors with two and with five independent parameters. Furthermore, the speeds cannot be superimposed on the molecular alignments, at least in the general case, and there are various types of singularities, which does nothing to simplify matters. But that does not necessarily prevent the results from being readable and even reproducible.

Another process of diffusion with hydrodynamic aspects was remarkably illustrated by Leduc, and D'Arcy Thompson highlighted its great interest [39]. One or more small copper sulphate crystals are placed in a solution of potassium ferrocyanide. The sulphate dissolves and the Cu^{2+} ions in the presence of the $Fe(CN)_6^{4-}$ ions produce insoluble copper ferrocyanide, forming a precipitation membrane enveloping each copper sulphate crystal. As the crystals dissolve, the copper sulphate solution inside each envelope becomes more concentrated, and the osmotic pressure rises. The precipitation membranes are permeable to water and the envelopes swell until they split; but each crack is the starting point for a new precipitation membrane, so that it is either repaired or gives rise to a new envelope attached to the edge of the crack. The whole thing forms long sacs, often branched, with lateral buds here and there, rather like some kind of seaweed.

Other heavy metal salts were used, diversifying these precipitation membranes; they produced the first examples of molecular sieves, permeable to water and small molecules. Leduc increased the number of situations, by introducing gradients of sea salt or sugar into these media where he provoked osmotic growth. He obtained astonishing forms, often reproducible, reminiscent of mushrooms or seashells. These resemblances could be coincidental, but not necessarily, as they involve the participation of processes, some of which are present within cells or in the interstices that separate them.

14.4 Stabilisation and Reorganisation of Forms

Artificial osmotic vegetation is produced by the diffusion of ions, which form precipitation membranes when they react. We therefore have all the ingredients of reaction-diffusion, but this in no way corresponds to Turing's model, or to any of its experimental illustrations. There is neither activator nor inhibitor, but a hydrodynamics sustained by high osmotic gradients, which do not play any part in the usual reaction-diffusion process. Precipitation membranes are *stabilising structures*. They illustrate an essential aspect of morphogenesis, but these membranes are very different from the cytoskeletons, membrane junctions and extracellular matrices that stabilise cell and tissue systems. The stabilisation aspect is rarely mentioned in current models of morphogenesis, and yet this is an essential aspect.

Whether they stem from games of construction, water games or, more subtly, from phenomena of diffusion, differentiations usually need to be stabilised, as many cell and tissue structures are. We have seen some examples with liquid crystalline secretions, which stabilise to form extracellular matrices, which display the initial geometry of the liquid crystal, give or take a few reorganisations imposed by the cells. This stabilisation often consists in *sol-gel transitions*, either by chemical bridging between neighbouring polymers or by local micro-crystallisations, grouping polymers together into fine fibrils. Mineralisation also contributes to stabilisation, by forming composite solids such as those found in skeletal tissues. Cell membranes are also liquid crystalline, and largely remain so, but they are stabilised at the level of various junctions, involving two-dimensional arrangements of proteins.

Each morphogenesis of a tissue or an organ includes a programme of the different elements of stabilisation that are indispensable to its subsequent functioning, in such a way that the young embryo develops in an almost fluid state, the stabilisation of which only occurs gradually. This stabilisation will be undermined locally, at various times during the life of the individual, because of reorganisation necessitated by growth or maintenance, or repair following injury, or in other circumstances. This involves the destruction of tissue or *histolysis*, the construction of new tissue or *histogenesis* and other transformations, which are accomplished according to programmes determined by regulations solidly anchored in the genome.

The modifications of form that occur during development take place either progressively or during distinct phases. Growth occurs through repeated spurts, in children and adolescents, but there are infinite variations throughout the different species of the world, and D'Arcy Thompson devotes the longest chapter of his book to the topic. In arthropods, such as crustaceans or insects, the shell cannot grow larger and it has to be changed during the moult – an often perilous operation, not least because of the predators willing to take advantage of the situation. D'Arcy Thompson showed that spurts of growth correspond above all to momentary accelerations, which do not really alter the gradual nature of the change in form. These transformations even appear to be differentiable, lending themselves to tensorial analysis [4]. D'Arcy Thompson extended this point of view to the study of evolution, provided that it was kept to sufficiently restricted groups of species. For example, he considered it pointless to look for a continuous sequence of transformations leading from the cuttlefish to the beetle. Development, like evolution, takes place in a gradual context, from which incremental spurts are not excluded.

It is probably this *theory of transformations* that readers of D'Arcy Thompson find the most striking, with its curvilinear modifications observed in the evolution of an organ, or in the overall morphology of the individual. As regards development, a moult is sometimes accompanied by a transformation, as in the case of the hatching of a butterfly, but the moult and the metamorphosis are prepared over a long time by a series of gradual histolyses, histogeneses and reorganisations. We only observe the result of these processes, the sole outward manifestation of the moult, over a short period of time, and this is what may give us an impression of great discontinuity. Evolution also appears to contain exceptions to the principle of gradualism, and we speak of missing links in the genealogy of current or extinct forms; these exceptions are never certain, for many reasons, some of which are analogous to those we have suggested on the subject of metamorphoses.

14.5 The Problem of Strong Local Curvature and New Prospects

Over the last twenty or thirty years, new knowledge has been acquired in the field of cell biology, the consequences of which for morphogenesis do not really seem to have been discussed, although they concern a precise problem, already considered by D'Arcy Thompson [39]. This is the problem of very high variations in the curvature of the membrane of certain cells, in very precise regions, for example at the level where cilia and flagella are attached to the membrane, a phenomenon that was later confirmed by numerous electronic microscope studies. Laplace's law states that the product of the mean curvature multiplied by the surface tension corresponds to the difference in pressure between the two sides of a film at the limit of two fluids. But it would take considerable differences in osmotic pressure between two neighbouring zones in the cytoplasm to account for these variations in curvature, and that would be difficult to explain. These morphologies have been ascribed to the effects of the cytoskeleton, and its attachments to membrane proteins, which is certainly true for a large number of the cases observed. Comparable deformations may be generated by links to the extracellular matrix.

However, it also appears that certain proteins present highly organised distributions within membranes, creating boundary lines that demarcate membrane compartments that probably have different compositions in terms of phospholipids or other molecules [1]. This could strongly modify the spontaneous curvatures from one domain to the next. These alignments of proteins would be capable of forming molecular filters, in one dimension, enabling the differences in composition to be maintained. Using freeze-etching, it is possible to observe several such protein belts, located at the base of microvilli, cilia or flagella. The cylindrical form of the membranes means that the hydrophilic layer on the extracellular side occupies slightly more space than its counterpart on the intracellular side, and the intermediate hydrophobic layer must occupy a surface area that is exactly the average of the two corresponding hydrophilic surfaces. This requires rigorous and constant chemical control of the composition of the whole length of the cilia or microvilli. The idea of spontaneous curvature of the membranes is one of the essential aspects at the origin of cell form. The tips of the cilia or microvilli are more or less hemispherical and they also contain proteins capable of influencing spontaneous curvature. There are other examples of macromolecules aligned along well-marked backbone dihedral angles, delimiting well-defined membrane domains, notably in pulmonary surfactant, or in pores of the nuclear envelope [1, 18].

Another new concept in the generation of cell forms is that of the rotary motor. Motor phenomena have long been known to exist in muscles, and also in many cells, as they contain the essential ingredients of muscle fibrils, namely filaments of actin and myosin, in a less compact distribution, that of the cytoskeleton. These muscles motors function by translation, and they are also involved in another type of translation with microtubules and transport processes. On the contrary, the first biological rotary motor was discovered at the base of bacterial flagella, with complex proteins constituting a stator at the level of the membrane bilayer and a rotor supporting the flagellum [17, 27]. The rotation is driven by the energy liberated by the hydrolysis of ATP, and at the same time a turbine structure is formed at the interface between the stator and the rotor, drawing H^+ ions into the cytoplasm of the bacteria from outside. Very similar structures, devoid of flagella, constitute H^+ ion pumps on the inner membrane of mitochondria, displaying similar ATP-ase activity. These "proton pumps" are mainly located along the mitochondrial crests, in the most curved areas. It has been demonstrated that there really is rotation in these pumps, by using fluorescent markers on an actin filament attached to the rotor, thanks to subtle nanotechnologies developed in the field of immunology [34]. This ATP-ase can function in reverse, becoming ATP-synthetases, when they are inserted, with the right orientation, into membranes separating two compartments maintained at different pH. Ion pumps are not uncommon inside cells, in the membranes of the endoplasmic reticulum, and many of them could constitute similar motors [1].

These two types of system, rotary motors and boundary proteins delimiting membrane compartments, play a role in what is now called *membrane and cytoplasm microfluidics*. The proteins that compartmentalise membranes along their planes increase their viscosity and could even stabilise at the level of certain cell junctions. Rotary motors allow the transport of ions that modify the state of bonds with water, with possible hydrodynamic effects, adding to those produced by the cytoskeleton. These membrane ultrastructures and rotary motors could therefore play a decisive role in morphogenesis.

14.6 Particular and General Morphogenetic Theories

The above paragraphs suggest that *thousands of mechanisms* are at work in morphogenesis: games of construction, water games, stabilisations, motors, and many other systems that we have not mentioned, the subject is so vast. Form appears at every scale of the structuration of matter, and at diverse levels of organisation of living matter: from the scale of molecules to the scale of populations, like schools of fish or flocks of birds in flight, ecosystems themselves, but there are also intermediate levels, notably that of intracellular organelles, cells, tissues and organs, and the individual.

Among these levels of organisation, two occupy a particularly important place in the thinking of D'Arcy Thompson: cells and individuals, because they are living units in their own right, possessing autonomy, with their capacity for growth and reproduction, notably in in vitro culture in the case of cells [39]. The two levels differ in multicellular organisms, but they coincide in unicellular beings, such as bacteria, certain algae, protozoa, etc.

It would take a treatise of several volumes to account for all the mechanisms of synthesis of molecules, from the smallest to the largest, organic or not. The same is true for other mechanisms that produce forms at higher scales, and we can be grateful that D'Arcy Thompson gave a measured presentation of them, with suitable comparisons between physical and biological examples. No phrase could express this point of view better than that written by Feynman, in his physics lectures [19]:

> Whatever life can do, atoms can do it too.

Another merit of D'Arcy Thompson is to have also shown us that beyond the countless particular mechanisms, there is still room to introduce more general ideas, as shown by his theory of transformations. But the question of the general principles of morphogenesis looms even larger now, to understand the synergy and the programming of these mechanisms in the development of the most complex living systems. Today, three approach paths are proposed: *genetics*, *symmetry breaking* and new ideas about the *viability* of systems.

14.6.1 The Direct or Indirect Role of the Genome in Morphogenesis

The immense repertoire of biosyntheses has become accessible down to its finest details, thanks to the progress in *genetics* and the new tools developed in that field. This has led many researchers to believe that work on this sole molecular level should suffice to explain all the processes of development, at all scales of biological organisation. But biochemical and genetic techniques will not be sufficient, although they remain essential at the scale of the individual, or even at those vaster scales involved in the collective behaviour of species or the patient construction of ecosystems.

The *direct* role of the genome is clearly apparent in the synthesis of proteins, which has been clarified and analysed with a high degree of precision. The transcription followed by the translation of genes produces, in particular, enzymes, the catalysts of metabolic reactions and notably of biosyntheses. But the role of genes is only *indirect* in most metabolic reactions, as they can be reproduced in vitro, in the presence of enzymes and in the absence of nucleic acids [1, 17, 21].

Other genes are *directly* at the origin of proteins capable of self-assembling into supramolecular structures, like the constituents of the cytoskeleton, but their role remains *indirect*, because we can reproduce these assemblies in vitro, in the absence of all the other cellular machinery, and notably in the absence of nucleic acids. Only the physico-chemical properties of the proteins that come into play in these constructions.

But the indirect role of genes remains, because mutations at their level can modify these proteins to the point where they can no longer assemble. In the normal situation, the genes allow the fabrication of subunits of assembly with the appropriate form and chemistry, but sometimes they require help from chaperones or enzymes (Sections 14.1.1 and 14.1.2).

Other biological assemblies resemble the mechanisms of formation of liquid crystalline phases, notably for cell membranes, and similar mechanisms are involved with non-biological amphiphilic bodies when they form fluid bilayers. For cell membranes, the form partly depends on the proteins they contain, and this allows for their control by the genome. But we must also cite the example of red blood cells. In humans, they are small cells, 7 μm in diameter, usually with a biconcave form, and reduced to a membrane, a cytoskeleton and a lot of haemoglobin; they have lost their nuclei and mitochondria, as if to rid themselves of nucleic acids. This biconcave form is maintained by physical mechanisms, involving the cytoskeleton and the liquid crystal characteristics of the cell membrane. Before the loss of the nucleus, the mitochondria and other organelles, these cells play a part in the synthesis of a large proportion of the molecules involved, but the biconcave form of red blood cells only depends very indirectly on genes, because the same biconcave form can be observed in bilayer vesicles, prepared in vitro from non-biological amphiphilic compounds [29].

At higher levels of organisation, D'Arcy Thompson underlined, in his own way, the direct and indirect roles played by genes and other very different factors of physical chemistry, a distinction that he already felt to be too lacking in the minds of his colleagues:

> It would, I dare say, be a gross exaggeration to see in every bone nothing more than a resultant of immediate and direct physical or mechanical conditions; for to do so would be to deny the existence, in this connection, of a principle of heredity ... But I maintain that it is no less an exaggeration if we tend to neglect these direct physical and mechanical modes

of causation altogether, and to see in the characters of a bone merely the results of variation and of heredity ...

Genetics has enabled us to identify numerous genes that intervene in the first stages of development, and at the present time, this is the domain of morphogenesis the most studied and in which the most advances are being made. However, these genes are only indirectly involved in morphogenesis, and this means, as we pointed out earlier, that we have yet to explore the interactions between the products of these genes – that is to say the proteins or other direct consequences of their expression – and the morphogenetic processes situated further downstream in the realisation of the phenotype. In the past, this was the subject of a separate discipline, *physiological genetics*, now hardly ever mentioned. In other words, it remains for us to study how the functioning of genes or their direct products (proteins, RNA etc.) fit into the physical chemistry – in the strict sense of the term – of morphogenesis. Genetics and physical chemistry provide two approaches to morphogenesis, which often remain separate when they should be combined. We have recently written about the example of bio-mineralisations, where this combined approach is being practised [10]. The nucleation and growth of mineral crystals in vitro gives rise to countless original forms, especially in the presence of natural or synthetic polymers, and without the least use of cellular machinery. By comparison with in vivo mineralisations, we can distinguish between purely physical processes and those in which the cellular machinery intervenes. This shows us that life knows how to take advantage of what the physical world can do, thus saving a substantial amount of work for the cells, which intervene more subtly by controlling the crystalline assemblies, their defects and their textures, and by making certain adjustments.

14.6.2 Symmetry Breaking and Differentiation

When the fertilised egg divides, it produces two identical cells, then four and so on ... but differences appear, sometimes very early on in the process, after the first stages of segmentation, even if it is only in the very unequal sizes of the daughter cells. These differentiations intervene at various stages in development, and populations of identical cells, with the same mother cells, cease to resemble each other after a few more divisions. It is as if a figure was made of two symmetrical parties and one of them has just changed, breaking the symmetry. The same is true of the cells of an embryo, when a specialisation into a particular tissue occurs in one part of the embryo but not in the rest. A population of cells that was *homogeneous* becomes *heterogeneous*. These cell differentiations are programmed in the development and they also determine perceptible modifications in the form.

Symmetry breaking also exists in non-biological systems, where they have mainly been studied. Many of these purely physical or chemical systems produce forms, as we have seen, and each morphogenesis may be accompanied by symmetry breaking, because every new aspect of the form must change something in what already existed. Thus, the problem of morphogenesis can be expressed as follows:

How can heterogeneity be brought into being in a homogeneous system? It would be difficult to think of more general terms for approaching morphogenesis, but some interesting solutions have been found.

Physicists, of course, proposed phase transitions, which break certain symmetries, such as they have observed them in the nucleation and growth of crystals and in liquid crystals. Chemists owe their response to the mathematician Turing, as we explained above, with the reaction-diffusion principle. This principle is very original, because diffusion is known above all for equalising concentrations of substances in solution. The possibility of obtaining localised zones with a higher concentration of one of the reagents therefore provoked incredulity, but chemists have found numerous demonstrative examples [26]. The response of biologists has often been that they do not face the problem in the same terms, because every cell comes from another cell and that the very material of cells is very finely heterogeneous, at every stage of development from the egg. In fact, there is no reason why a phase transition or reaction-diffusion, which breaks certain symmetries, should not also occur in an already heterogeneous medium, and this could be the case for biological morphogenesis. But the recognition of a reaction-diffusion process should also allow to identify the activator and inhibitor compounds, i.e. the morphogens, involved. Up until now, results concerning these morphogens have been almost non-existent or at best uncertain. Phase transitions that break symmetries are abundant [36, 37], notably with the various assemblies mentioned above.

Cells themselves take part in assemblies by forming tissues, a little like macromolecules when they make up diverse supramolecular structures, but in this case governed by rules that are more subtle. As we have already indicated, assembled cells often present polyhedral contours, reminiscent of soap bubbles in foam. This leads to precise geometries during the first stages in the segmentation of the egg. Subsequently, many cells join together to form epithelia, in other words lamina composed of one or more layers of cells, producing polygonal contours in the plane of the layers, either because the cells are very flattened, or because they are prismatic while being stretched perpendicular to the plane direction of the epithelium. The forms of these epithelia obey various geometric laws and partly depend on the frequencies of the number of sides of these polygonal cells, according to whether it is more or less than 6. Considerations about the symmetries and polarity of the cells are also involved, as well as their dimensions, their forms, the dynamics of cell division and the selective distribution of the orientation of mitotic spindles. The control of genes plays a role, notably by membrane proteins and especially those of the belt-like cell junctions and those of the basolateral and apical membranes (see Chap. 4).

It should also be remembered that each breaking of symmetry entails an *instability*, although the reverse is not true, since instabilities also appear in transitions without any change in symmetry, such as the transition from isotropic liquid to vapour or the reverse. Instabilities also appear during most chemical reactions. Symmetry breaking has been studied in detail in physics, but works on the subject are scarcer in biology, and I have attempted to discuss the essential aspects on several occasions [5, 6]. One of the unresolved problems is the following. We know that embryogenesis and subsequent development are highly programmed, and that regulation can take place in the event of an incident. This programming appears systematically to avoid the instabilities connected to symmetry breaking, whereas this is not the case over the course of evolution, which is blind and unprogrammed.

In other words, how can *ontogeny* avoid these instabilities, which are, on the contrary, essential to *phylogeny*, in generating the diversification of species and accelerating their separation? We can provide some idea of the answer with two examples from ontogeny. The activation energy required to carry out a biochemical reaction is considerably reduced when one of the reagents, the substrate, binds with an enzyme, increasing the probability and therefore the speed of the reaction. The energy of the substrate-enzyme system has a maximum level which, although reduced, nevertheless still exists. At the peak of this barrier of potential, the system is in unstable equilibrium, and depending on the fluctuations, it can just as easily return to its initial state as it can perform the reaction. But in living systems, this sort of hesitation is statistically swept aside for most reactions, because they take place far from equilibrium, in a direction imposed by the metabolism, as the use of a substrate requires its almost immediate replacement. Our second example is that of biomineralisations [10]. During the nucleation of a crystal, its orientation may be a matter of chance, but this is not so in biology, because the extracellular matrix already exists, constituting an ordered system and imposing constraints, related to phenomena of epitaxy or of another nature, such as we have particularly studied in crustacean shells. We could give many more examples, but in any case this *avoidance of instability* appears to be general in morphogenesis.

Among the first theoretical approaches to embryogenesis, we should cite that of Waddington [40], who compared the successive differentiations in embryo cells to the adventures of marbles travelling over a landscape of ridges and valleys. When a marble comes to a bifurcation in the valley it is rolling down, it continues down one of the branches without really making a choice, and so on. The contour lines of this landscape form saddle-type singularities at each bifurcation in the valley, and the descent can only be programmed by introducing a mechanism that allows to decide whether the marble takes the left- or right-hand valley, in other words a sort of switch system.

These conceptions were taken up by René Thom, who discussed them with Waddington (see Chap. 16) and added gradual changes in the landscape, so that a marble trapped in a basin can leave it if one of the sides disappears [37]. The marble then continues down one of the available valleys until it reaches another basin of attraction, more or less distant. These metaphors about the landscape of potential are well-known and can aid intuition in the search for models, but two problems remain: firstly that of activating the switches, to remove hesitation in the choice of the path to follow, and secondly that of programming these switches in a coordinated manner. To activate them, we can consider the mechanisms for avoiding instability and in particular the two examples described above, metabolic flow and pre-formed matrices. They could operate within a genetic context, and it would be useful to draw up a list of examples, so as to examine their general properties. Programming such a series of differentiating switches would also involve the genome. This can be imagined with the mechanisms of gene expression and repression, which could function in a similar way to certain networks of *Boolean automata* [38], and this is currently a very active domain of research.

14.6.3 New Prospects in Morphogenesis and the Concept of Viability

The general principles and particular mechanisms considered by D'Arcy Thompson and his successors are not sufficient to account for the functioning of a system as complex as a living being. We need to know all the interactions and the permissible latitudes beyond which the functioning is endangered. Viability is one of the thorny questions of biology. One sole mutation may be enough to stop a bacterium from being viable. The quantity of information in its genome has hardly changed, but its survival is jeopardised [30]. On the other hand, many similar changes in other parts of the genome will have no such dreadful effect.

Up until now, our approach has consisted in dissecting a global process of morphogenesis, as if to detach various elementary processes and then reproduce them in vitro. We do not know of many experimental studies in which these elementary mechanisms are made to coexist. This is a pity, because life is complicated right down to its most microscopic levels. The synergy of mechanisms is an essential aspect of life, but the example of the bacterium corresponds to a limit situation, where nothing can compensate for some of its mutations.

The concept of *viability kernel* designates the set of conditions that must be satisfied to ensure the existence of at least one viable trajectory. Outside this kernel, there is no salvation: sooner or later the system will fail (see Chap. 17). The study of these kernels should provide us with detailed information on the ingredients of survival, and in particular on the structures of synergy, a subject that is present throughout physiology. D'Arcy Thompson considered this problem implicitly, in discussing evolution, when it has entailed substantial changes in body plan. This concerns the appearance of the main zoological groups, from protozoa to invertebrates, with species as original as sea urchins, or again the origin of quadrupeds or birds. The palaeontological links that would allow such transitions are not only absent, but difficult to imagine. On this subject, D'Arcy Thompson evoked, on the last page of his book, alternatives of *physicomathematical possibility* that might have occurred repeatedly, to enable such changes. This is another way of defining viability kernels in extreme cases, and of conceiving that new functionalities can appear in complex systems, even though their evolution is strictly blind. The complexity diminishes, from the moment that we see an appropriate solution to a problem of survival, but up until now this has always been an understanding that we acquire *a posteriori*. Unlike evolution, the development and life of the individual are programmed, leaving a certain degree of freedom, but with adjustment mechanisms ready to intervene in the event of deviation from the path to be followed. The viability kernel of each individual lasts at least until the passing of their age of reproduction. If we succeed in exploring it, the structure of these kernels will be most revealing about the internal consistency of morphogenetic mechanisms in living beings.

References

1. Alberts B., Bray D., Lewis J., Raff M., Roberts K., and Watson J. D. (1998) *Molecular biology of the cell*, 2nd edition, Garland Publishing (New York).
2. Bloch K. (1952) Biological synthesis of cholesterol, *Harvey Lect.* **47**, 68.
3. Bookstein F.L. (1977) The study of shape transformations after d'Arcy Thompson, *Math. Biosci.* **34**, 177–219.
4. Bookstein F.L. (1996) Biometrics, biomathematics and the morphometric synthesis, *Bull. Math. Biol.* **58**, 313–365.
5. Bouligand Y. (1981) Symétries et brisures de symétrie en biologie, in *Symmetries and Broken Symmetries in Condensed Matter Physics*, edited by N. Boccara, IDSET (Paris), pp. 131–140, in French.
6. Bouligand Y. (1985) Brisures de symétrie et morphogenèse biologique, La Vie des Sciences, *Comptes Rendus Acad. Sci. Paris* **2**, 121–140, in French.
7. Bouligand Y. (1994) Champs de directeurs en morphogenèse biologique, in *Organisation et Processus dans les Sytèmes Biologiques*, edited by F. Gros and J. Friedel, Académie des Sciences, Actes de Colloques, pp. 199–201.
8. Bouligand Y. (1996) D'Arcy Thompson, in *Dictionnaire du Darwinisme et de l'Évolution*, edited by P. Tort, P.U.F. (Paris), pp. 4271–4278, in French.
9. Bouligand Y. (2004) D'Arcy Thompson et la logique des formes, special issue "Les Formes de la Vie" , *Pour la Science*, pp. 4–8, in French.
10. Bouligand Y. (2004) The renewal of ideas about biomineralisations. *Comptes Rendus Acad. Sci. Paris, Palevol* **3**, 617–628.
11. Bouligand Y. and Lepescheux L. (1998) La théorie des transformations. Les travaux de D'Arcy Thompson continuent d'irriguer la recherche, *La Recherche* **305**, 31–33, in French.
12. Bukau B. (ed.) (1999) *Molecular Chaperones and Folding Catalysts*, Harwood Academic (Amsterdam).
13. Carlier M.-F. and Pantaloni D. (1988) Binding of phosphate to F-ADP-actin and role of F-ADP-Pi-actin in ATP-actin polymerization, *J. Biol. Chem.* **263**, 817–825.
14. Carlier M.-F., Melki R., Pantaloni D., Hill T.L., and Chen Y. (1987) Synchronous oscillations in microtubule polymerization, *Proc. Natl. Acad. Sci. USA* **84**, 5257–5261.
15. Chapeville F. and Haenni A.-L.(1974) *Biosynthèse des Protéines, Traduction Génétique*, Hermann (Paris), in French.
16. Erickson H.P. and O'Brien E.T. (1992) Microtubule dynamic instability and GTP hydrolysis, *Annu. Rev. Biophys. Biomol. Struct.* **21**, 145–166.
17. Favard P. and Bouligand Y. (1981) La phénoménologie des autoassemblages biologiques, in *La Morphogenèse, de la Biologie aux Mathématiques*, edited by Y. Bouligand, Maloine (Paris), pp. 101–113, in French.
18. Fawcett D.W. (1981) *The Cell*, 2nd edition, Saunders (Philadelphia).
19. Feynman R.P. (1963) *The Feynman lecture on physics*, vol. 1 (Mainly mechanics, radiation, and heat), Addison-Wesley (Reading MA).
20. Gould S. J. (1971) D'Arcy Thompson and the science of form, *New Literary History* **2**, 229–258.
21. Gros F. and Grünberg-Manago M. (1974) *Biosynthèse des Acides Nucléiques. Réplication et Transcription*, Hermann (Paris), in French.
22. Haeckel E. (1999) Crystal Souls, translated from German by A.L. Mac Kay, *Forma* **14**, 1–146.
23. Houssay F. (1912) *Forme, Puissance et Stabilité des Poissons*, Hermann (Paris), in French.
24. Kagan H. (1985) Chiralité en chimie, La Vie des Sciences, *Comptes Rendus Acad. Sci. Paris* **2**, 141–156, in French.

25. Kellenberger E. (1980) Control mechanisms governing protein-protein interactions in assemblies, *Endeavour*, New Series **4**, 2–13.
26. Kepper P. de, Dulos E., Wit A. de, Dewel G., and Borckmans P. (1998) Taches, rayures et labyrinthes, *La Recherche* **305**, 84–89, in French.
27. Laüger P. (1977) Ion transport and rotation of bacterial flagella, *Nature* **268**, 360–362.
28. Lehmann O. (1908) *Flüssige Kristalle und die Theorien des Lebens*, J. A. Barth (Leipzig), in German.
29. Lipowsky R. (1991) The conformation of membranes, *Nature* **349**, 475–481.
30. Lwoff A. (1968) *L'Ordre Biologique*, Marabout Université (Paris), in French.
31. Meinhardt, H. (1982) *Models of biological pattern formation*, Academic Press (New York).
32. Miller S.L. and Orgel L.E. (1974) *The Origins of Life on Earth*, Prentice Hall (Englewood Cliffs NJ).
33. Needham J. (1936) *Order and Life*, Cambridge University Press (Cambridge).
34. Sambongi Y., Iko Y., Tanabe M., Omote H., Iwamoto-Kihara A., Wada Y., and Futai M. (1999) Mechanical rotation of gamma c subunit oligomer in ATPsynthase(F0F1): direct observation, *Science* **286**, 1722–1724.
35. Stevens P. (1974) *Patterns in Nature*, Penguin Books (New York).
36. Thom R. (1972) *Stabilité Structurelle et Morphogenèse. Essai d'une Théorie Générale des Modèles*, Benjamin (Reading MA), in French. English translation: *Structural stability and morphogenesis* (1975), Benjamin (Reading MA).
37. Thom R. (1980) *Modèles Mathématiques de la Morphogenèse*, Christian Bourgois (Paris), in French. English translation: *Mathematical models of morphogenesis* (1983), John Wiley and sons (Chichester).
38. Thomas R. (1994) Boucles de rétroaction, multistationnarité et différenciation, dans *Organisation et Processus dans les Systèmes Biologiques*, edited by J. Friedel and F. Gros, Académie des Sciences, Actes de Colloques, pp. 155–159, in French.
39. Thompson, D'Arcy W. (1917) *On Growth and Form,* Cambridge University Press (Cambdrige), 2nd edition in 1942, reprinted in 1952, 1959, 1963, 1968, 1972 ...
40. Waddington C.H. (1932) Experiments on the development of chick and duck embryos, cultivated in vitro, *Phil. Trans. Roy. Soc.* **179 B**, 221.
41. Witkowski N. (1998) D'Arcy Thompson, fantôme de la biologie, des outils mathématiques et physiques pour expliquer les formes du vivant, *La Recherche* **305**, 27–30, in French.

Chapter 15
The Morphogenetic Models of René Thom

Jean Petitot

At the end of the 1960s, René Thom was the first scientist to develop a general mathematical theory of morphogenetic processes. This chapter presents the fundamental principles of that theory.

15.1 General Content of the Model

Let S be a system satisfying the following hypotheses:

a. There exists an *internal process* (usually unobservable) X which defines the internal states that system S can occupy in a stable manner, and the number of these states is finite;
b. the internal states of S are in competition with each other and mutually determine each other, and the choice of one of them as the actual state makes the others virtual;
c. there therefore exists an *instance of selection* I which, on the basis of criteria specific to the system, selects the actual state from among the possible internal states;
d. the system S is controlled by a certain number of control parameters varying within a space W which we call, to distinguish it from the internal process X, the *external space* (or *control space* or *substrate space*) of S.

We also assume that the control is *continuous*, in the sense that the internal process X is a process X_w which depends continuously on the value w of the control. This process varies when w varies in W and when it is deformed it also deforms the structure of the internal states and their relations of mutual determination. We denote $\mathcal{X}$ the space of possible internal processes X. If the above hypotheses are verified, the

J. Petitot (✉)
Ecole des Hautes Études en Sciences Sociales, CREA (École Polytechnique), Paris, France
e-mail: jean.petitot@polytechnique.edu

P. Bourgine, A. Lesne (eds.), *Morphogenesis*, DOI 10.1007/978-3-642-13174-5_15,

system S will be described firstly by the (continuous) field $\sigma : W \to \mathcal{X}$ associating $w \in W$ with the process X_w and then by the instance of selection I.

The standard example is that of the thermodynamic phenomena of phase transitions. In this case, the system S is the thermodynamic system considered, the internal states are the thermodynamic phases (solid, liquid, gas), the instance of selection I is provided by the principle of free energy minimisation and the control parameters are, for example, pressure and temperature. As for the internal process X_w, indescribable because of its complexity, it is the process of molecular dynamics. When the control parameters cross certain specific values, known as critical values,[1] they present *phase transitions*, i.e. discontinuities in their observable qualities and abrupt transformations of their internal state. The critical values constitute a subset K of W, determining the *phase diagram*: K partitions W into domains corresponding to the different phases that S can present. In other words, it categorises it and externalises therein the competition between internal states, in the form of a system of discontinuities.

This is a direct consequence of hypotheses (a)–(d). A system $S = (W, \mathcal{X}, \sigma, I)$ manifests itself phenomenologically by the observable qualities $q^1, \ldots, q^n$ expressing its internal state. In other words, the internal process X_w is externalized in perceptible qualities q^i_w. When the control w varies continuously, the actual internal state varies continuously (hypothesis (d)) and the qualities q^i_w therefore vary equally. But phenomenologically speaking, a continuous variation is no more than a form of qualitative invariance. It is therefore not significant. So René Thom denoted by *regular point* of W a value w of the control such that the observable qualities q^i_w vary continuously – and therefore remain stable – throughout a neighbourhood U of w (this obviously presupposes that we have defined the concept of neighbourhood on W, i.e. a topology). By definition, the regular points constitute an open set R_W of W, the open set of quality stability. Then let K_W be the closed set complementary to R_W in W. By definition, the points of K_W are the values w of the control such that at least one observable quality q^i_w suffers a discontinuity. These are critical values, crossing which the system S presents critical behaviour. They are also called *catastrophic values*, the closed set K_W being called the *catastrophic set* of S. The K_W are also called *external morphologies*.

As René Thom often stressed, this concept of morphology is purely phenomenological. But it is closely connected to the mathematical concept of *bifurcation*. Let us suppose that the control w follows a path γ in W. Let A_w be the actual internal state initially selected by I. During the deformation of X_w along γ – and therefore, under hypothesis (d), of the structure of A_w and under hypothesis (b), of the relations of mutual determination it has with the virtual states B_w, C_w, etc. – when A_w crosses a critical value, it no longer satisfies the criteria of selection imposed by I under

[1] Here, the term *critical value* is related to bifurcation theory (and, in the rest of this chapter, to catastrophe theory), and not to the language of thermodynamics. As a general rule, these critical values do not correspond to a critical point in the thermodynamic sense (a particular point where the distinction between the different phases disappears and the phase transition becomes continuous, the singularity manifesting itself in thermodynamic derivatives, of free energy, for example).

hypothesis (c). The system therefore spontaneously bifurcates from A_w towards another actual (but hitherto virtual) state B_w. This catastrophic transition of internal state manifests itself by a discontinuity in some of the observable qualities q_w^i. In other words, it is the destabilisation (relative to the instance I) of actual internal states under the variation of the control which induces, in the external space W, a set of qualitative discontinuities K_W. In the right cases, the set K_W will constitute a system of interfaces, analogous to a phase diagram, partitioning the external space W into domains, each of which corresponds to a zone in W where one of the internal states is dominant.

15.2 Morphodynamics and Structural Stability

Thomian morphodynamics is based on the possibility of specifying the general model in mathematical terms (see [2] and [3]). The first specification consists in assuming that, with regard to their nature, the internal processes X_w constitute a space $\mathcal{X}$ equipped with a natural *topology* $\mathcal{T}$ significant for the type of process studied. This means that we can tell when two internal processes X and Y are neighbours and we can therefore rigorously define the continuity of the field $\sigma : W \to \mathcal{X}$. By moving in $\mathcal{X}$ we can therefore deform its elements X.

We then assume that we can define the *qualitative type* of the processes X. The qualitative type is a relation of equivalence (generally defined by the action of a group G on $\mathcal{X}$), which is a weak, qualitative identity. Let $\tilde{X}$ be the class of equivalence of X for the qualitative type (i.e. the orbit of X under the action of G). We seek to characterise that which remains invariant when X varies in $\tilde{X}$ (i.e. varies for a constant qualitative type) by means of discrete information, for example the values of a finite number of invariants. At the level of the invariants, the variation in a class of equivalence $\tilde{X}$ is reduced to the identity. Therefore, there is only qualitative variation when a deformation in $\mathcal{X}$ causes a change in the class of equivalence. The variation is manifested by a discontinuity in the value of certain invariants and we find the "right" situation of the general model.

Compared to a standard approach, which would consist in studying, for each physical system of type $S = (W, \mathcal{X}, \sigma, I)$, the internal processes X_w as isolated entities, morphodynamics introduces a double shift in perspective. Firstly, it takes as its object of study not only the processes X_w but also the parameterized families $(X_w)_{w \in W}$, by focusing on the geometry of the catastrophic sets K_W induced in the external spaces W by the destabilisation of the internal states defined by X_w. Secondly, and above all, it considers these families as the image of fields $\sigma : W \to \mathcal{X}$ sending the external space W (which is generally a part of the standard space $\mathbf{R}^n$ with n dimensions) into the generalised space $\mathcal{X}$.

Now, from the moment that we possess, for a space $\mathcal{X}$, a topology $\mathcal{T}$ and a relation of equivalence defining the qualitative type, we can define a concept of structural stability. Let $X \in \mathcal{X}$. We say that X is *structurally stable* if all Y close enough to X (in the sense of $\mathcal{T}$) are equivalent to X. The process X is therefore structurally stable if its qualitative type resists small perturbations, or if the class $\tilde{X}$ is open (in the sense of $\mathcal{T}$) in $\mathcal{X}$. Let $K_{\mathcal{X}}$ then be the closed subset of $\mathcal{X}$ composed of the structurally unstable $X \in \mathcal{X}$. $K_{\mathcal{X}}$ is an intrinsic catastrophic set, canonically associated with $\mathcal{X}$. It is categorised by a discriminating morphology, which classifies the qualitative types of its structurally stable elements.

Let $\sigma : W \to \mathcal{X}$ be the characteristic field of a system $S = (W, \mathcal{X}, \sigma, I)$. Let $K'_W = \sigma^{-1}(K_{\mathcal{X}} \cap \sigma(W))$ be the trace of $K_{\mathcal{X}}$ on W through the intermediary of σ. The hypothesis of morphodynamic modelling is that the catastrophic set K_W of S can be deduced from K'_W on the basis of the instance of selection I. This means that a value w of the control belongs to K_W (i.e. is a critical value) if and only if the situation in w is correlated in the manner regulated by I with a situation belonging to K'_W. The external morphology is essentially the apparent outline on the substrate space of the internal dynamics. It is therefore the analysis (at the same time local and global) of the intrinsic catastrophic sets $K_{\mathcal{X}}$ that lies at the heart of this theory.

If we introduce the additional hypothesis that a field σ can only concretely exist if it is itself structurally stable, we see that such a constraint drastically limits the complexity that K'_W can present. In the right cases, we can even obtain a classification of the local structures of K'_W and therefore of the local external morphologies. The theory thus brings to light purely mathematical constraints acting on the domain of morphogenetic phenomena.

15.3 The Theory of Dynamical Systems

The main mathematical specification of the general model consists in postulating that the internal process X is a *differentiable dynamical system* on a differentiable manifold M of internal parameters characteristic of the system S considered. We call the space M the internal space (to distinguish it from the external space W). A dynamical system X on M consists in associating with each point x of M a tangent vector $X(x)$ of M at x, a vector varying differentiably with x. X is therefore a vector field differentiable on M, in other words, in terms of local coordinates $x_1, \ldots, x_n$, a system of ordinary differential equations:

$$\frac{\mathrm{d}x_i}{\mathrm{d}t} = f_i(x_1, \ldots, x_n)$$

where the f_i (components of the field) are differentiable functions of x_j and where t is the time parameter. Given such a field, integrating it consists in finding, in M, differentiable curves parameterised by time (i.e. differentiable applications $\gamma : \mathbf{R} \to M$, $t \mapsto \gamma(t) = (x_1(t), \ldots, x_n(t))$ which admit at each point for velocity vector $\mathrm{d}x/\mathrm{d}t = \mathrm{d}\gamma(t)/\mathrm{d}t$ the field vector $X(x) = X(\gamma(t))$. We say that the field is a dynamical system if we can integrate the trajectories over an infinite time (i.e. if the trajectories do not leave M); if one and only one trajectory passes through each point (the principle of determinism: the initial condition $x(0)$ at time $t = 0$ univocally determines the future evolution $x(t)$ for $t > 0$ and the past evolution $x(t)$ for $t < 0$); and if the trajectories vary differentiably according to the initial conditions.

Then let $f_t : M \to M$ be the application which assigns to every point x of M the point at time t of the trajectory emanating from x at time $t = 0$. It is easy to see that f_t is a diffeomorphism of M (an automorphism of its differentiable manifold structure) and that the application $t \mapsto f_t$ of the additive group $\mathbf{R}$ in the group of diffeomorphisms of M is a morphism of the groups. f is called the *flow* of the dynamical system X. It is the integral version of the vector field X.

René Thom suggested that the models $S = (W, \mathcal{X}, \sigma, I)$ where the internal processes X_w are dynamical systems should be called *metabolic models*. Their internal states need to be defined. The basic idea is to introduce a difference between fast and slow dynamics, in other words between two timescales, one internal and fast, the other external and slow. In the internal space, the fast internal dynamics creates attractors that specify the local phenomenological quality of the substrate. The slow external dynamics operates in the substrate space W. As we assume that the internal dynamics of the evolution of instantaneous states is infinitely fast compared to the external dynamics of evolution in the external spaces W (a condition known as *adiabaticity*), the only significant states are the asymptotic states (for $t \to +\infty$) defined by the X_w, i.e. the limit regimes.

Now, the analysis of these asymptotic states has proved to be unexpectedly and formidably difficult. The complexity of a general dynamical system is prodigious. Firstly, the ideal determinism – which is mathematical – does not in any way entail determinism in the physical sense of the term (in the sense of "predictability"). An initial condition can only be defined approximately. It is not represented by a point x_0 in M but by a small domain U that "thickens" x_0. For determinism to be physical, the trajectories emanating from U must form a tube that "thickens" the trajectory γ emanating from x_0. This means that the trajectory γ is stable with respect to small perturbations of its initial condition. A physically deterministic dynamical system is therefore a dynamical system (by definition ideally deterministic) which has stable trajectories.

There is no reason why this should generally be the case. There are even dynamical systems (for example geodesic systems in Riemann manifolds with negative curvature) presenting the property that all their trajectories are unstable, and presenting it in a structurally stable way. As Vladimir Arnold observed ([1], p. 314–315, our translation):

> The possibility of structurally stable systems with complicated movements, each of which is in itself exponentially unstable, is one of the most important discoveries to be made in differential equation theory in recent years. [...] In the past, it was assumed that systems of generic differential equations could only contain simple, stable limit regimes: positions of equilibrium and cycles. If the system was more complicated (conservative, for example), it was accepted that under the effect of a weak modification in the equations (for example by taking into account small, non-conservative perturbations) the complicated movements "break down" into simple movements. Now we know that this is not the case, and that in the functional space of vector fields, there exist domains composed of fields where the phase curves [the trajectories] are more complex. The conclusions to be drawn from this affect a large number of phenomena in which deterministic objects have "stochastic" behaviour.

Physical indeterminism (chaos, chance, randomness, etc.) is therefore perfectly compatible with mathematical determinism. As Thom pointed out ([4], p. 124):

> What we call "laws of chance" are in fact no more than properties of the most general deterministic system.

Let us return to the specification of the general model. In terms of dynamical systems, the internal states of S are the attractors of X_w. The very tricky concept of *attractor* generalises the concept of stable equilibrium point. Intuitively, it is a stable asymptotic regime, a closed set A, X - invariant and indecomposable for these two properties (i.e. minimal), which attracts (i.e. captures asymptotically) all the trajectories emanating from the points of one of its neighbourhoods. The largest neighbourhood of A having this property is called the *basin of attraction* of A, denoted $B(A)$. In simple cases, the attractors have a simple topological structure (attractive point or attractive cycle), they are finite in number and their basins of attraction are "good" domains (of simple form) separated by separatrices. But this description is too optimistic, because the attractors may be infinite in number, their basins of attraction may be inextricable intermingled, and the attractors may have a very complicated topology (strange attractors).

On an attractor, the trajectories of a dynamical system present *recurrence*. Intuitively, the recurrence of a trajectory γ means that if $x \in \gamma$, then γ passes arbitrarily close to x again after an arbitrarily long time and so γ returns infinitely often close to its initial position. The trivial cases of recurrence are the fixed points of X (the points where X equals zero, i.e. the trajectories reduced to a point) and the cycles of X (closed trajectories). But there generally exists non-trivial recurrence. If γ is a complicated recurrent trajectory and A is its topological closure, then A is a whole domain of M (a non-empty closed set) where γ is dense.

Whatever we may make of these difficulties, Thom assumed in his morphodynamic models that for almost all initial conditions $x_0 \in M$ ("almost all" and not all because we must take into account the separatrices between basins), the trajectory emanating from x_0 is captured asymptotically by an attractor A_w of the internal dynamics X_w. This corresponds to a hypothesis of local equilibrium: the fast internal dynamics drives the system towards a stable asymptotic regime corresponding to an internal state.

Once these various hypotheses have been established, the general model becomes a mathematical programme: general structure of dynamical systems (qualitative dynamics or global analysis); geometric characterisation of structurally stable dynamical systems and their attractors; analysis of the ergodic properties of strange attractors; analysis of possible causes of instability; analysis of the deformations (perturbations) of structurally unstable systems; study of the geometry (which can be extremely complex) of catastrophic sets $K_{\mathcal{X}}$; etc. This programme, which we could call the Thom-Smale programme, is an extension of that of Poincaré and Birkhoff. It is in fact the programme of modern qualitative dynamics.

But if the Thom-Smale programme is of immense scope, it is also of immense difficulty. That is why Thom proposed to simplify it.

15.4 The Theory of Singularities and "Elementary" Morphogenetic Models

As the complexity of general dynamical systems is too great to master, we can start by carrying out a rough study, of a thermodynamic nature. This consists in ignoring the fine structure (the complicated topology) of attractors. This step is all the more necessary since the empirical catastrophic sets K_W are usually much simpler than those induced by the bifurcations of general dynamical systems. The aim is therefore to understand how systems can be internally chaotic (stochasticity of the attractors defining internal states) and externally ordered (simplicity of observable morphologies).

The idea is to apply to general systems that which can be observed to occur in the case of gradient systems, namely the existence of gradient lines and level manifolds. To do so, we use the fact that, if A is an attractor of a dynamical system X on a manifold M, we can build, on the basin $B(A)$ of A, a positive function f (called a *Lyapunov function*) which decreases strictly along the trajectories in $B(A) - A$ and vanishes on A. This function is a sort of local entropy, expressing the fact that over time, $B(A)$ contracts on A analogously to a gradient system. But it does not allow us to say anything about the internal structure of the attractor.

The next step is to retain, out of all the bifurcations of the attractors, only those that are associated with their Lyapunov function. This reduction is similar to thermodynamic averaging. It corresponds to a change in the level of observation, from the fine level described by X_w to the rougher level described by f_w. It is analogous to the reduction that is performed in Landau's mean-field theory of phase transitions. Thom gives the following justification for it ([5], p. 521, our translation):

> Personally, I like to think that what plays a role, is not the concept – too fine – of attractor, but a class of equivalence of attractors, equivalent because encapsulated in the level manifold of a Lyapunov function (a quasi-potential), provided that the attractor avoids implosions of an exceptional nature. I believe that this may be the path to follow to find a satisfactory mathematical definition of the concept of stationary asymptotic regime of a dynamics. From this perspective, the fine internal structure of the attractor is of little importance: the only thing that matters is the Lyapunov function that encloses it in one of its level manifolds. But we can consider that only the structure of the tube enclosing the attractor is phenomenologically important, and we thus obtain a problem that is similar to elementary catastrophe theory.

"Elementary" morphodynamic models consist in reducing quasi-potentials – the Lyapunov functions – to the gradient systems derived from potentials. We assume that the internal dynamics X_w is in fact the gradient dynamics associated with a differentiable potential function $f_w\colon M \to \mathbf{R}$. The internal states determined by f_w are then its minima (if f is equated with an energy, this principle is that of the energy minimisation of the system). In Thomian terminology, this sort of system is called a *static model*. Mathematically, the theory of static models is therefore an integral part of the bifurcation theory of potential functions. Now, for potentials, there exists a simple characterisation of structural stability (*Morse theorem*). Under the hypothesis that the manifold M is compact, $f : M \to \mathbf{R}$ is stable if and only if:

i. its critical points (i.e. its minima, maxima and saddle points) are non-degenerate, in other words they are not fusions of several minima, maxima or saddle points; and
ii. its critical values (i.e. the values $f(x)$ for critical x) are distinct.

There are therefore two causes of structural instability:

i. the degeneracy of critical points, corresponding to what are called *bifurcation* catastrophes;
ii. the equality of two critical values, corresponding to what are called *conflict* catastrophes.

Each of these two very distinct types of catastrophe has a corresponding type of instance of selection I, which Thom called conventions:

i. the *convention of perfect delay*, according to which the system S remains in an internal state (a minimum of f_w) as long as that state exists: there can only be catastrophe when a minimum disappears through fusion with another critical point (bifurcation);
ii. the *Maxwell convention*, according to which the system S always occupies the absolute minimum of f_w: there can only be catastrophe when another minimum becomes the absolute minimum (conflict).

15.5 The Principles of Morphodynamic Models

As we have seen, morphodynamic models receive a natural interpretation within the context of systems theory. In this setting, the space W is a control space and the phenomena we seek to account for are of the critical type. Most of the rigorous and accurate physical applications of morphodynamics are of this type: diffraction catastrophes and wave-front dislocations in wave optics (with their consequences for semi-classic approximation in quantum mechanics); the theory of phase transitions and phenomena of spontaneous symmetry breaking in ordered media; stability of defects in mesomorphic phases; bifurcation (buckling) of elastic structures; constrained differential equations, singular perturbations and chaotic solutions (feedback-induced chaos); theory of shock waves; analysis of singularities in variational systems; regime changes in hydrodynamics, chemical kinetics and thermodynamics (dissipative structures and spontaneous self-organisation of matter, etc.); strange attractors, deterministic chaos and routes to turbulence; etc.

In these rigorous and accurate applications, we know the internal dynamics of the system explicitly, one way or another. We therefore seek to derive mathematically the catastrophic sets K_W from our explicit knowledge of the fields $\sigma : W \to \mathcal{X}$ and, quite naturally, we postulate that the internal process X_w causally generates the external morphology K_W. Analysis of various physical examples shows that often, a "flesh" of fine-scale complex processes (renormalisation groups, oscillating integrals, etc.) is grafted onto a "skeleton" of medium-scale singularities. This allows us to speak with a certain degree of precision about the morphological infrastructures of certain classes of physical phenomena. Now, these infrastructures are phenomenologically dominant. We therefore possess – for the first time – a link between the mathematical formalisms of physical objectivity and the phenomenology of the manifestation.

15.6 The Models of Morphogenesis

The junction between physical models and morphological schemes is made by considering that the control space W is the spatio-temporal extension of a material substrate. Consequently, the models describe the qualitative variation of perceptible qualities that can be observed in the substrate. This is the case for the models proposed by Thom for embryogenesis. They are based on two guiding ideas.

The first is that the attractors of internal dynamics define local metabolic regimes on the substrate (whence the name of metabolic models) and that, since these regimes are controlled by the spatio-temporal extension of the substrate, their catastrophes (made elementary by thermodynamic averaging) manifest themselves as differentiations of qualities in the substrate, in other words as processes of morphogenesis.

The second idea, more speculative, is that it is possible to interpret the topology of attractors defining local regimes in terms of their functional significance within the global regulation of the organism.

References

1. Arnold V. (1978) *Mathematical methods of classical mechanics*, Mir (Moscow).
2. Thom R. (1980) *Modèles mathématiques de la morphogenèse*, Christian Bourgois (Paris), in French. English translation: *Mathematical models of morphogenesis* (1983), John Wiley and sons (Chichester).
3. Thom R. (1972) *Stabilité structurelle et Morphogenèse*, Benjamin (New York), Édiscience (Paris), in French. English translation: *Structural stability and morphogenesis* (1975), Benjamin (Reading MA).
4. Thom R. (1980) Halte au hasard, silence au bruit, *Le Débat* **3**, 119–132, Paris (Gallimard), in French. Commented english edition: Stop Chance! Silence Noise! *SubStance* **40** (1983).
5. Thom R. (1990) *Apologie du Logos*, Hachette (Paris).
   Collected works of René Thom are available on CD-Rom at the Institut des Hautes Études Scientifiques, 35 route de Chartres, 91440 Bures-sur-Yvette.

Chapter 16
Morphogenesis, Structural Stability and Epigenetic Landscape

Sara Franceschelli

This chapter offers a commentary on the correspondence exchanged between René Thom and Conrad Hal Waddington in 1967, concerning the interpretation in terms of catastrophe theory of the concepts of epigenetic landscape and chreod, introduced by Conrad Hal Waddington since the 1940s.

It is intended to provide some elements of reflection on the difficulties encountered in exchanges between a mathematician and a biologist on the subject of the mathematization – in this case by means of a "dynamical systems" approach – of a theoretical question in biology, expressed in images by the compound metaphor of epigenetic landscape. One interpretation of the disagreement between the two scholars is based on the difference between their mathematical cultures, making it difficult to establish a shared dictionary (this was René Thom's view). But another aspect must be taken into account to understand the difficulties of dialogue between the two scientists: the choice of variables (and the timescale specific to each of them) used to construct a mathematical model of the epigenetic landscape.

16.1 The Correspondence

Catastrophe theory, originating in René Thom's research into the topology and differential analysis of the problem of structural stability, was conceived by its author as a mathematical theory of morphogenesis.

The paper "A dynamic theory of morphogenesis", written by Thom in 1966 and published in 1968 in *Towards a Theoretical Biology I*, under the direction of Conrad Hal Waddington [7], is generally considered to mark the birth of catastrophe theory. In it, René Thom argued that embryology, and in particular the concepts of epigenetic landscape and chreod invented by Waddington, is both a source and a field of application of catastrophe theory. To illustrate his argument, he chose as an example of morphogenesis the model of cell differentiation developed by Max Delbrück in

S. Franceschelli (✉)
École Normale Supérieure des Lettres et Sciences, Lyon, France
e-mail: sfrances@ens-lsh.fr

P. Bourgine, A. Lesne (eds.), *Morphogenesis*, DOI 10.1007/978-3-642-13174-5_16,

1949 [2], and showed how it could be applied in the context of catastrophe theory. This choice sparked off a correspondence between Waddington and Thom, some of which was published in 1980 as an annex to the French version of the paper in the collection *Modèles mathématiques de la morphogenèse* (five letters, dated between 25 January and 23 February 1967) [8]. This correspondence reveals disagreements between the two authors about the relevance of the example chosen by Thom to the question of cell differentiation in the context of development.

Waddington made two criticisms of the first version of Thom's paper. The first concerned the paternity of the biochemical interpretation of cell differentiation, which he claimed for himself. He questioned the following phrase in Thom's text ([8], p. 23, our translation):

> The biochemical interpretation (due to Delbrück and Szilard) of cell differentiation.

The second criticism was more subtle. It concerned the use (by Delbrück and Szilard) of steady states, instead of time-extended chreods, when treating the question of cell differentiation. According to Waddington, the following phrase in Thom's text was problematical ([8], p. 23, our translation):

> All cell specialisation being — according to the idea of Delbrück and Szilard — characterised by a stable regime of the metabolism, that is to say an attractor A of the local biochemical dynamics.

This suggested that Thom supported the idea of a description in terms of steady states. For Waddington, the expression *"stable regime"* was synonymous with *steady state*, and he therefore asked Thom to change it. On the subject of the first point, he wrote ([8], p. 23):

> I had stated the main point as early as 1939.

And concerning the second ([8], p. 23):

> I got it right, and spoke of alternatives between time-extended chreods (though I did not yet call them that), whereas Delbrück and Szilard had the simpler and basically inadequate idea in the context of development of an alternative between steady states.

In response to these remarks, Thom proposed the following changes ([8], p. 24, our translation):

> This idea of interpreting cell differentiation in terms of "a stable regime of the metabolism", i.e. of an attractor of the biochemical kinetics, is often attributed to Delbrück and Szilard. In fact it was stated — under its local form, which is the only correct one — in C.H. Waddington, *Introduction to Modern Genetics*, 1939.

And secondly ([8], p. 24, our translation):

> All cell specialisation being characterised by a stable regime of the local metabolism.

But concerning his second, more conceptually important criticism, Waddington preferred ([8], p. 25):

> ... by a stable but evolving regime of the local metabolism.

For Waddington, the specification "local state", by implicit contrast to "global state", was not explanatory. He felt that the distinction should be made between a stable regime, invariant over time (which he believed to be the case in the Delbrück model), and a regime that is ([8], p. 25):

> at any moment, stable, but which changes progressively as time passes.

Thom then suggested ([8], p. 33, our translation):

> ... an attractor of the biochemical kinetics tangent to the point considered.

This response shows how Thom was seeking the best expressions, in mathematical terms ("local form", "kinetics tangent to the point considered"), to meet Waddington's requirements. The next instalment in the correspondence shows that Waddington did not find this wording completely satisfactory; but this is nevertheless what Thom kept in the final version of his article.

One interpretation of this terminological disagreement between the two scholars is based, on first analysis, on the difference between their scientific cultures, especially in the field of mathematics. They suggested this idea themselves. Thus, René Thom presented this correspondence as a typical example of difficulties in understanding, or even of misunderstanding, between mathematicians and biologists, caused by a difference in the requirements of mathematical rigour with which the concepts are defined. For René Thom, it is the property of structural stability that provides the link between his catastrophe theory and the concept of chreod introduced by Waddington: a chreod is none other than those parts of the domain of parameter space for which a process is structurally stable. Consequently, according to Thom, the property of *homeorhesis*[1] invoked by Waddington follows quite naturally from the very definition of chreod. According to Waddington, the disagreement arose from the fact that Thom, as a mathematician, did not appreciate the particular value of the time variable in the analysis of development, wrongly considering it to be a variable that could be equated, for example, to concentrations[2] ([8], pp. 33, 34):

> Delbrück in 1949 was talking about the alternatives of driving round and round the Place de la Concorde, or round and round the Étoile ; and that is only a degenerate case of what I had been talking about in 1940, which is the alternative of taking the bus from the Aérogare des Invalides to the Aéroport Orly or the Aéroport Le Bourget. The only way to eliminate this difference between Delbrück and myself would be if you are so "pure" a mathematician that you acknowledge no difference between a dimension devoted to a material variable, such as concentration, and one devoted to time. But this is a level of abstractness at which mathematics looses touch with the real-world problems of biologists.

---

[1] Waddington coined the term *homeorhesis* to describe the property of stability (in modern parlance we would call it robustness) of the processes of development when subject to perturbations. For Waddington, it was essential not to confuse this property of equilibrium along a trajectory of development (which he called the *chreod*) with the property of homeostasis, indicating an equilibrium around an unchanging state.

[2] As we shall see in Sect. 16.2, to interpret in mathematical terms the problem of cell differentiation as presented in the Delbrück model, Thom wrote a differential law in which the variables are concentrations.

To clarify the terms of this disagreement, it will be useful to examine the underlying issue: Delbrück's model.

16.2 Delbrück's Model

This model was introduced by Max Delbrück at a conference on genetics held in Paris in 1949, during which a proof was presented for the heritability of certain phenotypic traits in paramecia and other ciliates over a large, but finite, number of generations [2]. This could be taken as evidence against the chromosomal transmission of these traits. G.H. Beale then advanced the hypothesis of the existence of populations of cytoplasmic genes, or plasmagenes, that would be transmitted through a finite number of cell divisions before disappearing, and which would be responsible for the existence of these traits. Delbrück wanted to show that the same observations could be explained without resorting to genes or plasmagenes. To this end, he put forward the following argument ([2], p. 33, our translation):

> ...many systems in flow equilibrium are capable of several different equilibria under identical conditions. They can move from one equilibrium state to another under the influence of transient perturbations.

Delbrück thus proposed a model of interacting metabolic pathways, as illustrated in Fig. 16.1, but without explicitly writing the associated equations. A_1, A_2, B_1, B_2 represent different types of enzymes within the cell (represented by the circle). a_1 and b_1 are substances in the environment, while a_2 and b_2 are intermediate metabolites produced from a_1 and b_1 under the influence of A_1 and B_1. They are, in turn, the substrates of enzymes A_2 and B_2, which transform them into a_3 and b_3, which are waste products. In a constant environment, this model always remains in a stable state. But at this point, Delbrück added the hypothesis that there exist mutual interactions between the two series of enzymatic reactions (shown in the diagram as

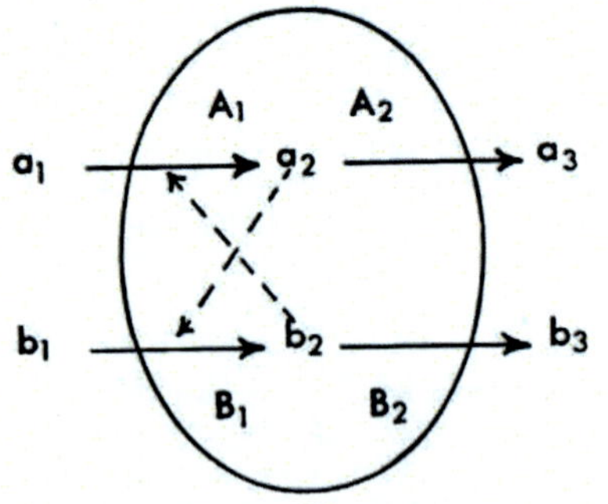

Fig. 16.1 Delbrück's model. A_1, A_2, B_1, B_2 are different types of enzymes inside the cell (which is represented by the circle). a_1 and b_1 are substances in the environment. a_2 and b_2 are intermediate metabolites produced from a_1 and b_1 under the influence of A_1 and B_1. They are in turn the substrates of enzymes A_2 and B_2, which transform them into a_3 and b_3, waste products. After [2]

dotted arrows). Now there exist three possible equilibria for the same conditions of the environment: two stable states and one unstable state ([2], p. 33, our translation):

> To sum up, our cell model is capable of existing in two functionally different states of flow equilibrium, without that entailing any change in the properties of the genes, plasmagenes, enzymes or any other structural units; transition from one state to another can be provoked by transient modifications in the conditions of the environment.

Commenting on this model and the role it has played in the development of developmental biology, Evelyn Fox-Keller observed that it has undergone a series of semantic shifts according to the contexts in which it has been considered. She also suggested that it has acted more as a metaphor than as a model [3]. To illustrate, but purely qualitatively, an application of catastrophe theory, Thom wrote an explicit system of differential equations inspired by this model/metaphor, introducing k chemical substances $s_1, s_2, \dots s_k$ with respective concentrations $c_1, c_2, \dots c_k$:

$$\frac{\mathrm{d}c_i}{\mathrm{d}t} = X_i(c_1, \dots c_k) \qquad \text{(differential law for concentration variations).} \qquad (16.1)$$

He then extended the model by taking into account its spatial extension, through the introduction of a system of coordinates (x) over the domain U occupied by the system:

$$\frac{\partial c_i(x_i, t)}{\partial t} = X_i(c_i, x, t) + k\Delta c\,. \qquad (16.2)$$

where the Laplacian term $k\Delta c$, assumed to be small compared to X, can be neglected.

16.3 Structural Stability and Morphogenetic Field

In the very broad sense of the term adopted by Thom, "morphogenesis" describes any process that creates (or destroys) forms, without specifying anything about either the nature (material or immaterial) of the substrate of the forms considered, or the nature of the forces causing these changes. The key concept allowing Thom to link his view of morphogenesis in terms of catastrophe theory with Waddington's concepts of chreod and epigenetic landscape is structural stability. The idea is that a function F is *structurally stable* if, for a sufficiently small perturbation of that function, the perturbed function $G = F + \delta F$ keeps the same topological form as the initial function F.

Andronov and Pontrjagin had formulated a more technical definition in 1935 [1]. They asked what properties a dynamical system (a model) must have to correspond to a physical system. One cannot take into account all the factors that influence a physical system; moreover, there is nothing to guarantee that the factors considered will remain perfectly constant during the evolution of the system. Whence the following concept of structural stability ([9], p. 48):

For a dynamical system defined by the vector field X on the manifold M, we say that this system (M, X) is structurally stable if, for all fields X' topologically close enough to X, there exists a homeomorphism h_x of M to itself, which transforms any trajectory of X into a trajectory of (M, X'). In other words, the total decomposition of M into orbits does not change topological type when X is perturbed into X'.

According to Thom, a morphogenetic field on an open set U of space-time resides in a pre-existing "universal model", of which the particular process under study is a copy. Such a process will unfold in accordance with the universal model given *a priori* and will therefore be structurally stable.

16.4 Epigenetic Landscape: A Mental Picture, a Metaphor . . . of What?

In *The strategy of the genes* [12], Waddington explicitly called for a mathematization of the processes of development on the basis of their geometrical rather than algebraic properties. In the chapter "The cybernetics of development", he argued that the processes of development cannot be modelled in terms of alternatives between several steady states, as they were, according to him, in Delbrück's model. He knew the property of sensitive dependence on initial conditions, which he called "the exaggeration of initial differences" ([12], p. 16). He drew this knowledge from the biomathematical work of Lotka [5], Kostizin [4] and others. And in an appendix to his book, he included a treatment of autocatalytic reactions written by Kacser. According to Needham [6], Waddington had been searching since the 1930s for a way to represent the course of embryonic determination in terms of a succession of choices between unstable equilibria, in the tradition of embryology research ([6], p. 58 *et seq.*). The analysis of different types of equilibrium in the study of living matter had already been taken into consideration by Lotka in the domains of epidemiology and population dynamics, where he succeeded in plotting the integral curves for systems with two variables. Needham ([6], p. 61), proposed a three-dimensional plaster model to represent these curves. The similarity to the hilly contours of the epigenetic landscape is obvious. Waddington knew these studies and he knew Lotka's results. And it is possible that he drew on them for the mental pictures he evoked with his landscapes. The thing that remained difficult to achieve, and which Waddington did not in fact achieve, as he never got any further than proposing metaphors, was to transfer the techniques of mathematization (and of the study of equilibria) from the biology of populations to the domain of development.

By breaking down the compound metaphor into its three significant aspects, we shall now see which variables Waddington believed ought to be taken into account.

i. Cell differentiation (Figs. 16.2 and 16.3)

Figure 16.2 is the first pictorial representation of the epigenetic landscape. It dates back to 1940. It is a landscape with a river flowing towards the sea (on the horizon, under the clouds) and branching into different valleys, at the ends of which we can

THE EPIGENETIC LANDSCAPE

From a drawing by JOHN PIPER

Looking down the main valley towards the sea. As the river flows away into the mountains it passes a hanging valley, and then two branch valleys, on its left bank. In the distance the sides of the valleys are steeper and more canyon-like. (See p. 91.)

Fig. 16.2 Waddington's epigenetic landscape. This river flowing towards the sea (on the horizon), and the valleys that form along its sides, was Waddington's first pictorial representation of the epigenetic landscape. From [10]

imagine the different products of cell differentiation. In 1957, Waddington proposed a more explicit image of the epigenetic landscape (Fig. 16.3). The interpretation of Fig. 16.3 as a metaphor of cell differentiation is based on the concept of sensitivity to initial conditions ([12], p. 16 *et seq.*). At the end of the hilly landscape, we must imagine the different tissues or organs produced by differentiation. The initial position of the marble at the top of the hill represents one of the different cytoplasmic states occupying the different regions of the egg. Waddington thus showed that he had well understood the property of sensitivity to initial conditions, which he cited as being responsible for progressive cell differentiation ([12], p. 29):

> Or we could represent the various different initial conditions by imagining various degrees of bias on the balls which are to run across the surface.

ii. Robustness, chreod, homeorhesis

The different paths that the ball may take are stable pathways of development, or chreods – guarantees of the robustness of the process. If we adopt the hypothesis

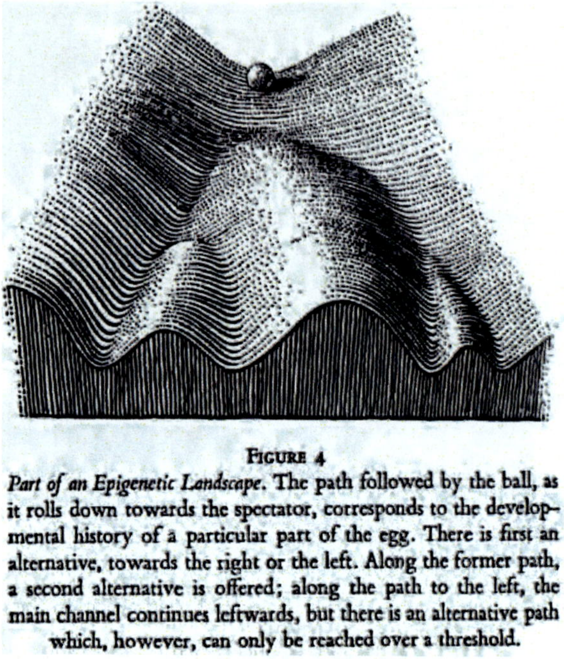

Fig. 16.3 In 1957 Waddington gave this representation of the epigenetic landscape, in the form of a hilly landscape down which a ball is rolling. The path followed by the ball corresponds to the history of the development of a given part of the egg. From [12], p. 29

that Lotka's work on different types of equilibrium gave Waddington the idea for this image of the epigenetic landscape, then Waddington's conception marks a shift in thinking, at the level of the chreods. Whereas in Lotka's work the hilly plaster model was no more than a three-dimensional representation of integral curves, and therefore a consequence of the study of equilibria, Waddington focused directly on the chreods, believing that he was thereby "inventing" a new type of equilibrium, specific to the study of living matter, because it could guarantee an equilibrium that evolves over the course of time, along a stable path of development. From this point of view, Thom seems to be right in affirming that Waddington did not fully appreciate the scope of a dynamical systems approach. But my thesis is that the reasons for their mutual misunderstanding go far beyond this. What I find even more remarkable in Waddington's ambition, which was certainly very high, was the idea of combining this image, already suggesting two metaphorical interpretations, with a third image, which Thom did not explicitly take into account in his modelling.

iii. Influence of gene interactions on the process of (epigenetic) development

For Waddington, epigenetics studies the effect of causal relations originating in the genes on the genotype-phenotype transition [11]. Waddington wanted to express the idea that it is the genes and their interactions that determine the form of the epige-

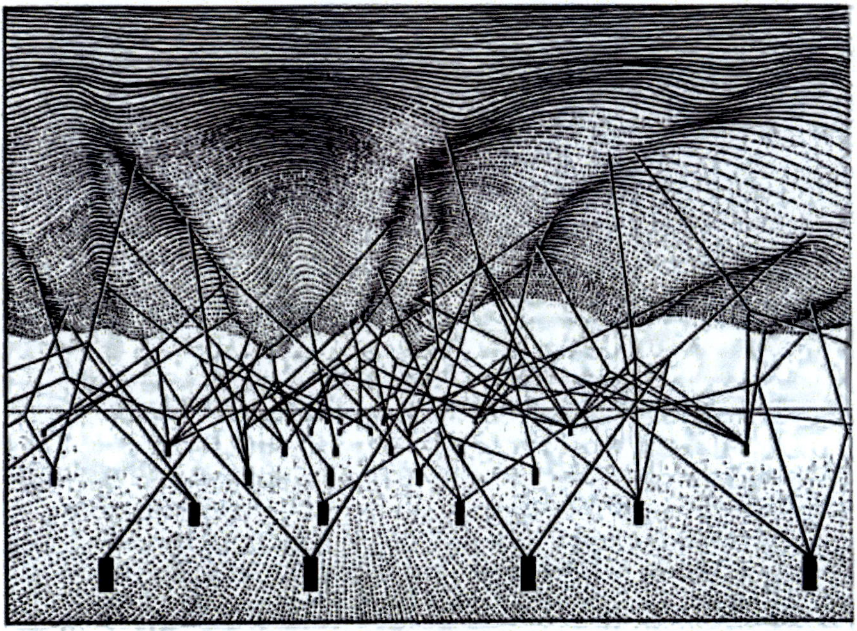

FIGURE 5

The complex system of interactions underlying the epigenetic landscape. The pegs in the ground represent genes; the strings leading from them the chemical tendencies which the genes produce. The modelling of the epigenetic landscape, which slopes down from above one's head towards the distance, is controlled by the pull of these numerous guy-ropes which are ultimately anchored to the genes.

Fig. 16.4 In the same work ([12], p. 36), Waddington adds this image, representing the system of interactions woven between the genes underlying the epigenetic landscape

netic landscape (Fig. 16.4). For him, the surface of the epigenetic landscape, which governs the course of cell differentiation, can be seen, if we change the timescale, as a metaphor for the resultant of gene interactions ([12], p. 34, 35):

> It is important to realise that the comparatively simple orderliness of the epigenetic landscape – its restricted number of valleys with their branching point and characteristic contours – is a property of higher order dependent on an underlying network of interactions which is vastly more complicated. The cells proceeding along any development pathway must have a metabolism of some corresponding complexity. [...] But genetics still gives us more insight into the real complexity of apparently simple epigenetic processes than does biochemistry. [...] Since each gene must be regarded as a distinct chemical entity, the path of development as it is observed by the anatomist must be viewed as the resultant of all the very numerous processes in which these genes are involved in the cells concerned.

It was this desire to represent phenomena taking place at different scales that prevented Waddington from being completely satisfied with Thom's proposition ... Of course, as Waddington made very clear, these were only mental pictures. But if we are prepared to reason in terms of images, what Thom proposed did not account for the process of cell differentiation, within Waddington's metaphorical context. At most, it can account for the genesis of chreods!

16.5 Interpretations

To sum up, this commentary brings to light two possible interpretations, which are not mutually exclusive, for the disagreement between the two scholars.

The first is based on the difference between the scientific cultures of the two authors, especially their requirements in terms of mathematical rigour. This amounts to saying that Waddington did not have enough mathematical knowledge to understand that the terminology used by Thom simply gave explicit expression to the properties implicit in the concept of chreod, of which the characteristic and essential trait is that it represents the domain (in parameter space) of a structurally stable process, a trait from which all its other traits ensue. In addition, and again from Thom's point of view, Waddington did not understand that if Delbrück's model was expressed in a very general form, in mathematical terms, that did not mean that it could be reduced to a choice between alternative steady states.

The second is based on the existence of a theoretical problem, underlying the compound metaphor formed by these different metaphorical images of the epigenetic landscape, which Thom failed to take into account: how to model mathematically two processes taking place over different timescales (one slow, the timescale of evolution, acting at the level of gene interaction, and one fast, the timescale of development, acting at the level of the different cytoplasmic states in different regions of the egg). Such modelling would involve writing a dynamical system with time-dependent variables. Or describing the dynamics on a network (underlying the landscape, which would be an emergent property of the network). The variables of this dynamics would be the different cytoplasmic states, and the nodes of the network would be the genes ...

References

1. Andronov A.A. and Pontryagin L.S. (1937) Coarse Systems, *Dokl. Akad. Nau. SSSR* **14**, 247.
2. Delbrück M. (1949) *Unités biologiques douées de continuité génétique*, Éditions du CNRS (Paris), pp. 33–35, in French. Reprinted in [8].
3. Fox-Keller E. (2002) *Making Sense of Life. Explaining Biological Development with Models, Metaphors and Machines*, Harvard University Press (Cambridge MA).
4. Kostitzin V.A. (1937) *Biologie mathématique*, Colin (Paris), in French. English translation: *Mathematical biology* (1939), Harrap (London).
5. Lotka A.J. (1925) *Elements of Physical Biolology*, William & Wilkins (Baltimore).
6. Needham J. (1936) *Order and Life*, MIT Press (Cambridge MA).
7. Thom R. (1968) Une théorie dynamique de la morphogenèse, in *Towards a theoretical biology I*, edited by C.H. Waddington, University of Edinburgh Press, pp. 152–166. Reprinted in [8].

8. Thom R. (1980) *Modèles mathématiques de la morphogenèse*, Christian Bourgois (Paris), in French. English translation: *Mathematical models of morphogenesis* (1983), John Wiley and sons (Chichester).
9. Thom R. (1980) Mathématique, pp. 37–56 in [8].
10. Waddington C.H. (1940) *Organisers and genes*, Cambridge University Press (Cambridge).
11. Waddington C.H. (1942) The epigenotype, *Endeavour* **1**, 18–20.
12. Waddington C.H. (1957) *The strategy of the genes*, Allens & Unwin (London).

Chapter 17
Morphological and Mutational Analysis: Tools for the Study of Morphogenesis

Jean-Pierre Aubin and Annick Lesne

17.1 Objectives

The study of morphogenesis deals with the evolution of forms, which are essentially sets, in the mathematical sense of the term. Consequently, their evolution, their control (visual, for example), their analysis, their processing (when the forms are images) or their optimisation naturally require an intrinsic analysis *at the level of sets*. These "specifications" have driven the recent development of set-valued (or multi-valued) analysis[1] [7, 34] and the even more recent development of morphological and mutational analysis [5, 9], which will be the subject of this chapter.

The word morphology was coined by Johann Wolfgang von Goethe (1749–1832), who also deployed his many talents in the field of biology. In an essay [21] published in 1790, he proposed a bold unifying hypothesis, according to which most of the main plant forms had evolved from one archetypal plant (Urpflanze). Goethe foreshadowed the work [41] of the founder of morphogenesis, D'Arcy Thompson (1860–1948). He was bitterly disappointed that his scientific work attracted so little attention and was not taken seriously (except by Isidore Geoffroy Saint-Hilaire). He complained that:

> [...] The public [...] expects someone who has distinguished himself in one field [...] no to abandon it, and even less to venture into another domain that has no relation. When somebody attempts this experiment, no recognition will be accorded to him; indeed, even if the task is well done, he will receive no praise.

In his *Maximen und Reflexionen*, he added, on the subject of mathematics:

> Mathematicians are like Frenchmen: whatever you say to them they translate into their own language, and forthwith it is something entirely different.

J.-P. Aubin (✉)
CREA and the LASTRE (Applied Controlled Tychastic Systems Laboratory), Paris, France
e-mail: aubin.jp@gmail.com

[1] In contrast to the "pointwise" analysis of *single-valued maps*, set-valued analysis deals with continuity, with the differential and integral calculus of *(set-valued or multi-valued) maps*, assigning a (possibly empty) subset in the final space to every element in the initial space.

P. Bourgine, A. Lesne (eds.), *Morphogenesis*, DOI 10.1007/978-3-642-13174-5_17,

Morphogenesis is a challenge for mathematicians because the dynamical systems they are used to studying the evolution of vectors, or functions, governed by differential equations or partial derivative equations, whereas the purpose of morphogenesis is to study the evolution[2] of forms, in other words of subsets of one kind or another, *which are not necessarily regular*. Indeed, since Husserl, forms have been more elaborate structures than sets: they are composed of sets known as *substrates* on which are defined *perceptive functions* describing their perceptible qualities; in addition they are equipped with morphological accidents, describing the singularities of the perceptive functions. We refer readers interested in the programme of "*morphodynamics*" – inspired by the theory of singularities and catastrophes and dedicated to these structures that develop over space and time – to the works of Thom [40, 39], Petitot [30–32] and their collaborators, and to Chap. 15 that summarises them.[3] Nor is this the place to describe in detail the countless works on mathematical and computer processing of images. Other tools have been forged in the field of functional analysis, *which all, in one way or another, represent shapes by functions*. This tradition has been maintained up to the present day. We refer readers to the excellent work [29]. These various methods of representing a set in terms of a function (characteristic function, support function, indicators, gauges, usual or signed distances, level functions, etc.), each more ingenious than the other, allow to use the properties of functions, the available applications and all the many results accumulated in analysis and in geometry, but at a price: *these techniques require of the sets a mathematical regularity that biological forms do not possess* (typically properties of boundary differentiability).

Moreover, these theories were not designed for studying sets, but functions. That is the essential reason why, as is so often the case, it is better to confront the difficulty head-on. To do so, we must return to the origin of dynamical systems and:

1. choose a new approach by defining a concept of "velocity" for a set which *moves and mutates*, "*grows and multiplies*";
2. then, armed with this concept of velocity, to give a meaning to sorts of differential equations (called *morphological equations*) governing the evolution of sets, their movement, their deformation, but also their growth (expansion): during an infinitesimal evolution, not only can each element of the shape be "shifted" to another point in space, but the shape may be "expanded" if some of its elements each possess several successors; this is where the set-valued character mentioned in the introduction appears.

Moreover, these evolutions of sets must respect constraints of confinement, or what are called "geometric" constraints; they have to reach targets in finite times; they

[2] In the mathematical sense of time-dependent function $t \mapsto x(t)$, not in the biological sense of "Darwinian evolution".

[3] For readers familiar with viability theory, it is interesting to note that this programme could incorporate work on the principle of inertia in viability theory, which itself generates discontinuities in parameters, with viability niches playing the role of attractors. But that is not the subject of the present text, which focuses on the *evolution* of sets.

also have to "co-evolve" with their elements, which are governed individually by other dynamics ("non-local" problems in mathematical language). We must therefore follow almost systematically the reverse strategy to that of functional analysis: *by characterising the functions and their maps in terms of their graphs, we shall consider them too as sets*, and from this perspective, study the operations on sets, the evolution of sets, the hyperspaces (spaces whose elements are subsets of another space), set-valued maps (assigning a set to a point), shape maps (assigning a point to a set) and maps that assign one set to another.

Morphological analysis has thus been conceived to enrich the mathematical toolkit of morphogenesis, by providing a structure, *mutational analysis*, that encompasses and incorporates into one same framework the differential calculus in all metric spaces (vector spaces, hyperspaces of forms, and other examples, equipped with the *mutational structures* described in Sect. 17.6. *We have absolutely no wish to make a clean sweep of the past*: it would be a shame to deprive ourselves of the efficient tools of functional analysis (and it would be contrary to Darwinian "ideogenesis"!); we suggest quite simply to complete it by these new instruments of set-valued and morphological analysis.

17.2 Motivations

A large number of domains have supplied explicit motivations for morphological analysis. The following list, far from exhaustive, is intended to give some idea of the diversity of existing or potential applications.

17.2.1 Problems of Co-Viability

Problems of co-viability are at the very origin of mutational and morphological analysis. The purpose of viability theory is to study the equations governing evolutions $t \mapsto x(t)$ (where x represents the state of the system) which are "viable", i.e. which remain in an environment $K \subset X$ described by what are called the viability constraints [2, 9]. By extension, "co-viability" means that the joint evolution $t \mapsto x(t)$ and $t \mapsto K(t)$, in a time-dependent environment, is viable in the sense that:

$$\forall t \geq 0, \quad x(t) \in K(t) \qquad \textit{(property of co-viability or joint viability)} \quad (17.1)$$

Firstly, it was assumed that the evolutions $t \mapsto K(t)$ were given (for an intuitive reason, they were called *tubes*, the axis of the tube corresponding to the time axis and its section at time t corresponding to the set $K(t) \subset X$, see Fig. 17.1). At the beginning of the 1980s, the characterisation of their viability led to the introduction of "graphical derivatives" of shape maps in [1], inspired by Fermat, but using the concept of *contingent cone* introduced by Georges Bouligand and Francesco Severi at the beginning of the 1930s.

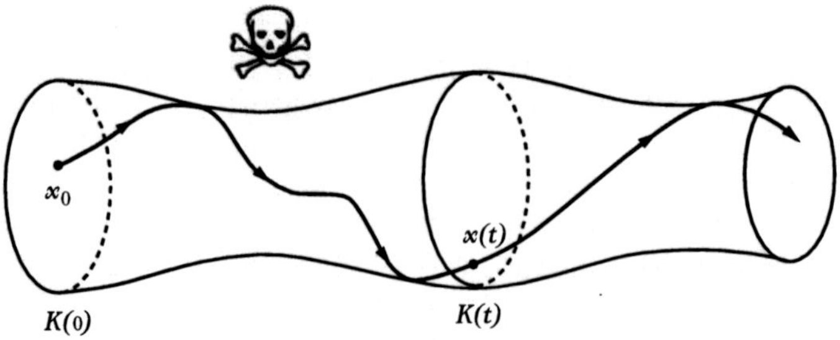

Fig. 17.1 Graph of a viable evolution in a tube. *Tube* is the nickname given to set-valued maps $K : t \in [0, T] \mapsto X$

The contingent cone $T_L(x)$ to the set $L \subset X$ at a point $x \in L$ is the set of directions $v \in X$ such that there exist real sequences h_n, with $h_n \to 0$ and $h_n > 0$, and sequences v_n in X, with $v_n \to v$, such that $x + h_n v_n \in L$ for all $n \geq 0$. Obviously, $T_L(x) = X$ for any point x in the interior of L, and its property of being a cone means that if $v \in T_L(x)$, then $\lambda v \in T_L(x)$ for any positive real λ.

The "graphical derivative" of a set-valued map $t \mapsto K(t)$ is defined at a point $(t, x) \in \text{Graph}(K)$ of its *graph* as being the *set-valued map* $u \in \mathbf{R} \mapsto DK(t, x)(u) \subset X$ *of which the graph is the "contingent cone" to the graph Graph*(K) *of the map at point* (t, x). This graphical derivative is a generalisation of the concept of usual derivative, but different from that of derivative in the sense of Laurent Schwartz distributions. Their parallel history, similarities and differences are described in [6].

By definition, the graph Graph(K) of the map $t \mapsto K(t)$ is the set of points (t, x) with $t \in \mathbf{R}_+$ and $x \in K(t)$; it is therefore a subset of $\mathbf{R}^+ \times X$; the graph of its graphical derivative is a subset of $\mathbf{R} \times X$.

As an illustration of the concept of contingent cone, let us consider the example of a rectangle: $K_{rect}(t) = \{(x, y), |x| \leq a, |y| \leq b\}$. The boundary of the rectangle presents four angular points where it is non-differentiable. At these points, where the tangent is not defined, the contingent cone is reduced to the quadrant associated with the directions $v = (v_x, v_y)$ pointing into the rectangle, for example $(v_x \leq 0, v_y \leq 0)$ for the corner (a, b). For the other points on the boundary of K_{rect}, the contingent cone is a half-plane, for example $v_x \leq 0$ for the points on the right-hand edge, or $v_y \leq 0$ for the points on the upper edge. Lastly, at any point in the interior of the rectangle, it is the whole plane. The reader will verify that the contingent cone to a circle K_{circ} with centre 0 and radius R at a point $\mathbf{r}$ on its circumference ($r = R$) is the half-plane containing all the directions $\mathbf{v}$ such that $\mathbf{v}.\mathbf{r} \leq 0$.

Concerning the graphical derivative, the first thing to note is that the graphical derivative of a constant tube $K(t) \equiv K$ (therefore of the graph Graph$(K) = \mathbf{R}^+ \times K$) is directly linked to the contingent cone to the set K: $DK(t, x)(u) \equiv T_K(x)$, because in this case $T_{\text{Graph}(K)} = \mathbf{R} \times T_K(x)$. To illustrate the concept in a non-trivial situation, let us take the rectangle again, but now we let the lengths of the edges vary over time: $K_{rect}(t) = \{(x, y), |x| \leq a(t), |y| \leq b(t)\}$ The graphical derivative $DK_{rect}(t, a(t), b(t))(u)$ of the tube $t \mapsto K_{rect}(t)$ at the point $(t, a(t), b(t))$ is the set of points (v, w) such that $v \leq ua'(t)$ and $w \leq ub'(t)$. On the other hand, $DK_{rect}(t, x, y)(u)$ is trivially equal to the plane at any point in the interior of the rectangle. Lastly, if (x, y) belongs to an edge, for example the right-hand edge, then the constraint on w disappears, while v remains constrained by $v \leq ua'(t)$. In the example of the circle $K_{circ}(t)$ with centre 0 and radius $R(t)$, now variable, the graphical derivative $DK_{circ}(t, \mathbf{r}(t))(u)$ at a point $\mathbf{r}(t)$ on the circumference (i.e. $(t) = R(t)$ at time t) is the set of points $\mathbf{v}$ such that $\mathbf{v}.\mathbf{r}(t) < uR(t)R'(t)$ or $\mathbf{v} = uR'(t)\mathbf{r}(t)/R(t)$.

However, biological and economic considerations led us not only to characterise and control the evolutions governed by controlled viable systems in a given tube $K(t)$, but also to "endogenise" the evolution of the tube $K(t)$ to obtain a globally autonomous tube. The question soon arose of how to describe the evolution of tubes by means of a "morphological equation", with the aim of studying and characterising the property of co-viability (17.1) of the joint evolution of a vector and an environment, the one governed by a differential equation, the other by a "morphological equation" of the form:

$$\begin{cases} i) \ x'(t) = f(x(t), K(t)) \\ ii) \ \overset{\circ}{K}(t) \ni g(x(t), K(t)) \end{cases} \quad \textit{(morphological equation)} \qquad (17.2)$$

To give significance to the (17.2(ii)), it was necessary to design a concept of velocity (which we had written in anticipation by introducing the notation $\overset{\circ}{K}(t)$), and this will be the main subject of the present chapter. We shall see in Sect. 17.4 that the graphical derivatives of tubes are unfortunately inadequate for defining the velocity of tubes. This led to the development of "differential calculus" in the metric space of non-empty compact sets, as a means to treat these morphological equations: the concept of *mutation of a tube*, introduced at the beginning of the 1990s in [3], allows to define a concept of set velocity endowed with the same properties as vector velocity.

Beyond problems of co-viability, in [22] Anne Gorre studied more general problems involving tubes $K(t)$, $L(t)$, $M(t)$ evolving according to a system of morphological equations that satisfy at all times relations of the form

$$\forall t \geq 0, \quad \begin{cases} i) \ L(t) \cap M(t) \neq \emptyset & \textit{(property of intersectability)} \\ ii) \ K(t) \subset L(t) \cap M(t) & \textit{(property of confinement)} \end{cases} \qquad (17.3)$$

17.2.2 Biological Morphogenesis

Biological morphogenesis and other domains in biology provide a vast range of problems that can be studied by morphological analysis, whenever the dynamics governing the evolution of biological forms depends not only on individual elements, but also on the environments in which they operate. In physiology, for example, we can create a mathematical metaphor for "postural dynamics" when $K(t)$ describes the convex envelope of the feet at time t and $x(t)$ the centre of gravity of the body, which must remain within this set at all times [11]. Environmental problems and numerous domains in ecology, dealing with the evolution and interactions of populations, also offer new themes of research (see Sect. 17.8 on the embryogenesis of zebrafish).

17.2.3 Image Processing

Georges Matheron worked on image processing in the context of *mathematical morphology*, to design algorithms for image processing based on the algebraic operations of Minkowski [25, 26, 35, 36]. Black and white images are essentially subsets (of "pixels") of two- or three-dimensional vector spaces $\mathbf{R}^n$ and, in the case of digital images, of "grids" $\mathbf{Z}^n$. Suitable tools were developed to be valid in both cases, as the images are processed by computers which only operate on discrete sets.

Grey and colour images are single-valued maps that assign intensities of grey and of the three primary colours respectively to each pixel. By characterising these maps in terms of their *epigraphs*[4] and their *graphs*, grey and colour images are in turn characterised by subsets of vector spaces. Moreover, these maps are not regular, since an image represented by a continuous grey scale function would be fuzzy! Image processing has been – and still is – at the root of many problems dealing naturally with sets and their evolution. We refer readers to the book [28] for a presentation of the Grenander-Mumford-Shah approach.

17.2.4 Shape Optimisation

Shape optimisation is concerned with mechanics (design and construction of industrial structures) and the optimal control of distributed systems. Many of the problems in this domain are formulated as the minimisation of functionals over a class of sets under "geometric" constraints. To be able to apply Fermat's rule[5] and to carry out sensitivity analysis, Jean Céa and Jean-Paul Zolesio introduced the concept of the *shape gradient* of a *shape function* (function assigning a number or a vector to a set), which was to be the prototype for the concept of map mutation [16, 17, 38].

17.2.5 Dynamic Economics

Dynamic (or evolutionary) economics is also a rich source of motivations. In the Arrow-Debreu setting, production technologies are often represented as sets, in fact

[4] The epigraph of a function f is the set of points (x, y) above its graph, i.e. such that $y \geq f(x)$.
[5] Fermat's rule states that the gradient of a function is zero at all extrema of the function.

graphs of input-output set-valued maps. Keeping this general description of production processes inherited from static theory, it is possible to endogenise the evolution of production processes by means of mutational equations, where the property of co-viability expressing the requirement that total consumption must at all times be supplied by the production processes themselves in co-evolution. Coalitions of economic agents, diffusion of technologies and problems of migration provide other avenues of research to which morphological analysis could be applied [4].

17.2.6 Front Propagation

Front propagation problems are concerned with the study of sets

$$K(t) \; := \; \{x \in X \mid a(t,x) = \text{constant}\} \tag{17.4}$$

defined as level sets of a function $(t,x) \mapsto a(t,x)$, when this is a solution to a partial derivative equation (for example, reaction-diffusion equations, Hamilton-Jacobi equations (level sets), etc.) [13–15, 29, 37]. However, this approach requires regularity in the sets $K(t)$ that is missing from certain single-valued maps. As we shall see, no condition of regularity is required when the sets $K(t)$ are solutions to a mutational equation.

17.2.7 Visual Robotics

Visual robotics deals with systems controlled by mechanisms which, in the final analysis, act through feedback on shapes, images or environments, i.e. sets. In such problems, feedback allows to *act on the images or shapes* to respect viability constraints, to reach objectives or to minimise an intertemporal criterion. Mutational analysis allows us to obtain feedback and Lyapunov functions that are shape functions [19, 20].

17.2.8 Interval Analysis

Interval analysis is a domain of *numerical analysis* that studies the evolution of intervals containing the solution to a differential equation (see [27] for an introduction to this subject). This domain falls naturally into the framework of mutational and morphological analysis.

17.3 The Genesis of Morphological Analysis

Paradoxically, as is so often the case in history, research into hyperspaces (spaces of subsets of a given space) started at the same time as set theory, at the dawn of the twentieth century. In 1902, Painlevé defined the concepts of upper and lower limits of sets (called Kuratowski limits after the mathematician who presented them in his famous work) and in 1907, Pompeiu, a student of Painlevé, introduced the distance over the hyperspace of non-empty compact sets of a metric space (called the Hausdorff distance after that scientist used them in his no-less famous book), defined by

$$\mathbb{d}(L, M) := \max\left(\sup_{y\in L}\inf_{z\in M} d(y,z), \sup_{z\in M}\inf_{y\in L} d(y,z)\right) \quad \left(\begin{array}{c}\textit{Pompeiu-Hausdorff}\\ \textit{distance}\end{array}\right) \tag{17.5}$$

However, this set-valued approach was neglected for nearly half a century, before rising from the ashes under the pressures of an increasing number of problems after the Second World War. This neglect was due to the fact that the "pointwise approach", which considers the set-valued (multi-valued) maps of X in Y as singled-valued maps of X in the hyperspace $\mathbf{P}(Y)$ of all the parts of Y, had been chosen by Bourbaki. It then became the predominant approach in analysis, to the detriment of the "old" graphical approach that characterised single- and set-valued maps in terms of their graphs, a point of view going back to Pierre de Fermat and René Descartes. And so, for perfectly sound reasons, functional analysis became the norm, especially when questions of differentiability emerged. Numerous astute techniques have been used to circumvent the obstacles raised by the mathematical treatment of sets.

Why did it take us so long to resurrect set-valued analysis from the purgatory to which it had been condemned … by Bourbaki among others?

The loss of the paradise of differentiability was the punishment for those who wished to explore the purgatory of set-valued analysis, depriving the sinner of the grace of differential and integral calculus. Another of the many reasons for this hiatus in the history of set-valued analysis lies in the fact that hyperspaces only inherit part of the properties of the underlying space (of which the elements of the hyperspace are subsets). In particular, the families of sets of a vector space lose its linear structure, and only possess the much poorer algebraic properties of lattices, dioids, or exotic algebra (max-plus), which have nevertheless been exploited in mathematical morphology. Likewise, the topological properties of the underlying space are not easily or naturally transferred to the various hyperspaces, except in the case of non-empty compact spaces (the book [10] presents an exhaustive study of topologies on hyperspaces).

17.4 From Shape Optimisation to Set-Valued Analysis

In the meantime, the pressure to find a way to define the velocity of tubes was growing ever stronger, leading Jean-Paul Zolesio in 1976 to define tube velocities as vector fields, a definition that has now become a particular case of the concept of *mutation* of a set-valued map [43]. The concept of mutations of maps was in fact strangely hidden behind the concept of the *shape gradient* of shape functions.

Let X and Y be two *vector spaces* of finite dimension (and therefore naturally equipped with a topology) and let $\mathbf{K}(X)$ and $\mathbf{K}(Y)$ be the families of non-empty compact subsets of X and of Y respectively,which are metric spaces for the Pompeiu-Hausdorff distance (17.5). We shall first consider a *shape map* V mapping the subsets $K \subset X$ of a vector space X into vectors $V(K) \in Y$ of Y: it therefore assigns a point in Y to a subset of X. The underlying idea is to generalise the usual differential quotient

$$\frac{U(x+hv)-U(x)}{h} \qquad (x \in X)$$

of a single-valued map $U : X \mapsto Y$ constructed by following the half-lines $x + hv$ to obtain the differential quotient

$$\frac{V(\vartheta_\varphi(h,K))-V(K)}{h} \qquad (K \subset X)$$

where $\varphi : X \mapsto X$ is a Lipschitz map, $\vartheta_\varphi(h,x) \in X$ denoting the value at time h of the solution to the differential equation $z' = \varphi(z)$ starting from $z(0) = x$ at time 0 and $\vartheta_\varphi(h,K) := \{\vartheta_\varphi(h,x)\}_{x\in K} \subset X$ *the reachable set* at time h from K by the evolution generated in X by the map φ. Note that $\vartheta_\varphi(0,x) = x$ and therefore $\vartheta_\varphi(0,K) = K$.

These "half-curves" $h \mapsto \vartheta_\varphi(h,K)$, associated with the Lipschitz map φ, play the role of the half-lines involved in the definition of the usual directional derivative of a function, allowing us to transpose the concept of directional derivative of a single-valued map of variable $x \in X$ to the shape map of variable $K \subset X$: if the limit

$$\overset{\circ}{V}(K)\varphi := \lim_{h\to 0+} \frac{V(\vartheta_\varphi(h,K))-V(K)}{h} \qquad \textit{(shape derivative)} \qquad (17.6)$$

exists, it is called the *directional shape derivative of* V *at* K in the "direction" φ. If it is also linear and continuous on the space of Lipschitz maps, then it defines a vector distribution that was baptised the *shape gradient* by Céa and Zolesio. But we shall depart from the path marked out by those two authors, to focus on the adaptation of this idea to set-valued maps that assign a subset of Y to each point in X.

In other words, the "half-curve" $h \mapsto \vartheta_\varphi(h,K)$ plays the same role in $\mathbf{K}(X)$ as the half-line $h \mapsto x + hu$ plays in X. Both of them can be considered as *transitions*, assigning to each element ($x \in X$ in one case, $K \in \mathbf{K}(X)$ in the other) a neighbouring element in a given "direction", described by a vector v in the case of vector spaces and by a Lipschitz map φ in the case of hyperspaces. In particular, this allows us to define infinitesimal variations of the "points" K in $\mathbf{K}(X)$, making natural the definition (17.6) of the shape derivative $\overset{\circ}{V}(K)\varphi \in Y$ of the shape map V in the direction φ. The definition can thus be reformulated as follows: *the transition* $V(K) + h\overset{\circ}{V}(K)\varphi$ *of the image* $V(K)$ (transition in Y) and *the image* $V(\vartheta_\varphi(h,K))$ *of the transition* $\vartheta_\varphi(h,K)$ (transition in $\mathbf{K}(X)$) are equivalent in the sense that

$$\lim_{h\to 0+} \frac{d[V(\vartheta_\varphi(h,K)), V(K) + h\overset{\circ}{V}(K)\varphi]}{h} = 0 \qquad \textit{(shape derivative)} \tag{17.7}$$

where d is the usual distance[6] in the vector space Y. This formula appears as the generalised application to shape maps of the first-order Taylor formula (written $U(x+hv) - U(x) - hU'(x)v = hr(h)$ with $\lim_{h\to 0+} r(h) = 0$ for usual maps). Since the space $\mathbf{K}(X)$ is only a metric space, without linear structure, replacing the half-lines by half-curves in the definition of the differential quotients is after all a reasonable strategy. For this example of metric space, these "half-curves" ϑ_φ are transitions that allow us to define the analogue to a differential calculus of shape maps. This reworking of the concept of directional shape derivative *only uses the concepts of transitions in metric space* $\mathbf{K}(X)$ *and in vector space* Y. When we compare this shape derivative to the directional derivative of the single-valued map $U : X \mapsto Y$ involving transitions in X and in Y, the only "privilege" that X derives from being a vector space is having "de luxe" transitions of the form $x + hv$ associated with vectors v belonging to the same vector space X.

Because a set-valued map goes from X to $\mathbf{K}(Y)$, while a shape map goes from $\mathbf{K}(X)$ to Y, we only have to "inverse" the generalisation made in the definition of shape derivatives to obtain *mutations* (the relevant analogues to derivatives of set-valued maps, exchanging the initial space $\mathbf{K}(X)$ in X and the final space Y in $\mathbf{K}(Y)$. As this strategy gave good results for shape maps, it should do the same for set-valued maps $F : X \mapsto \mathbf{K}(Y)$, and this is effectively the case.

For that purpose, we endow the vector space X with transitions $h \mapsto x + hv$ and the metric space $\mathbf{K}(Y)$ with transitions $h \mapsto \vartheta_\varphi(h, K)$ (with $K \subset Y$) associated with Lipschitz maps φ of Y in Y (defining the transitions on the metric space $\mathbf{K}(Y)$ of the non-empty compact sets K of the vector space Y). *We thus compare the transition* $\vartheta_\varphi(h, F(x)) \subset Y$ *of the image* $F(x)$ *and the image* $F(x + hv)$ *under* F *of the transition* $x + hv$. *It is then tempting to say that the map* φ *from* Y *to* Y *(or the associated transition* ϑ_φ*) belongs to the mutation of* F *at* x *in the direction* v if

$$\lim_{h\to 0+} \frac{d\!\!l[F(x+hv), \vartheta_\varphi(h, F(x))]}{h} = 0 \quad \begin{pmatrix}\textit{mutation of a set-valued map}\\ \textit{between vector spaces}\end{pmatrix} \tag{17.8}$$

and to denote *the set* $\overset{\circ}{F}(x)(v)$ *of these maps* φ (or the transitions ϑ_φ generated by them) *as being the mutation of* F *at* x *in the direction* v. The Lipschitz map φ from Y to Y plays the role of a "directional derivative of the set-valued map F at x in the direction v", chosen in the mutation $\overset{\circ}{F}(x)(v)$ (which can be a set containing one, several or no Lipschitz maps φ). In other words, the elements φ of the mutation

[6] Note that since Y is a vector space, the difference $V(\vartheta_\varphi(h,K)) - V(K) - h\overset{\circ}{V}(K)\varphi$ has a meaning and (17.7) could be written $\lim_{h\to 0+}(1/h)\|V(\vartheta_\varphi(h,K)) - V(K) - h\overset{\circ}{V}(K)\varphi\| = 0$; the expression (17.7) is preferred here because it will lead us more directly to the generalisation (17.8) allowing to define the mutations.

$\overset{\circ}{F}(x)(v)$ determine the "correct way" to transport the set $F(x)$ to approximate to the first order the evolution $h \mapsto (x + hv)$ in the direction v from x. We should stress that the mutations are *pointwise mutations*, in the sense that they are defined at each point x of the domain of the set-valued map F (x is fixed in their definition (17.8)). They are single-valued maps assigning a set of Lipschitz maps of Y in Y (or of the transitions it generates from $F(x)$) to each direction v (or to the transition $x \mapsto x+hv$ it generates *from each* x). Mutations differ from *graphical derivatives,*[7] *which are* ... "*graphical*" in the sense that they are defined at each point of the *graph* of the set-valued map F as mappings of X in Y.

It is therefore indispensable to introduce a new term, "mutation", to avoid adding to the confusion, see Fig. 17.2. The last two decades have seen the parallel development of a multitude of approaches, turning the analysis of shapes and images into a scientific Tower of Babel, threatening to break apart the profound unity of concepts to which mathematics aspires. It has therefore been reassuring to prove that mutations are linked to graphical derivatives by a formula as simple as it is important, restoring unity to a domain inhabited by so many definitions that they threatened to make a "multiverse" of the universe bequeathed to us by Pierre de Fermat, Isaac Newton and Gottfried Leibniz. In the next section, we shall explain this formula in detail for the case of tubes $t \mapsto K(t) \subset X$.

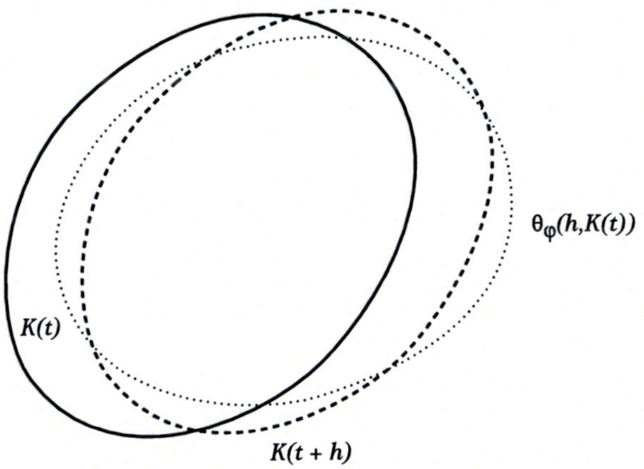

Fig. 17.2 We can interpret the definition of the mutation $\overset{\circ}{K}(t)$ by saying that if $\varphi \in \overset{\circ}{K}(t)(1)$, then the transition $\vartheta_\varphi(h, K(t))$ of the image $K(t)$ and the image $K(t+h)$ of the transition $t+h$ are "equivalent" to the first order, in the sense that $\lim_{h \to 0+} \frac{d(K(t+h), \vartheta_\varphi(h, K(t)))}{h} = 0$

[7] We should be quite clear about this point, so as to fully grasp the difference between graphical derivatives and mutations, remembering that the graphical derivative of F at $(x, y) \in \text{Graph}(F)$ (i.e. such that $y \in F(x)$) is the set-valued map $u \in X \mapsto DF(x, y)(u) = \{v \in Y \text{ such that } (u, v) \in T_{\text{Graph}(F)}(x, y)\} \subset Y$ (see Sect. 17.2.1).

17.5 Velocities of Tubes as Mutations

Thanks to the concept of mutation introduced in the last section, we are now in a position to define the *velocity of the tube* $t \mapsto K(t)$ at t (destined for use in (17.2)) as being the mutation $\overset{\circ}{K}(t) := \overset{\circ}{K}(t)(1)$ at time t in the direction 1.

When the tube is defined as a level set $K(t) := \{x \in X \mid a(t,x) = 0\}$ of a function a with real, regular and non-degenerate values, the Lipschitz map $\varphi(t)$ defined by

$$x \mapsto \varphi(t)(x) \quad := \; -\frac{a'_t(t,x)}{\|a'_x(t,x)\|} \frac{a'_x(t,x)}{\|a'_x(t,x)\|} \tag{17.9}$$

belongs to the mutation $\overset{\circ}{K}(t)$ of $K(t)$. Its restriction to $K(t)$

$$\forall\, x \in K(t), \;\; \varphi(t)(x) \;\; = \; -\frac{a'_t(t,x)}{\|a'_x(t,x)\|} \;\; \times \; (\text{unit normal to } K(t) \text{ at } x) \tag{17.10}$$

is the *normal velocity* of $K(t)$ in differential geometry. When it is valid, the level set approach thus appears as a particular case of the morphological analysis.

Let us stress once again that the graphical derivative of the tube is defined at each point (t,x) of the graph of K (i.e. such that $x \in K(t)$), which does not allow to define a velocity of the tube at time t; but which does allow the concept of mutation, which is a pointwise concept. It is possible to prove the formula

$$\forall\, (t,x) \in \text{Graph}(K), \;\; DK(t,x)(1) \; = \; \varphi(x) + T_{K(t)} \subset X \tag{17.11}$$

which links the graphical derivative $DK(t,x)(1)$ of the tube in the direction 1 to any transition generated by a Lipschitz map $\varphi \in \overset{\circ}{K}(t)$ drawn from the mutation of K at t in the direction 1.

17.6 Mutational Analysis

The choice of a Lipschitz single-valued map as a natural candidate for defining the directional derivative of a set-valued map might have seemed counter-intuitive, without this detour by way of the shape derivatives of shape maps. But once this particular case had been elucidated, we noticed that the theorem proofs never used the explicit properties of the Pompeiu-Hausdorff distance (17.5) on the space of non-empty compact sets of a vector space, but only the three axioms of a distance. In fact, the use of this specific distance proved to be an unnecessary hindrance. It turned out that *linearity is not really indispensable for designing a differential calculus* (although it does greatly simplify the definition and study of derivatives). It is easier and more useful to develop these ideas within the context of metric spaces, beyond the sole hyperspace $\mathbf{K}(X)$. Other examples can be found in the works [8, 18, 24, 33].

In this way, we can replace the linear structure of vector space by a *mutational structure on a metric space* X: this structure is described by a space of *transitions*[8] $(h, x) \mapsto \vartheta(h, x)$ (single-valued maps of $\mathbf{R}^+ \times X \mapsto X$) satisfying a small number of axioms, turning the metric space X into a mutational space. The rigorous axiomatic definitions of *transitions* on a metric space and of the *structure of mutational space* are too intricate to be reproduced here, especially since they do not help to understand the ideas. If two metric spaces E and E' are thus equipped with a structure of mutational space and if $f : E \mapsto E'$ is a map of E in E', we say that the transition $\tau \in \mathring{f}(x)(\vartheta)$ (in the mutational space E') is a *mutation of* f *in* x *in the direction of the transition* ϑ (in the mutational space E) if *the value* τ *of the transition* $\tau(h, f(x)) \in E'$ *of the image* $f(x)$ *and the image* $f(\vartheta(h, x))$ *by* f *of the value* $\vartheta(h, x) \in E$ *of the transition* ϑ *are equivalent* in the sense that (d being the distance in the metric space E)

$$\lim_{h \to 0+} \frac{d[f(\vartheta(h, x)), \tau(h, f(x))]}{h} = 0 \quad \begin{pmatrix} \textit{mutation of a single-valued map} \\ \textit{between metric spaces} \end{pmatrix} \tag{17.12}$$

Note that the richer the transition space of a mutational structure, the more chances there are that the mutation of a mapping to one point will be non-empty (the situation is analogous to the choice of topology to make families of maps continuous). In the particular case where $E = \mathbf{R}$, we define a *velocity* of the evolution $t \mapsto x(t)$ at t as the mutation $\mathring{x}(t) := \mathring{x}(t)(1)$ at time t in the direction 1, containing the transitions $\vartheta(., x(t))$ such that:

$$\lim_{h \to 0+} \frac{d[x(t + h), \vartheta(h, x(t))]}{h} = 0 \tag{17.13}$$

As we had observed in the case of the velocity of a tube (Sect. 17.5), and for the same reason, we have $\mathring{x}(t)(\lambda) = \lambda \, \mathring{x}(t)(1)$ for all $\lambda > 0$.

When f is a map of the mutational space E in its transition space, the concept of speed allows us to define the *mutational equation*

$$\mathring{x}(t) \ni f(x(t)) \qquad \textit{(mutational equation)} \tag{17.14}$$

governing the evolution of $x(t)$ in the metric space E.

It is also possible to apply the concept of *tangent transition* (inherited from the original concept introduced by Bouligand and Severi) to a subset of a mutational

---

8 When the metric space is no longer that of the non-empty compact sets of a vector space, we must work directly at the level of the transitions: they can no longer be associated with Lipschitz maps as they could in Sect. 17.5.

space, and adapt some useful geometric concepts to metric spaces: a transition is ϑ *tangent to a subset* $K \subset E$ at a point $x \in K$ if

$$\liminf_{h\to 0+} \frac{d(\vartheta(h,x),K)}{h} = 0 \qquad \textit{(tangent transition)} \tag{17.15}$$

With the concepts of mutation and tangent transition at our disposal, we can adapt the concepts of set-valued analysis to mutational spaces – an often routine and consequently tedious task, but not always easy. The theorems of Cauchy-Lipschitz and Nagumo concerning differential equations, the concepts of viability kernel and capture basins can be adapted to the case of mutational equations. It is also possible to define and characterise Lyapunov functions, and even, by using the concept of viability kernel, to construct optimal exponential Lyapunov functions and use them to adapt the *Montagnes Russes algorithm* to the case of metric spaces, to obtain the global constrained minimum of a function [23]. The theorem of inverse functions (derived from the theorem of implicit functions), allowing to locally inverse a shape map V whenever its shape derivative is surjective, has been adapted in [19, 20], as have Fermat's rule and numerous applications, to the constrained optimisation of functions defined on metric spaces. The centre manifold theorem is one of the theorems that transposes correctly. The same is true of systems of first-order "partial mutational equations".

17.7 Morphological Equations

We end this account with a more technical section, with the aim of convincing the reader that this new structure offers, in return for a greater degree of abstraction, an only partially-tapped gold mine of interesting results, the first of which have been presented in [5] and subsequent articles.

Morphological analysis is specifically devoted to the conception of mutational structures on *hyperspaces* (families of subsets of a space) and the conception of *mutations of hypermaps*, which map from one hyperspace to another, and therefore include set-valued and shape maps. We shall limit ourselves to the space $\mathbf{K}(X)$ of non-empty compact subsets of a vector space X of finite dimension equipped with the Pompeiu-Hausdorff distance (17.5). We shall now introduce several examples of transitions.

"Structuring transitions" are inspired by Matheron's *mathematical morphology*: they are the "half-curves" $\vartheta_B(h,K) := K + hB$, where B is a compact convex set considered as a *structuring element* (the sum of subsets is well-defined because we are in a vector space). The real number h appears as a scale factor.

"Shape transitions" are derived from *shape optimisation*: as we have seen, they are associated with Lipschitz maps.

We can then classify these two examples of transitions under the same more general term of "*morphological transitions*" associated with Lipschitz set-valued maps (and no longer solely with single-valued maps) $\Phi : X \mapsto X$ with compact convex images, as follows: $\vartheta_\Phi(h, x)$ denotes the set of values at time h of solutions to the differential inclusion[9] $z' \in \Phi(z)$ starting from $z(0) = x$ at time 0 and $\vartheta_\Phi(h, K) := \{\vartheta_\Phi(h, x)\}_{x \in K}$ the *reachable set* at time h for Φ from K. For example, the formula (17.11) linking graphical derivatives and mutations can be extended to transitions generated by such set-valued maps Φ, more general than the single-valued maps φ:

$$\forall\ (t, x) \in \text{Graph}(K),\quad DK(t, x)(1) \;=\; \Phi(x) + T_{K(t)}(x) \;\subset\; X \tag{17.16}$$

We now possess the mathematical tools to study morphogenesis, in the sense of the evolution of forms governed by a morphological equation, thanks to the concept of velocity defined as the mutation $\overset{\circ}{K}(t)$ of the tube $t \mapsto K(t)$ at t in the direction 1. We shall simply cite the main results; the interested reader will find an exposition of the theorems mentioned, their context and extensions, in [5].

- When f assigns the compact subsets K to Lipschitz set-valued maps $x \mapsto f(K)(x)$ defining morphological transitions on $\mathbf{K}(X)$, this concept of velocity allows us to define the *morphological equation*

$$\overset{\circ}{K}(t) \ni f[K(t)] \qquad \textit{(morphological equation)} \tag{17.17}$$

  governing the evolution of the sets $K(t)$, in the same way as the differential equation $x'(t) = g[x(t)]$ governs the evolution of the vectors $x(t)$. It states that at each moment, the second term of the morphological equation belongs to the mutation $\overset{\circ}{K}(t)$. We can prove that a set K is an equilibrium of this equation (in the sense of stationary evolution) if and only if K is both invariant and negatively viable by the Lipschitz set-valued map $x \mapsto f(K)(x)$.
- *Interval equations* are examples of morphological equations of the type

$$\overset{\circ}{K}(t) \ni f[V(K(t))] \tag{17.18}$$

  where V is a shape map of $\mathbf{K}(X)$ to X and where $f : X \mapsto X$. Problems of visual control and other problems using feedback involving shape functions come into this category.
- The *Morphological Nagumo theorem* allows to characterise the dynamics governing shape evolutions, satisfying at all times the "geometric constraints", such as those accounting for problems of confinement.
- The theory of *Lyapunov functions* can also be adapted to the case of morphological equations, so as to enable us to study the asymptotic stability of tubes converging to a given shape.

[9] Note than $\Phi(z)$ is now a subset of X and no longer a single point. The extension to this situation of the differential equation $z' = \varphi(z)$ (involving a single-valued map φ) is a *differential inclusion*: $z'(t) \in \Phi[z(t)]$ for all $t \geq 0$.

- The morphological centre manifold theorem provides solutions that are shape maps $K \mapsto u(K)$ which *track* the solutions $t \mapsto K(t)$ of a given morphological equation in the sense that $\forall\, t \geq 0,\;\; x(t) = u[K(t)]$, when $t \mapsto x(t)$ is governed by a given differential equation. *Such a shape map $K \mapsto u(K)$ provides a way of "summing up" in terms of "vector characteristics" sets evolving under a morphological equation.* Along the same lines, it is possible to prove the adaptation of the viability theorem to the case of co-evolution: *the system (17.2) supplies the evolutions $t \mapsto x(t)$ and $t \mapsto K(t)$ satisfying the property of "co-viability" (17.1) $\forall\, t \geq 0,\;\; x(t) \in K(t)$ if and only if the dynamics f and g satisfy*

$$\forall\, K \in \mathbf{K}(x),\; \forall\, x \in K,\;\; f(x, K) \in g(x, K)(x) + T_K(x) \tag{17.19}$$

where $T_K(x)$ denotes the *contingent cone* to K at $x \in K$.
- Anne Gorre has proved that the system of morphological equations

$$\begin{cases} i) \;\; \overset{\circ}{L}(t) \ni g(L(t), M(t)) \\ ii) \;\; \overset{\circ}{M}(t) \ni h(L(t), M(t)) \end{cases} \tag{17.20}$$

governs evolutions possessing the *property of intersectability* (17.3(i)) if and only if, for each intersecting pair (L, M) of non-empty compact subsets,

$$\exists\, x \in L \cap M, \text{ such that } (f(L, M)(x, x) - g(L, M)(x, x)) \cap P_M^L(x) \neq \emptyset \tag{17.21}$$

where $P_M^L(x)$ denotes the *Bouligand paratingent cone* to (L, M) at $x \in L \cap M$: the paratingent cone is the subset of directions $v \in X$ such that there exist sequences $h_n \to 0$ with $h_n > 0$, $x_n \in L$ converging to x and $v_n \to v$ such that $x_n + h_n v_n \in M$ for all $n \geq 0$. In the same way, she characterised the systems of morphological equations

$$\begin{cases} i) \;\; \overset{\circ}{K}(t) \ni f(K(t), L(t), M(t)) \\ i) \;\; \overset{\circ}{L}(t) \ni g(K(t), L(t), M(t)) \\ ii) \;\; \overset{\circ}{M}(t) \ni h(K(t), L(t), M(t)) \end{cases} \tag{17.22}$$

governing evolutions satisfying the *property of confinement* (17.3(ii)) with the help of the same tools. *So the two cones – contingent and paratingent – introduced more than seventy years ago by Georges Bouligand have found profound applications in a domain – morphogenesis – that his son Yves Bouligand studied from a biological perspective* (see [12] for example, and the author's contributions to the present book, Chaps. 4 and 14).

17.8 Embryogenesis of the Zebrafish

In collaboration with Nadine Peyriéras, Alexandra Fronville has used mutational inclusions (in discrete time) to explain the first stages in the embryogenesis of tissues composed of cells constrained by the confinement of cellular tissues and their non-interpenetrability. The behaviour of a cell can be understood within the context of its interactions with its neighbours in a cellular tissue (cell populations). The co-evolution of tissues and their constituent cells is sufficient, on its own, to limit the generation of biological forms. The aim of Alexandra Fronville's work is to see how far the mathematical metaphor of mutational inclusions can support this hypothesis. Her results, Fig. 17.3, are to be compared to the three-dimensional image analysis of live zebrafish embryos (*Danio rerio*) obtained by Nadine Peyriéras and her colleagues, see Chap. 9. This tiny fish has the great advantage (for researchers) of being transparent, thus lending itself to experimentation with in vivo observation. They have recorded remarkable transitions in cell behaviour, filmed thanks to experimental feats that command the greatest admiration. The early stages of segmentation lead to a multiplication and diversification in the cells of the embryo. The first cell divisions are synchronous, of constant volume, and leave the cells little freedom of movement. To explain at least the first stages in embryogenesis, before

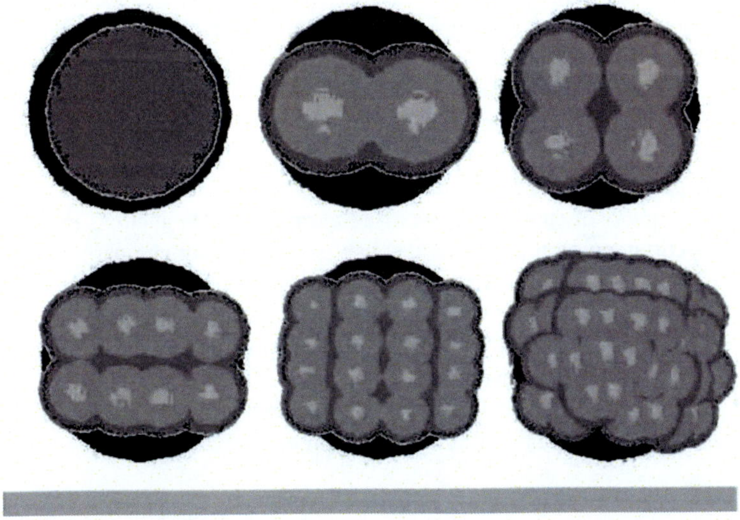

Fig. 17.3 This figure is obtained by discrete morphological inclusion under the constraints of confinement (above the yolk) and non-interpenetrability (*in virtuo*, to use the expression suggested by Jacques Tisseau [42], or *in silico*, derived from the Latin *in silicio* (in silicon) to describe computer simulations, or *in cognitivo* to denote activities involving mathematical metaphors. It is to be compared with snapshots of the divisions of the zebrafish embryo presented in Chap. 9. The interest of this comparison lies not so much in the similarities as in the reasoning and results underlying the morphological equations.

the biological mechanisms of cell differentiation come into play, supplanting the simple mechanisms of growth, Alexandra Fronville has developed a morphogenetic dynamics. At each step, the cells in the cellular tissue have the choice between:

1. dividing and placing the daughter cell in an available site inside the tissue and in the available space in which it is confined;
2. doing nothing;
3. dying of old age within a fixed time span;
4. committing suicide by apoptosis when the viability is threatened.

A visiting order of the cells in the tissue and a visiting order of the four choices listed above, considered as regulons of the discrete morphogenetic inclusion, are fixed at each step. These visiting orders can evolve from one step to the next, according to the nature of the cellular tissue, according to the very spirit of morphological equations. While awaiting more precise information about this mechanism, the visiting orders are *chosen arbitrarily, but not randomly*, in order to analyse their influence on the morphogenetic evolution. At this stage of the study, the constraints on viability are purely physical: a new cell cannot take the place of an existing cell in the cellular tissue, and this tissue must remain confined within a given space (for example, the cells cannot penetrate into the yolk). It is remarkable to observe that during the first divisions, this very simple scenario is sufficient to reproduce the form of the tissue of the zebrafish embryo.

This work by Alexandra Fronville invites the use of viability theory to study the control mechanisms necessary to the maintenance of viability. Firstly, without apoptosis, the morphological viability kernel is empty. Knowing this viability kernel, we can deduce a correspondence of morphological control that specifies which cells must die in order for the tissue to survive, study how the principle of inertia comes into play and search for the heavy evolutions in this context. These first mathematical metaphors of embryogenesis using morphological equations explain these morphogenetic dynamics, even outrageously simplified, by comparing them with the constraints of non-interpenetrability, the emergence of one or another shape that nature has produced (by *invagination*, or *ingression*, for example) in order to understand their diversity and plasticity without resorting to simulations. These forms are complex to grasp for the human brain, which paradoxically knows how to construct simple geometric forms that nature has not been able to produce (except through the intervention of the human mind): circles, rectangles, polyhedra, etc.

References

1. Aubin J.-P. (1981) Contingent derivatives of set-valued maps and existence of solutions to nonlinear inclusions and differential inclusions, *Advances in Mathematics*, Supplementary studies, edited by L. Nachbin, pp. 160–232.
2. Aubin J.-P. (1991) *Viability Theory*, Birkhäuser (Boston).
3. Aubin J.-P. (1993) Mutational Equations in Metric Spaces, *Set-Valued Analysis* **1**, 3–46.
4. Aubin J.-P. (1997) *Dynamic Economic Theory: A Viability Approach*, Springer (New York).

5. Aubin J.-P. (2000) *Mutational and morphological analysis: Tools for shape regulation and morphogenesis*, Birkhäuser (Boston).
6. Aubin J.-P. (2000) *Applied functional analysis*, 2nd edition, Wiley Interscience (New York).
7. Aubin J.-P. and Frankowska H. (1990) *Set-Valued Analysis*, Birkhäuser (Boston).
8. Aubin J.-P. and Murillo-Hernandez J. A. (2004) *Morphological equations and sweeping processes*, Advances in Mechanics and Mathematics, Kluwer Academic Publisher (Dordrecht).
9. Aubin J.-P., Bayen A., and Saint-Pierre P. (2011) Viability Theory, New directions, Springer (New York).
10. Beer G. (1993) *Topologies on closed and closed convex sets*, Kluwer Academic Publisher (Dordrecht).
11. Bouisset S. and Maton B. (1997) *Muscles, posture et mouvement*, Hermann (Paris), in French.
12. Bouligand Y. (1980) *La morphogenèse : de la biologie aux mathématiques*, Maloine (Paris).
13. Cardaliaguet P. (1999) *Contributions à la théorie du contrôle, aux jeux différentiels, au calcul des variations et aux propagations de front*, Habilitation à Diriger des Recherches, Université de Paris-Dauphine, in French.
14. Cardaliaguet P. (2000) On front propagation problems with nonlocal terms, *Adv. Diff. Eq.* **5**, 213–268.
15. Cardaliaguet P. (2000) Front propagation problems with nonlocal terms. II. *J. Math. Anal. Appl.* **260**, 572–601.
16. Delfour M. and Zolesio J.-P. (2001) *Shapes and geometries: Analysis, differential calculus and optimization*, SIAM (Philadelphia).
17. Delfour M. and Zolesio J.-P. (2000) *Intrinsic differential geometry and theory of thin shells*, Quaderni, Scuola Normale Superiore (Pisa).
18. Demongeot J., Kulesa P., and Murray J.D. (1996) Compact set-valued flows: Applications in biological modelling, *Acta Biotheoretica* **44**, 349–358.
19. Doyen L. (1993) *Optimisation, évolution et contrôle de formes*, Thèse de l'Université de Paris-Dauphine.
20. Doyen L. (1995) Mutational equations for shapes and vision-based control, *J. Math. Imaging Vis.* **5**, 99–109.
21. Goethe J. von (1790) *Versuch, die Metamorphose der Pflanzen zu erklären*, C.W. Ettinger (Gotha), in German. English translation: Essay in Elucidation of the Metamorphosis of Plants, in *Goethe's Botany* by A. Arber (1946), The Chronica Botanica Company (Waltham MA).
22. Gorre A. (1996) *Evolution de tubes opérables gouvernée par des équations mutationnelles*, Thèse de l'Université de Paris-Dauphine, in French.
23. Gorre A. (1996) The "Montagnes Russes" algorithm in mutationnal spaces, in *Parametric Optimization IV*, Springer (Berlin).
24. Lorenz, T. (2004) *First-order geometric evolutions and semilinear evolution equations: A common mutational approach*, Thèse de l'Université de Heidelberg.
25. Matheron G. (1975) *Random sets and integral geometry*, Wiley (New York).
26. Mattioli J. and Schmitt M. (1993) *Morphologie mathématique*, Masson (Paris), in French.
27. Moore R.E. (1979) *Methods and applications of interval analysis*, SIAM Studies in Applied Mathematics (Philadelphia).
28. Morel J.-M and Solimini (1994) *Variational Methods in Image Segmentation*, Birkhäuser (Boston).
29. Murray J.D (2002) *Mathematical biology*, 3rd edition, Springer (Berlin).
30. Petitot J. (1980) "Forme", Encyclopédie Universalis (Paris), in French.
31. Petitot J. (1985) *Morphogenèse du sens*, Presses Universitaires de France (Paris), in French.
32. Petitot J. (1992) *La physique du sens*, Éditions du CNRS (Paris), in French.
33. Pichard K. (2002) *Équations différentielles dans les espaces métriques. Applications à l'évolution de domaines*, Thèse de l'Université de Pau, in French.
34. Rockafellar R.T. and Wets R. (1997) *Variational Analysis*, Springer (Berlin).
35. Serra J. (1982) *Image Analysis and Mathematical Morphology*, Academic Press (New York).

36. Serra J. (1988) *Image Analysis and Mathematical Morphology, Volume 2: Theoretical Advances*, Academic Press (New York).
37. Sethian J.A. (1996) *Level set methods*, Cambridge University Press (Cambridge).
38. Sokolowski J. and Zolesio J.-P. (1992) *Introduction to shape optimization*, Springer (Berlin).
39. Thom R. (1972) *Stabilité structurelle et Morphogenèse*, Benjamin (New York), Édiscience (Paris), in French. English translation: *Structural stability and morphogenesis* (1975), Benjamin (Reading MA).
40. Thom R. (1980) *Modèles mathématiques de la morphogenèse*, Christian Bourgois (Paris), in French. English translation: *Mathematical models of morphogenesis* (1983), John Wiley and sons (Chichester).
41. Thompson D'Arcy W. (1917) *On growth and form*, Cambridge University Press (Cambridge).
42. Tisseau J. (2001) *Virtual reality – in virtuo autonomy*, Accreditation to direct research, University of Rennes I.
43. Zolesio J.-P. (1976) Existence de vitesses convergentes, *Comptes Rendus de l'Académie des Sciences, Paris* **283**, 855–858, in French.
44. Zolezio J.-P. (1979) *Identification de domaines par déformations*, Thèse de Doctorat d'Etat, Université de Nice, in French.

Chapter 18
Computer Morphogenesis

Jean-Louis Giavitto and Antoine Spicher

18.1 Explaining Living Matter by Understanding Development

18.1.1 The Animal-Machine

In 1739, Jacques de Vaucanson (1709–1782) presented a celebrated automaton to the French Academy of Sciences. It was called the *Canard Digérateur* (Digesting Duck, Fig. 18.1), a masterpiece of anatomical simulation, with more than four hundred moving parts reproducing the main vital functions (respiration, digestion, locomotion): the animal flapped its wings, ate grain and defecated (the grain being digested by dissolution, according to the inventor).

In making these "mobile anatomies", Jacques de Vaucanson was almost certainly influenced by the biomechanistic philosophy of René Descartes (1596–1650), who reduced the organs of the human body to parts in a machine "designed by God". Indeed, Descartes believed that one can understand life by comparing it to a machine: that one can explain the main bodily functions – digestion, locomotion, respiration, but also memory and imagination – as if they were produced by an automaton, like a clock designed to show the time simply by the layout of its wheels and counterweights. But when René Descartes tried to convince Queen Christina of Sweden that animals were just another form of machine, she is said to have replied:

> Can machines reproduce?

Three centuries were to pass before her question received an answer. A hundred years later, the automata of Vaucanson were imitating the main physiological functions, but they still could not reproduce, and it was only with the publication of an article by John Von Neumann in 1951, *The General and Logical Theory of Automata*, that it was finally possible to believe that a machine could effectively build a copy of itself [34].

J.-L. Giavitto (✉)
University of Évry, Paris, France
e-mail: giavitto@ibisc.univ-evry.fr

P. Bourgine, A. Lesne (eds.), *Morphogenesis*, DOI 10.1007/978-3-642-13174-5_18,

Fig. 18.1 Vaucanson's duck. Voltaire described Vaucanson in these lines: *"While rival of the old Prometheus' fame, Vaucanson brings to man celestial flame. Boldly to copy nature's self aspires, And bodies animates with heavenly fires"*

To meet Queen Christina's objection, it is necessary to define precisely what we mean by "machine" and what we mean by "reproduction". For Von Neumann, who had a very functionalist approach to this question, mechanics can ultimately be reduced to a computer programme, and reproduction consists in duplicating this programme. This does not mean using a command in the computer's operating system to copy a file containing a programme, but ensuring that the functioning of the programme produces a complete and functional description of the programme itself. Fig. 18.2 shows an example of such a programme written in the programming language C: its execution produces a file containing the exact copy of its own code. This is called a *self-replicating* code.

Von Neumann's purpose was clearly to show that living processes can be reduced to mechanical processes, described by operations that can be performed autonomously, without the help of an "invisible mahout": to a machine, in other

```
#include<stdio.h>

main(){char*c="\\\"#include<stdio.h>%cmain() {char*c=%c%c%c%.
102s%cn%c;printf(c+2,c[102],c[1],*c,*c,c,*c,c[1]);exit(0);}
\n";printf(c+2,c[102],c[1],*c,*c,c,*c,c[1]);exit(0);}
```

Fig. 18.2 A self-replicating code. This programme is made up of two lines of code in the programming language C. The second line of the programme (starting with main) has been arbitrarily typeset over three lines to make it more legible words. And for Von Neumann, like Queen Christina, reproduction and development are a specific characteristic of living things. But for Von Neumann, this characteristic is just a particular property possessed by certain machines, not a quality that transcends physical processes, giving special status to biological ones. The existence of a machine, an automaton, capable of reproduction, is therefore a key factor in the age-old debate opposing the relative status of biology and physics.

This debate has not been easy to settle, reproduction being one of the most fundamental processes in the life of organisms and appearing to resist any physical explanation. Intuition suggests that if, as a result of its functioning, a machine A can produce a machine B, then A must contain, in one form or another, a complete description not only of B but also of the specific mechanisms instructing it how to use that description to actually produce (construct) B. This description must be internal to A, otherwise we would be dealing with a mechanism of copying rather than reproduction. We should therefore be able to define a certain measure of complexity and show that A is necessarily more complex than B. But in this case, our intuition leads us astray.

18.1.2 From Self-Reproduction to Development

Modern biologists may ask themselves the same questions as the philosophers and queens of past centuries, but today they seek to understand the mechanisms of reproduction by elucidating the *processes* leading from the germ cell to the complete organism: the aim is to understand, step by step, the *construction of an organism over the course of time*, through the multitude of local interactions of its constituent elements. In a word, *development*.

The elements that Von Neumann brought to the debate are very abstract: they are based on the description of a cellular automaton which reproduces, over the course of time, the configuration of a spatial subdomain in a neighbouring region. A cellular automaton can be described by a *predefined* network of sites, called cells, each cell possessing one of a finite set of states. The state of each cell is updated according to a predefined rule of evolution, which takes into account the state of the cell and the state of its neighbours at time t to calculate the state of the cell at time $t + 1$. The functioning of the automaton corresponds to the updating of the state of its cells at discrete time intervals (see Fig. 18.3).

We are a long way from the molecular mechanisms to which modern biologists wish to reduce biological phenomena. The existence of a self-replicating automaton suggests that there is no problem of principle in the existence of such a machine, but it tells us nothing about the "how" of biological processes. Nevertheless, the concepts of programme, code, automaton, memory and information have invaded biology and assumed an explanatory value, especially in developmental biology [19, 20]: biologists need models and metaphors to understand (i.e. to represent, analyse and interpret) the huge mass of experimental data they have collected. For example, the concept of *genetic code* plays a similar role in the living cell as the rule governing the evolution of states does in the Von Neumann automaton.

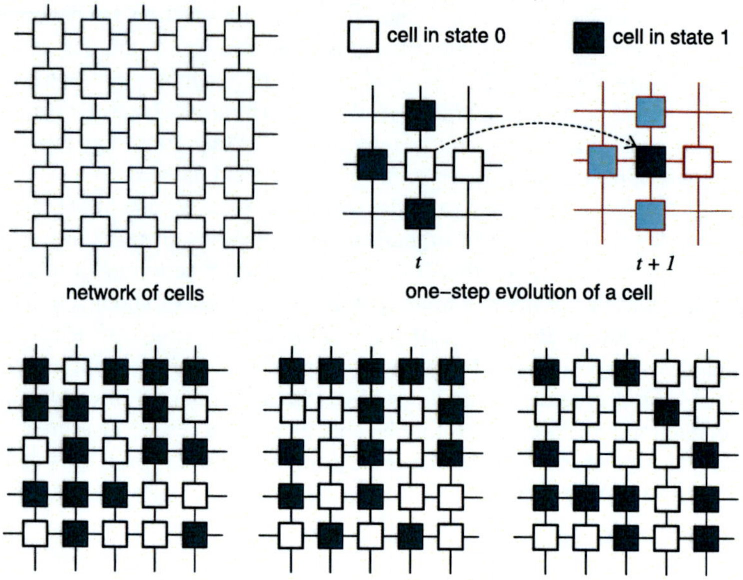

Fig. 18.3 A cellular automaton is a network of cells, each joined to its neighbours by links. Here the network is a rectangular grid. Each cell possesses a state (here either 0 or 1). The rule of evolution used here is: the state of a cell is the modulo-2 addition of the states of the neighbouring cells. An example of the evolution of one cell is shown top right. The three networks below show three successive stages in the evolution of the automaton. The rule is applied simultaneously to all the cells. Von Neumann's self-replicating automaton is a model of this type, where the rules of evolution lead to the reproduction of the initial configuration of a given region in an adjacent region

18.1.3 Development as a Dynamical System

The concept of *dynamical system* allows to formalise the idea of process of development. A dynamical system (DS) is characterised by observations that evolve over time. These observations are the *variables* of the system, and they are linked by certain relations. These variables account for relevant properties of the system (whether they be biological, physical, chemical, sociological, or other). At a given moment in time, they have a certain value, and the set of these values constitutes the *state of the system*. The set of all the possible states of a system constitutes its *state space* (or configuration space). For example, a falling stone is a system characterised by the variables *position* and *velocity* of the stone. These two variables are not independent: if we consider the position of the stone as a function of time, then its velocity is the derivative of that function.

The succession of system states over time is called a *trajectory*. A DS is a formal way of specifying how the system moves from one point in the configuration space (one state) to another point (the next state). This can be done directly, by a function (the function of evolution of the system), or indirectly, by giving constraints (equations) on the possible future state (which is not necessarily unique, if the system

Table 18.1 Three examples of formalism used to specify a dynamical system according to the continuous or discrete nature of the variables and of time. Iterated functions correspond to sequences $x_{n+1} = f^{n+1}(x_0) = f(x_n)$ for a given function f on $\mathbf{R}$. Many other formalisms have also been studied

C : Continuous D : Discrete	Differential Equation	Iterated Functions	Finite Automaton
Time	C	D	D
State	C	C	D

is not deterministic). A variety of mathematical formalisms correspond to this very general concept of dynamical system. For example, the variables can take continuous or discrete values. Likewise, the progression of time can be continuous or in discrete steps. Examples of formalisms corresponding to these cases are listed in Table 18.1.

In simple cases, the trajectory of a dynamical system can be expressed explicitly by an analytic function of time t. In the case of the falling stone, for example, the differential equations $\mathrm{d}x/\mathrm{d}t = v$ and $\mathrm{d}v/\mathrm{d}t = \mathbf{g}$ can be explicitly integrated to give the distance travelled by the stone as a function of time: $x = \mathbf{g}t^2/2$.

In more complex cases, an analytic equation giving the trajectory does not exist, and computer simulation is then a favoured approach for studying the trajectories of the system. In addition, instead of focusing on one particular trajectory, we can look at qualitative properties satisfied by all the possible trajectories, for example: "if we wait long enough, the system ends up in a well-defined state in which it then remains" or "if the trajectory passes through these states, it will never return". When there is no faster means of predicting properties than by observing or simulating them, we qualify them as *emergent properties*. Note that DS with very simple specifications can produce very complex trajectories (we sometimes speak of *chaotic behaviour*); moreover, calculating the trajectory of the system can be expensive in terms of computer time and require a vast amount of memory.

18.1.3.1 The Structure of States

Another important characteristic by means of which dynamical systems can be classified is the structure of states. In the example of the falling stone, the structure of a state is simple: it is a pair of vectors (velocity, position).

Very often, the structure of a state reflects the spatial organisation of the system. Let us take the example of the diffusion of heat in a volume. The distribution of the temperature has a structure, related to the spatial organisation of the volume. We can therefore define a scalar field assigning a temperature to each point. The evolution of this field follows a law of diffusion specified by a partial derivative equation. This links the temperature at time $t + dt$ of a point p to the values of the temperature field at p and in its neighbourhood at time t.

Very often, subsystems only interact if they are connected or physically close: we call this the property of *locality* (there is no action at a distance). The structure of a state then reflects this division into subsystems, and the function of evolution respects the property of locality. For the evolution of temperature in a volume, each state assigns a temperature to each point in the volume V and the state space is therefore the set of functions of from in to. The heat diffusion equation governing the evolution of the system indicates that the temperature of a point in V depends solely on the temperature of the neighbouring points.

18.1.3.2 Development as Trajectory of a Dynamical System

Above, we stated that the concept of *genetic code* has much in common with the rules specifying the evolution of cell state in Von Neumann automata. This is the concept underlying the "all-genetic" paradigm, according to which the complete evolution of the organism is coded in its genetic material, and every characteristic is uniquely determined by the genes. This viewpoint has been substantially challenged [2], in favour of a more flexible approach, reconciling the genetic and epigenetic viewpoints on development. Living systems may be dynamical, but they are also open systems, interacting with their environment. Development should therefore be regarded as a co-construction, depending on interactions both within the system and outside it (with the environment). Genetic material does not constitute a complete and sufficient description of any given organism, although it is indispensable. Cell machinery, for example, also plays a central role, as has been demonstrated experimentally by the technique of cloning in which a nucleus (i.e. the genetic material of a cell) is introduced into a germ cell.

However, the processes of morphogenesis involving the movement and reorganisation of matter are also characterised by a second property: the state space and its topology can also evolve over time.

Let us illustrate this idea by comparing it to the two examples described above. In the case of the falling stone, the velocity and the position of the stone change at each moment but the system is always adequately described by a pair of vectors. In this case, we say that the dynamical system has a *stable (static) structure*. The same is true for the evolution of the temperature in the volume V: V is fixed in advance and each state is always an element of $V \to \mathbf{R}$. In these two examples, the state space can be described adequately at the beginning of time, before the simulation; it corresponds to the space of the measurements of the system. The value of these measurements changes over time, but the data of the state space and its topology are not variables of the system and cannot evolve over the course of time.

Quite the opposite holds true for the processes of development: biological processes form highly structured and hierarchically organised dynamical systems, the spatial structure of which varies over time and must be calculated in conjunction with the state of the system. We call this type of system a *dynamical system with dynamical structure*, which we shall abbreviate to $(\mathrm{DS})^2$.

The fact that the very structure of a biological system is dynamical has been highlighted by several authors; we can cite, in different domains: the concept of *hypercycle* introduced by Eigen and Schuster in the study of autocatalytic networks [13], the theory of *autopoietic systems* formulated by Maturana and Varela [43], *systems of variable structure* developed in control theory by Itkis [30], or the concept of *biological organisation* introduced by Fontana and Buss to formalise and study the emergence of self-maintaining functional structures in a set of chemical reactions [18]. The objective of all of these works has been to grasp and formalise the idea of change in the structure of a system, change that is coupled with the evolution of the state of the system.

$(DS)^2$ are widespread in models of plant growth and more generally in developmental biology, in multiscale cell models, mechanisms of protein transport and compartmentalisation, etc. But they are also relevant in other domains, such as the modelling of mobile networks, Internet and the Web, the development of cities, traffic jams, self-assembly processes, autocatalytic networks in chemistry, semantic networks in learning, social behaviour, etc.

18.1.3.3 An Example

To illustrate the concept of $(DS)^2$, let us take the example of the development of an embryo. The initial state of the embryo is described by the state $s_0 \in \mathcal{S}$ of the germ cell (however complicated that description might be). After the first division, we have to describe the state with 2 cells, that is to say a new state $s_1 \in \mathcal{S} \times \mathcal{S}$. But when the number n of embryo cells becomes large enough, the state of the system can no longer be adequately described by an element of $\mathcal{S}^n$. This set only describes the state of each cell; it does not contain the spatial information necessary to describe the network of cells (their positioning in relation to each other). And yet this network is of prime importance, because it conditions the diffusion of signals (chemical, mechanical or electrical) between cells and therefore, in the end, their functioning. With each movement, division or death of a cell, the topology of this network changes. For example, during *gastrulation*, cells initially far apart become neighbours, enabling them to interact and changing their destiny (cell differentiation).

18.1.4 What Formalism for Dynamical Systems with Dynamical Structure?

Dynamical systems with dynamical structure are difficult to study because they are difficult to formalise. Let us return to the example of the embryo to illustrate this.

We have indicated that the position of each cell changes over time, making it difficult, for example, to specify the processes of diffusion between cells. One solution that comes immediately to mind is therefore to complete the state of a cell with information about its position, and to consider $\mathcal{T} = \mathcal{S} \times \mathbf{R}^3$ as a building block[1] allowing to construct the set:

[1] To simplify, we only take into account the position of each cell in $\mathbf{R}^3$, but we should also specify its form, which conditions its neighbourhood and its exchanges with other cells (for example the surface exchange area between two neighbouring cells, which conditions intermembrane flow).

$$\begin{aligned}\mathcal{T}^* &= \mathcal{T} \cup \mathcal{T}^2 \cup \cdots \cup \mathcal{T}^n \cup \dots \\ &= \mathcal{T} \cup \mathcal{T} \times \mathcal{T}^* .\end{aligned}$$

It is certainly possible to characterise an embryo as a point in this phase space, but that does not get us very far: $\mathcal{T}^*$ has very little intrinsic structure and does not provide much information about the possible trajectories of the systems. For example, the function of evolution will be very difficult to define and there is little chance that it will be continuous.

18.1.4.1 The Problem of Locality

The function of evolution will be difficult to define because specifying the position of each cell in terms of its coordinates $\mathbf{R}^3$ presupposes the definition of a global reference point. During the evolution of the embryo, the growth of a cell pushes away the neighbouring cells, which in turn push away their neighbours, until the position of every cell has been changed. Between two successive states, we therefore have to express the change in the position of each cell by a *global transformation* of coordinates. Because it must express globally the changes in each position, and because these changes are due to multiple concurrent local transformations, the expression of this transformation can be arbitrarily complex.

The origin of this problem lies in the *extrinsic and global* expression of the form of the system[2] and one solution is therefore to specify intrinsically the position of each cell, for example by including the distance from its neighbours in the state $s \in \mathcal{S}$ of each cell. In this case, the specification of changes in the position of a cell is local, but as the neighbourhood of each cell changes, we are again faced with the problem of a state space that changes over time.

18.1.4.2 The Problem of Continuity

Let us return to the example of the falling stone. The position and the velocity of the stone vary continuously. The state of the system therefore varies continuously over time and the trajectory of the system is a *continuous function* of time in the state space. This continuity allows to reason in terms of infinitesimal evolutions of the system and to write a differential equation characterising the trajectory. In more complicated cases, we obtain a partial derivative equation (when the state has a spatial structure) or a set of such equations when several different modes of functioning have to be taken into account (a finite and usually small number).

---

[2] In the approach described, the specification of the position of the cells uses a global reference point independent of the growing embryo. This reference point corresponds to the identification of points in the space surrounding the form, and not to a process intrinsic to the growing form: the laws governing the movement, division and death of cells would be the same if the embryo was developing within a toric volume (but the result could be different because the neighbourhoods of the cells would be different).

In the case of embryo development, this is no longer possible: as long as there is no movement,[3] division or death of cells, the state s belongs to a certain $\mathcal{T}^n$ and this evolution is continuous (assuming that the electric potentials, chemical concentrations, mechanical constraints, etc. evolve continuously). But the essential morphogenetic events (for example a cell division that changes the state from $\mathcal{T}^n$ to $\mathcal{T}^{n+1}$) are by nature discontinuous.[4]

18.1.4.3 Towards Other Solutions

The modelling and simulation of the evolution of a $(DS)^2$ are therefore particularly arduous: it is difficult to define the structure and the dynamics of the system at the same time, because one is dependent on the other. The example given above highlights the inadequacy of global and continuous formalisms (we want to express an evolution as a succession of discrete morphogenetic events corresponding to qualitative discontinuities and changes). However, it is still possible to describe these systems, with the laws of evolution often being informally described as a set of *local* transformations acting on an ordered set of discrete entities.

Faced with these difficulties, several researchers have suggested using *rewriting systems* to formalise this type of description.

18.2 Rewriting Systems

18.2.1 Introduction

Rewriting systems (RS) are among the formalisms that computer scientists have appropriated and developed, especially for modelling changes in the state of a process. A rewriting system is a mechanism allowing to define the replacement of one part of an object by another. The objects concerned are usually *terms* that can be represented by a tree, of which the inner nodes are operations and the leaf nodes are constants (see Fig. 18.4). An RS is defined by a set of rules, and a rule is a pair denoted $\alpha \to \beta$. A rule $\alpha \to \beta$ indicates how a sub-term α can be replaced by a term β.

18.2.1.1 An Example

Let us take the arithmetical expressions and the rule $0 + x \to x$. Intuitively, this rule specifies that any expression that can take the form "0 added to something

[3] Cell movement is sufficient to change the topology and therefore the interaction between cells.

[4] In the example we have been using, morphogenetic events are discontinuous because the modelling is done at cell level. We could have modelled the concentration of different molecules at each point in space, which might have avoided this problem of discontinuity (the movement of each molecule being *a priori* continuous). But this raises another problem: how do these concentrations represent the biological entities that interest us: cells, tissues, organs, etc.?

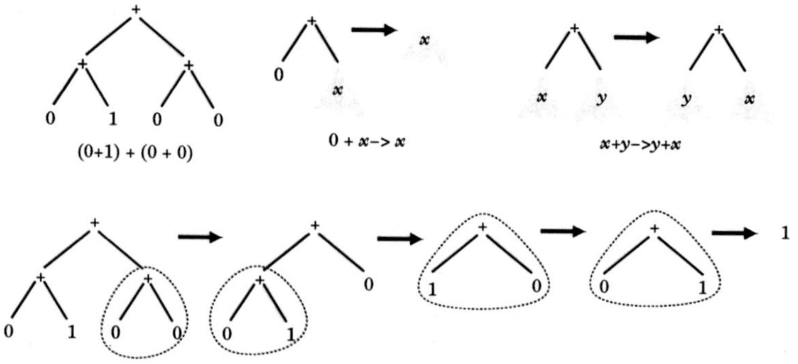

Fig. 18.4 Representation of the term $(0+1)+(0+0)$ and application of the two rules $0+x \to x$ and $x+y \to y+x$. At each reduction, the strategy here is to apply one rule at a time. The subtree filtered by the left side of the rule to be applied is circled by a dashed line. The applications are non-deterministic, in the sense that we could have chosen other applications at each step. For the first reduction, for example, we could have applied the same rule $0+x \to x$ to the left subtree of the root rather than the right subtree. We could also have chosen to apply the rule $x+y \to y+x$ to any of the three inner nodes (3 possibilities). The final term obtained is the constant 1, and this is a *normal form* for the two rules

denoted by x" can be rewritten more simply as "the thing denoted by x". Thus, the expression $e = 1+(0+3)$ can be rewritten as $e' = 1+3$ by applying the above rule to the sub-term $(0+3)$ of e. We also write $e \to e'$ to indicate that e can be rewritten as e' through one sole application of the rule.

The sequence $e \to e_1 \to \cdots \to e_n \to e'$ is called a *derivation* of e. We say that e is a *normal form* if there is no e' such that $e \to e'$.

18.2.1.2 RS and Decision Procedure in an Equational Theory

The original motivation behind RS was to provide a decision procedure in equational theories. In these theories, the aim is to prove automatically the equality of two complex terms solely by using predefined elementary equalities. The idea is to orientate the equations (for example, to orient the equality $0+x = x$ into a rule $0+x \to x$) and to use the rules obtained to derive the normal form e' of a term e. The normal form e' is equivalent to e (since each substitution transforms a subterm into an equivalent term) and can be interpreted as a simplification of e. Two terms e_1 and e_2 are then equivalent in the theory if they reduce to the same normal form e. For example, e_1 defined by $0+(1+3)$ is equivalent to e_2 defined by $1+(0+3)$, because e_1 and e_2 reduce to the same normal form e: $1+3$.

For this decision procedure always to succeed, there must exist a normal form for each expression (property of *normalisation*) and each expression must have one sole normal form (property of *confluence*). These two properties are not quite sufficient for the decision procedure to calculate automatically; at each step we must also choose a derivation, i.e. choose which subterm will be rewritten and by which rule: this is the *strategy* of rule application.

The theory of RS [9, 10] is mainly used in algebra and logic, but it can be applied in almost every branch of computing (from Petri networks to symbolic calculus, from the theory of demonstration to lambda calculus). One key result is that RS, considered as processes of calculation, are *Turing-complete* (any computational process, i.e. described by a Turing machine, can be formalised by an RS). The use of rules to transform a term is such a fundamental operation that several generic environments have been developed to define and apply RS (see, among others, the websites of the projects ELAN [16] and MAUDE [33]). The tools differ in the terms they take into account, the α patterns allowed on the left-hand side of a rule for selecting subterms, and the strategies of application that can be defined.

18.2.2 Rewriting Systems and the Simulation of Dynamical Systems

The above presentation suggests that a rule $\alpha \to \beta$ specifies a term β equivalent to (and simpler than) the term α. But we can interpret this rule as the result of a computation (the expression β is the result of evaluating the expression α) or as the evolution of a subsystem changing from state α to state β. RS can therefore be used to model and simulate DS:

- a state is represented by a term and the state of a subsystem is represented by a subterm;
- the evolution function is encoded by the rules of the RS in the following manner: the left side of the rule corresponds to a subsystem in which the elements *interact*, and the right side of the rule corresponds to the result of their interaction.

Thus, the derivation of a term s corresponds to a possible trajectory of a DS starting from the initial state s. A rewriting rule then corresponds to the specification of the evolution of a subsystem. A normal form corresponds to a fixed point in the trajectory (the system is in equilibrium and no evolution can take place).

18.2.2.1 An Example

For the development of the embryo, a rule $c \oplus i \to c'$ can be interpreted as a cell in the state c which, on receiving a signal i, evolves to the state c'; a rule $c \to c' \oplus c''$ represents a cell division; a rule $c \to \emptyset$ (c gives nothing) represents apoptosis; etc. [17, 23]. The idea is that the evolution of a biosystem is specified by rewriting rules of which the left side selects an entity in the system and the messages sent to it, and the right side describes the new state of the entity. The operator $\oplus$ which appears in the rule denotes the composition of local entities in a global system (in our example, the aggregation of cells in an embryo). The capacity to represent both the changes of state and the appearance and disappearance of cells within the same formalism makes RS good candidates for the modelling of $(\mathrm{DS})^2$.

18.2.2.2 Dealing with Time

One important factor in the modelling of a $(\mathrm{DS})^2$ is the treatment of time. The model of time favoured in RS is clearly an event-driven, atomic and discrete model: time passes when an evolution occurs somewhere in the system, the application of a rule corresponds to an event and specifies an atomic and instant change in the state of the system. The concept of duration is not taken into account (although it could be, within this formalism, by considering the start and the end of a time interval as events). The choice of a strategy of application provides a certain degree of control over the model of time: for example, a maximal parallel application of rules to change from one global state to another corresponds to synchronous dynamics, while the application of one sole rule corresponds to asynchronous dynamics.

18.2.2.3 Dealing with Space

A rule of the form $c \oplus i \rightarrow c'$ presupposes that a signal i produced by a certain cell will reach its target c somewhere else in the system. The operation $\oplus$ used to amalgamate the states of the subsystems and the messages of interaction into the state of a complete system must therefore express the spatial dependencies and functional organisation of the system studied.

The concept of rewriting has mainly been developed and studied for the rewriting of terms. These represent a severe restriction on RS, because their use requires the encoding of the highly organised structure of $(\mathrm{DS})^2$ in tree form. The possibility of defining rules of evolution depends on this encoding. This work demands a great deal of creativity and intuition. It is difficult to represent in a satisfactory manner the organisation of a biological system into molecules, compartments, cells, tissues, organs and individuals, and this has motivated an extension of the concept of rewriting to structures more sophisticated than terms (for example, we can define a concept of rewriting on a graph, see also [21, 22]).

Nevertheless, even when they are limited to trees, RS offer remarkable examples of modelling of $(\mathrm{DS})^2$, particularly in the biological domain. By playing on the properties of the operators, it is possible to model several types of organisation. In the following sections, we shall give examples where:

- the operation $\oplus$ is associative and commutative, which allows to model a "chemical soup";
- several operations can be considered simultaneously, as a means to introduce the idea of compartmentalisation;
- the operation $\oplus$ is simply associative, which allows to represent sequences and tree structures.

18.3 Multiset Rewriting and Chemical Modelling

The state of a chemical solution can be represented by a *multiset*: a set in which one element can appear several times, as in a chemical solution where several molecules of the same species are present at the same time. A multiset can be formalised by a formal sum in which the operator $\oplus$ is *associative* and *commutative*. For example:

$$(a \oplus b) \oplus (c \oplus b)$$

represents a multiset (e.g. a chemical solution) containing the elements (e.g. the molecules) a, b and c, where two copies of b appear. Since the operation $\oplus$ is associative, we can discard the brackets, and the property of commutativity allows us to reorganise the elements in this sum as we like:

$$(a \oplus b) \oplus (c \oplus b) = a \oplus b \oplus c \oplus b = a \oplus b \oplus b \oplus c = c \oplus b \oplus a \oplus b = \ldots$$

A multiset therefore corresponds to a tree in which associativity allows us to "flatten" the branches, and commutativity allows us to permute the leaves.

In a chemical solution, Brownian motion agitates the molecules, and after a sufficiently long time each molecule will have had the opportunity to meet and interact with any other molecule in the solution. Once we have represented the state of a chemical solution as a multiset, it is therefore easy to formulate the chemical reactions as rewriting rules on multisets. The associativity and commutativity of the operator $\oplus$ play the role of Brownian motion and allow to "group together" arbitrarily the elements of the multiset corresponding to a left side of the rule before that rule is applied. For example, the three rules:

$$r_1 : a \oplus a \rightarrow a \oplus a \oplus b \qquad r_2 : a \oplus b \rightarrow a \oplus b \oplus b \qquad r_3 : b \oplus b \rightarrow b \oplus b \oplus a$$

represent second-order catalytic reactions between type a and type b molecules (a collision between two molecules catalyses the formation of a third molecule, without consuming the first two). Thus, if a reaction r_1 occurs in state $a \oplus c \oplus a \oplus b$, the result will be the state $a \oplus c \oplus a \oplus b \oplus b$ where an extra b has been produced. Note that it is not necessary for the two a molecules to be side by side, because we can always rearrange the term to make it so.

Several chemical reactions can happen at the same time, *in parallel.* This corresponds to the simultaneous application of several rules to different molecules. The strategy of applying as many rules as possible at a given time step is called a *maximal parallel* application. Such a strategy is non-deterministic: on the multiset $a \oplus a \oplus b$ we can apply r_1 or r_2, but not both at the same time, due to a lack of resources. In this case, one of the rules is chosen at random. A reduction step is then repeated to simulate the evolution of the state of the chemical solution. Several approaches are possible, in terms of adjusting the strategy of rule application, to take into account the kinetics of chemical reactions [5, 24].

Note that in this approach, each molecule is explicitly represented and each interaction is explicitly treated: this is known as *agent-based simulation.* This approach can be compared to more classic approaches which represent the concentration of each chemical species rather than each molecule. Obviously, in this particular case, the agent-based approach is more costly in computing time and memory, but it allows to simulate finely the complex phenomena, such as fluctuations and correlations, that are beyond the reach of global approaches.

This abstract formalisation of chemical reactions constitutes a domain of research called *artificial chemistry* [11, 12], tackling problems ranging from the automatic generation of combustion reactions [6] to the study of mechanisms of self-organisation in the evolution of self-catalytic networks [18].

18.3.1 Some Examples of Application

18.3.1.1 A simple example of population growth

To illustrate multiset rewriting and its application to modelling, we shall look at an example of a biological nature: the multiplication of a unicellular organism in a test-tube. We assume that a cell exists in two forms, A and b: A represents a mature cell ready to divide and b a young cell that will evolve to form A. Each cell division of A produces one cell of type A and one cell of type b. These evolutions can be formalised by the two rules:

$$\begin{aligned} r_1 : &\quad A \longrightarrow A \oplus b \\ r_2 : &\quad b \longrightarrow A \end{aligned}$$

If the initial state of our test-tube is represented by $m_0 = A \oplus b \oplus b$, the first three evolutions give us:

$$\begin{aligned} m_0 \to A \oplus b \oplus A \oplus A \to A \oplus b \oplus A \oplus b \oplus A \oplus b \oplus A \to \\ \to A \oplus b \oplus A \oplus b \oplus A \oplus b \oplus A \oplus b \oplus A \oplus A \oplus A \to \dots \end{aligned}$$

Simulation of this process can be used to determine, for example, the ratio of forms A to forms b in the population after a given time. Moreover, as we mentioned earlier, we can test properties verified by all the processes satisfying these rules of evolution. For example, Fibonacci[5] proved that the ratio $\#A/\#b$ of the number of A to the number of b converges asymptotically towards the golden number, whatever the initial state.

18.3.1.2 Applications to the Modelling of Networks of Biological Interaction

The modelling of biological interaction networks (genetic control networks, signalling networks, metabolic cascades, etc.) is a relatively new domain of application of these techniques. In [17], Fisher and his co-authors proposed using the concept of multiset to represent proteins involved in a cascade of interactions in a signalling

[5] In 1602, Fibonacci studied the question of how fast a population of rabbits would grow under ideal conditions. Imagine that a pair of rabbits, one male and one female, are put in a field. These rabbits are capable of reproducing after one month, so that at the end of the second month, the female has given birth to another pair of rabbits. To simplify, we assume that the rabbits never die and that each female gives birth to a new pair, composed of a male and a female, every month starting from the second month. If we represent a newly-born pair by b and a mature pair by A, then the rule r_1 corresponds to the breeding of a new pair and the rule r_2 to the maturing of a young pair.

pathway. This avenue has been widely developed, in particular to take into account the different complexes that proteins can form [14, 15].

If multiset rewriting has been used to model signalling networks and metabolic pathways, we should not deduce that the cell can be compared to a test-tube containing a chemical soup. On the contrary, the cell is a spatially highly organised medium, with compartments, vesicles, cargos, membranes, etc., which allow to localise the different chemical species involved (for example the receptors are localised on the cell membrane, while the genes are located in the nucleus; other proteins are anchored and diffuse in membranes like the endoplasmic reticulum). Among other things, this localisation helps to make certain reactions much more efficient. Other phenomena, such as the extreme density of proteins in the intracellular medium, render the simple model of chemical soup simply inadequate. Taking into account this spatial organisation is one of the main challenges currently faced in the modelling of cellular processes [31, 42].

18.3.1.3 Heat Diffusion in a Bar

Above, we stated that the properties of associativity and commutativity allow us to deconstruct a term so that each element can interact with any other element, in the manner of molecules in a well-mixed chemical soup. But with the appropriate encoding, multiset rewriting can be "diverted", so as to take into account geometric information.

The process we want to model is the diffusion of a set of particles along a line. This problem also corresponds to the diffusion of heat in a thin rod, with each particle representing a quantum of heat. The line is discretized into a sequence of small intervals indexed by consecutive integers. Each interval contains a number of particles (possibly zero). At each time step, a particle can stay in the same interval or diffuse into the neighbouring interval (see Fig. 18.5). We can represent a state of the line by means of a multiset in which each number n represents a particle present in the interval numbered n. The evolution of the system is then specified by the following three rules:

$$\begin{aligned} r_1 : &\quad n \longrightarrow n \\ r_2 : &\quad n \longrightarrow n-1 \\ r_3 : &\quad n \longrightarrow n+1 \end{aligned}$$

where n is an integer and the operations $+$ and $-$ which appear on the right side are the usual arithmetic operations. The rule r_2 (respectively r_3) specifies the behaviour of a particle that diffuses into the interval on its left (respectively right) and the rule r_1 specifies a particle that remains in the same interval.

18.3.2 Păun Systems and Compartmentalisation

The above encoding allows us to deal with linear geometry. Other variations have been proposed to facilitate the representation of more complex biological structures,

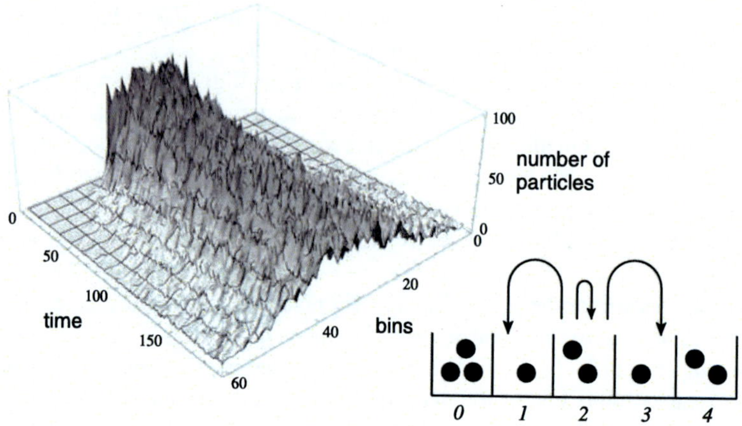

Fig. 18.5 Diffusion of particles in a thin rod. The left-hand figure shows the evolution over 160 time steps of 1500 particles initially distributed over 20 intervals in the middle of a bar discretized into 60 intervals. The diagram on the right specifies the behaviour of a particle (see text). The number of particles in an interval corresponds to the concentration of a chemical product or a quantity of heat. The multiset representing the state of the line (which is discretized into 5 intervals numbered from *0* to *4*) is given by: $0 \oplus 0 \oplus 0 \oplus 1 \oplus 2 \oplus 2 \oplus 3 \oplus 4 \oplus 4$

such as the nesting of membranes and compartments in a cell: the elements of a multiset can be molecules or other multisets, which can in turn contain molecules or other multisets.

This nesting is studied using the formalism of *Păun systems* (P systems) [36], in which the classical rewriting of multisets is extended by the concept of *membrane*. A membrane is a nesting of compartments represented, for example, by a Venn diagram[6] without intersections and with one sole superset: the *skin* of the system (see Fig. 18.6).

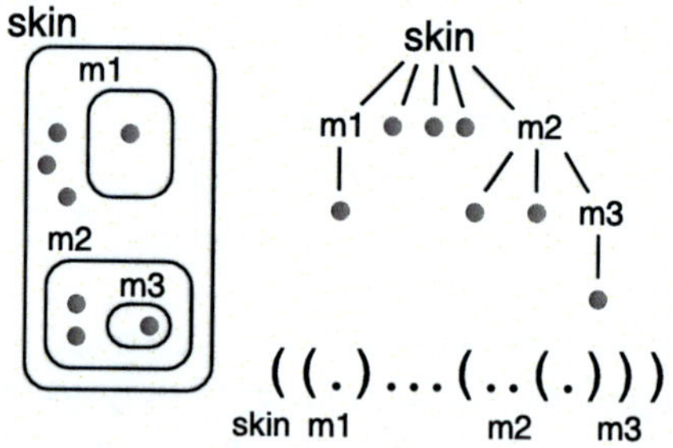

Fig. 18.6 Păun systems. Membrane nesting can be represented by a Venn diagram without intersections and with one sole superset, by a nesting tree or by a well-bracketed word

[6] Venn diagrams, invented by the English logician of the same name, are a means of visualising set operations by representing the sets as surfaces delimited by closed curves.

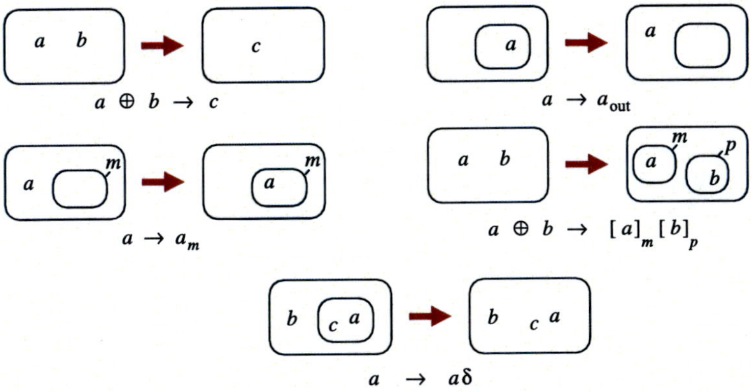

Fig. 18.7 Example of a rule of evolution in a Păun system. The symbol δ on the right side of a rule entails the dissolution of the enclosing membrane. The destination membranes are indicated by the suffix (the membranes are named). The enclosing membrane can always be referred to by the name "out"

Objects are placed in each region delimited by a membrane, and they then evolve according to diverse mechanisms: an object (or a multiset of objects) can change into other objects, but it can also cross a membrane or provoke the dissolution or creation of a membrane. Fig. 18.7 shows some examples of rules of evolution in P systems. Formally, such a system can be specified by using several operations $\oplus, \oplus', \oplus'', \ldots$, each corresponding to a certain membrane. These operations are associative and commutative, but they are not associative with each other (in order to keep the membranes separate).

18.3.3 In Parenthesis: The Application to Parallel Programming

The dialogue between computing and the other scientific disciplines is not all one-way. Here is an example. Inspired by the chemical metaphor, computer scientists have used multiset rewriting not only to simulate chemical reactions or biological processes, but also as a parallel programming language. The idea was first developed in the language GAMMA [3]. Here is a particularly elegant example of a parallel programme:

$$x \oplus y \;/\; (x \bmod y = 0) \longrightarrow y\,.$$

This rule specifies that the pair of numbers x, y must be replaced by y when the condition "y divides x" is satisfied (the condition is written after the / symbol). If we apply this rule as far as possible to the multiset composed of all the integers between 2 and n, we obtain a multiset in which the rule cannot be applied (because the condition is no longer satisfied) and which therefore contains all the prime numbers up to n.

In the above programme, there is no trace of artificial sequencing in the calculations: the rule can be applied in any order whatsoever. Note that the parallelism comes from the simultaneous application of rules and that the "unfolding" of the programme consists simply in repeating the application of the rules until a normal form (a fixed point) is obtained. These programmes are non-deterministic, unless the rewriting rules are confluent: in that case, when the programme ends, we do obtain a perfectly determined result, although the intermediate values calculated during the execution of the programme can differ (we speak of deterministic results despite a non-deterministic evaluation).

18.4 Lindenmayer Systems and the Growth of Linear Structures

In the previous section, we considered a process of rewriting on associative and commutative terms, allowing us to model a "chemical soup". In this section, we shall explore associative terms: these terms then correspond to sequences and we speak of *rewriting strings* (of symbols). Chomsky's work on formal grammars [7] marked the beginning of a long series of works on string rewriting, and these works have been at the origin of developments concerning syntax, semantics and formal languages in computing. Grammars are generative formalisms. In other words, they allow us to construct families of objects, by generating sets of *phrases*: a phrase is a sequence of symbols generated by successive rewritings. The set of phrases that can be generated is a *language*.

In 1968, the biologist Aristid Lindenmayer (1925–1989) introduced a new type of string rewriting to serve as the foundation for a formal theory of developmental biology [32]: Lindenmayer systems, more often abbreviated to L-systems. The main difference from Chomsky grammars lies in the strategy of rule application. In Chomsky grammars, only one rewriting is applied at a time, whereas in L-systems the rewritings take place in parallel, replacing all the symbols in a phrase at each step. Lindenmayer justified this strategy by analogy to cell development: all the cells in an organism divide independently and in parallel.

The objective of L-systems is to construct a complex object (like a plant) by successively replacing the different parts of a simpler object, by means of rewriting rules. Symbols are interpreted as components of a living organism, such as cells or organs, rather than words. L-systems have found numerous applications not only in the modelling of plant growth, but also in computer graphics, with the generation of fractal curves or virtual plants.

18.4.1 Growth of a Filamentous Structure

A simple example of L-system is the one that describes the growth of cyanobacteria *Anabaena Catenula*. These blue-green algae form filaments composed, for our example, of four types of cells, `G`, `g`, `D` and `d`, which can be interpreted as follows: `G` and D are mature cells capable of dividing; g and d are quiescent cells. In addition, the cells are polarised: D and d are polarised towards the right in the filament; G and g are polarised towards the left. When we examine a filament, we can see that the cells do not succeed each other in any old order. And the rewriting system presented below generates sequences very similar to those observed in nature.

The derivations on the right show that at each step all the symbols are rewritten in parallel according to the rules on the left. Numerous variations can be developed from this basic mechanism, with the aim of extending the expressivity of the formalism. One of the most important extensions involves attaching attributes to the symbols, for example a number representing a size, or the concentration of a chemical product. Fig. 18.8 illustrates the use of one such extension in a more realistic model of *Anabaena* growth. Instead of simply considering mature cells and quiescent cells, each cell possesses a size that grows over time. Furthermore, in a nitrogen-free medium, some cells become specialised: the heterocysts. Wilcox *et al.* [44] proposed that a cell differentiates into a heterocyst under the action of two chemical substances, an activator and an inhibitor, which diffuse in the filament and react with each other. This reaction-diffusion model allows to explain the appearance of an isolated heterocyst cell every n vegetative cells, as observed in nature, with a relatively constant n. Prusinkiewicz and Hammel [25] used an L-system to specify and simulate this system, thereby achieving the simulation of a reaction-diffusion in a growing medium (the filament), an important example of $(DS)^2$.

Fig. 18.8 Differentiation of heterocysts in an *Anabaena* filament. The *left-hand figure* represents the same diagram as the *right-hand one*, but seen from another angle. This graphical representation, called an extrusion in space-time, was introduced in [25]. In this diagram, times moves from the *top-left corner* to the *bottom-right corner*. Each "slice" represents the cells of a filament at a given moment in time. The height of each cell represents the concentration of activator, as does the shading of the cell (from *black to white*). The *black cells* are vegetative. Differentiation occurs when the concentration of activator rises above a threshold level

18.4.2 Development of a Branching Structure

It is easy to represent a branching structure by a string, by introducing two symbols that serve as "brackets". There are subtle differences between the rewriting of such strings and the direct rewriting of terms. String rewriting is the method used in L-systems to represent the branching structure of a plant and its rules of development. The example below is caricatural, and does not correspond to the growth of any real plant. But it does allow us to illustrate the power of this approach.

Let us assume that a plant is made up of two types of "branch": simple branches `b` and budded branches `B`. From one year to the next, budded branches lose their buds and become simple branches. We therefore have the rule `B` $\rightarrow$ `b`. A simple branch grows and produces a section of plant comprising an axis made up of three simple branches with two budded branches branching off it, 1/3 and 2/3 of the way up. This specification is expressed by the rule:

$$\mathtt{b} \longrightarrow \mathtt{b}\,\langle\, q\mathtt{B}\,\rangle\, \mathtt{b}\,\langle\, p\mathtt{B}\,\rangle\, \mathtt{b}\ .$$

In this rule, we use the brackets $\langle$ and $\rangle$ the represent the development of the lateral axes. We use the additional symbols p and q to indicate development on the left or the right of these axes.

At the beginning of the 1980s, P. Prusinkiewicz introduced a graphical interpretation of words produced by an L-system [38–40]. This interpretation is based on the concept of graphical turtle, as used in the LOGO programming language, for example. This allows to directly visualise the structure of objects described by a word generated by the L-system. Thus, we use the successive derivations of an L-system to represent the successive states of a developing plant, or to draw the successive curves that tend towards a fractal curve.

A state of the graphical turtle is the triplet (x, y, θ), where (x, y) represents the current position of the turtle in Cartesian coordinates and θ represents the orientation of the turtle. This orientation is interpreted as the angle between the body of the turtle and the horizontal axis. The turtle moves following commands represented by the symbols of a word.

- In our example, the symbol `b` corresponds to the command "move forward one length Δ". So if the current state was (x, y, θ), then after reading the symbol `b` it becomes $(x + \Delta \cos\theta, y + \Delta sin\theta, \theta)$.
- The symbol `B` is interpreted in the same way, except that after drawing the corresponding segment, we also draw a circle centred on the current position.
- The symbols $\langle$ and $\rangle$ save and restore the current state respectively. The position is saved in a stack. When the turtle meets the symbol $\rangle$, it "jumps" to the position corresponding to the open bracket.
- Finally, the symbol p (resp. q) increments (resp. decrements) the current angle θ by a predefined angle.

Using this graphical interpretation, the first three derivations of a simple branch are illustrated in Fig. 18.9.

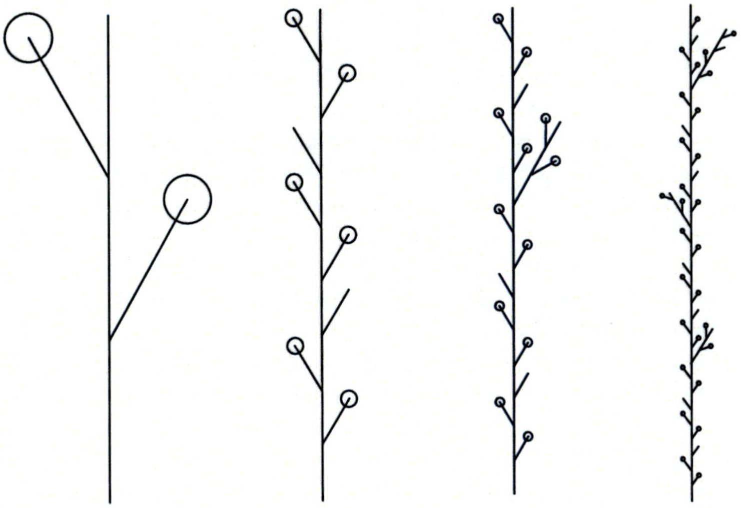

Fig. 18.9 Graphical representation of the first three derivations in a Lindenmayer system. The initial state is given by the word b ⟨ qB ⟩ b ⟨ pB ⟩ b and the rules of derivation are the two rules defined in the text. The scale of the representation of each "plant" is different

18.5 Beyond Linear Structures: Calculating a Form in Order to Understand It

L-systems have proved to be perfectly suited to the modelling of plant growth [37, 40]: they allow to define in a particularly compact and synthetic way the creation of the complex form of a plant and above all, in their recent extensions, to couple the process of form creation with the physical-chemical processes that take place within that form.

However, although L-systems are suitable for the representation of linear forms (filaments or trees), their use for the construction of more complex shapes (ordinary graphs, surfaces or volumes) depends on arbitrary encoding that rapidly becomes inextricably complex. Researchers are therefore trying to design more suitable formalisms. The import of this search for formalisms to specify the processes of development reaches far beyond the question of simulation, for two reasons: these formalisms could fill a conceptual vacuum in biology, and they could potentially have an enormous epistemological impact. What is more, their application could extend far beyond the domain of biology.

18.5.1 Simulation and Explanation

Drawing firstly on purely physical models (osmotic growth with Leduc, optimal forms with D'Arcy Thompson, reaction-diffusion processes with Turing, etc.), then purely genetic models (with concepts such as gene action or the genetic programme), the different formalisms proposed over the course of the last century to specify the processes of development have filled a conceptual vacuum and modified the perception of what has explanatory value for biologists [20].

As an example, advances in computing and the data produced in biology allow the simulation of certain processes of development with predictions that can then be empirically validated [8]. Very recently, for instance, several cell-level models [4, 41] of meristem development (the meristem being the growing tissue of the plant) have succeeded in reproducing characteristic phyllotactic patterns observed in nature and in linking them to the circulation of auxin (a plant hormone) in this tissue. The accumulation of auxin triggers the development of new organs, which modify the form of the meristem and consequently the flow of auxin: a marvellous example of $(DS)^2$.

However, no matter how predictive these simulations are, they can only have an explanatory value if they allow us to express the processes of development in a form that is intelligible to the human mind, so that we can analyse them and reason about them [29]. After all, what kind of understanding can we hope to derive from the simple observation of a succession of complex calculations? We might just as well observe these processes in nature, instead of reproducing them on a computer.

Computer morphogenesis allows us to define a formal framework, in which we can speak rigorously of genetic programme, memory, information, signal, interaction, environment, etc. and to relate these concepts to a completely mechanistic view of development processes. It introduces the concept of *computation* as an explanatory scheme in the modelling of development. But if the embryo can be deduced by computation from a description of the egg and its interactions with the environment, the embryo must be considered both as the result of a computation and as part of the computer that produces this result. This problem is studied in computer science (reflexive interpreters, meta-circular evaluators). The future will tell whether these concepts will enable us to grasp that most specific aspect of living beings: their development.

18.5.2 Giving Form to a Population of Autonomous Agents

The modelling of development processes is important for biologists, but it is also important for computer scientists, who are always looking for new computational models and for whom biology is clearly a great source of inspiration.

Computational models are constrained by the particularities of a material model or inspired by a metaphor of what a computation should be. Today, new material supports for computation are being studied. One celebrated example is the experiment that Adleman performed in 1994 [1], proving that a combinatorial problem[7] can be solved using DNA molecules in a test-tube. But other possibilities

[7] The problem he chose was the Hamiltonian path problem, consisting in determining whether a given graph contains a path that starts at the first vertex, ends at the last vertex, and passes exactly once through each remaining vertex.

are currently the subject of very active research, including using the growth of colonies of bacteria, the diffusion of chemical reagents or the self-assembly of biomolecules ... to compute. The programming of these new computational supports certainly raises some substantial problems, and is driving the development of new languages and algorithms to allow us to use an immense population of autonomous entities (biomolecules, viruses or cells) that interact locally and irregularly, to construct and develop a reliable computation (a form).

But the mechanisms offered by a programming language, or by new algorithms, can also be directly inspired by a biological metaphor without resorting to biological machines built using biotechnologies. For example, evolutionary algorithms are inspired by the mechanisms studied in evolution theory, even though they are executed on electronic machines like present-day computers.

In the same order of idea, formalisms providing a conceptual grasp of the mechanisms of development could well revitalise the concept of "programme", by suggesting new approaches in the development of very big software, notably in the specification of their architecture and the interconnection of their different parts, or by offering new mechanisms for hiding useless information, abstracting details or capitalising and reusing code. Computer scientists are actively seeking, for their software, properties usually attributed to living matter: autonomy, adaptability, self-repair, robustness, self-organisation. Clearly, the dialogue between computing and biology [26–28, 35], so ambiguous and so fertile, is not about to end.

References

1. Adleman, L. (1994) Molecular computation of solutions to combinatorial problems, *Science* **266**, 1021–1024.
2. Atlan H. (1999) *La Fin du "tout-génétique"? Vers de nouveaux paradigmes en biologie*, INRA Éditions (Paris).
3. Banatre J.-P., Coutant A., and Metayer D.L. (1988) A parallel machine for multiset transformation and its programming style, *Future Generation Computer Systems* **4**, 133–144.
4. Barbier de Reuille P., Bohn-Courseau I., Ljung K., Morin H., Carraro J., Godin C., and Traas J. (2006) Computer simulations reveal properties of the cell-cell signaling network at the shoot apex in Arabidopsis. *PNAS* **103**, 1627–1632.
5. Bournez O. and Hoyrup M. (2003) Rewriting logic and probabilities, in *14th Int. Conf. on Rewriting Techniques and Applications (RTA'03)*, Valencia, June 2003, Lecture Notes in Computer Science, vol. 2706, edited by R. Nieuwenhuis, Springer (Berlin), pp. 61–75.
6. Bournez O., Côme G.-M., Conraud V., Kirschner H., and Ibanescu L. (2003) A rule-based approach for automated generation of kinetic chemical mechanisms, in *14th Int. Conf. on Rewriting Techniques and Applications (RTA'03)*, Valencia, June 2003, Lecture Notes in Computer Science, vol. 2706, edited by R. Nieuwenhuis, Springer (Berlin), pp. 30–45.
7. Chomsky N. (ed.) (1957) *Syntactic structures*, Mouton & Co. (The Hague).
8. Coen E., Rolland-Lagan A.-G., Matthews M., Bangham J.A., and Prusinkiewicz P. (2004) The genetics of geometry, *PNAS* **101**, 4728–4735.
9. Dershowitz N. (1993) A Taste of Rewrite Systems, Lecture Notes in Computer Science, vol. 693, Springer-Verlag (Berlin), pp. 199–228.
10. Dershowitz N. and Jouannaud J.-P. (1990) Rewrite systems, in *Handbook of Theoretical Computer Science*, vol. B, Elsevier (Amsterdam), pp. 244–320.

11. Dittrich P. (2001) Artificial chemistry webpage, `ls11-www.cs.uni-dortmund.de/achem`
12. Dittrich P., Ziegler J., and Banzhaf W. (2001) Artificial chemistries — a review, *Artif. Life* **7**, 225–275.
13. Eigen M. and Schuster P. (1979) *The Hypercycle: A Principle of Natural Self-Organization*, Springer (Berlin).
14. Eker S., Knapp M., Laderoute K., Lincoln P., Meseguer J., and Sonmez J. (2002) Pathway logic: Symbolic analysis of biological signaling, in *Proceedings of the Pacific Symposium on Biocomputing*, January 2002, pp. 400–412.
15. Eker S., Knapp M., Laderoute K., Lincoln P., and Talcott C. (2002) Pathway logic: Executable models of biological networks, in *Fourth International Workshop on Rewriting Logic and Its Applications (WRLA'2002)*, Electronic Notes in Theoretical Computer Science, vol. 71, Elsevier (Amsterdam).
16. The Elan project. Elan home page, 2002. `http://www.loria.fr/equipes/protheo/SOFTWARES/ELAN/`
17. Fisher M., Malcolm G., and Paton R. (2000) Spatio-logical processes in intracellular signalling, *BioSystems* **55**, 83–92.
18. Fontana W. and Buss L. (1994) "The arrival of the fittest": Toward a theory of biological organization, *Bull. Math. Biol.* **56**, 1–64.
19. Fox Keller E. (1995) *Refiguring Life: Metaphors of Twentieth-century Biology*, Columbia University Press (New York).
20. Fox Keller E. (2002) *Making Sense of Life: Explaining Biological Development with Models, Metaphors, and Machines*, Harvard University Press (Cambridge MA).
21. Giavitto J.-L. (2003) Invited talk: Topological collections, transformations and their application to the modeling and the simulation of dynamical systems, in *14th Int. Conf. on Rewriting Techniques and Applications (RTA'03)*, Valencia, June 2003, Lecture Notes in Computer Science, vol. 2706, edited by R. Nieuwenhuis, Springer (Berlin), pp. 208–233.
22. Giavitto J.-L. and Michel O. (2002) The topological structures of membrane computing, *Fundamenta Informaticae* **49**, 107–129.
23. Giavitto J.-L. and Michel O. (2003) Modeling the topological organization of cellular processes, *BioSystems* **70**, 149–163.
24. Gillespie D.T. (1977) Exact stochastic simulation of coupled chemical reactions, *J. Phys. Chem.* **81**, 2340–2361.
25. Hammel M. and Prusinkiewicz P. (1996) Visualization of developmental processes by extrusion in space-time, in *Proceedings of Graphics Interface '96*, pp. 246–258.
26. Head T. (1987) Formal language theory and DNA: an analysis of the generative capacity of specific recombinant behaviors, *Bull. Math. Biology* **49**, 737–759.
27. Head T. (1992) Splicing schemes and DNA, in *Lindenmayer Systems: Impacts on Theoretical Computer Science, Computer Graphics, and Developmental Biology*, Springer (Berlin), pp. 371–383. Reprinted in *Nanobiology* **1**, 335–342 (1992).
28. INTERSTICE (web site presenting research activity in the domain of the science and techniques of information and communication), *Dossier sur la bio-informatique*, in French. `http://interstices.info/display.jsp?id=c_6474`
29. Israel G. (1996) *La mathématisation du réel*, Seuil (Paris), in French.
30. Itkis Y. (1976) *Control Systems of Variable Structure*, Wiley (New York).
31. Lemerle C., Di Ventura B., and Serrano L. (2005) Space as the final frontier in stochastic simulations of biological systems, *Minireview. FEBS Letters* **579**, 1789–1794.
32. Lindenmayer A. (1968) Mathematical models for cellular interaction in development, Parts I and II, *J. Theor. Biol.* **18**, 280–315.
33. The Maude project, Maude home page, 2002. `http://maude.csl.sri.com/`
34. von Neumann, J. (1966) *Theory of Self-Reproducing Automata*, Univ. of Illinois Press (Urbana-Champaign).
35. Paton R. (ed.) (1994) *Computing With Biological Metaphors*, Chapman & Hall (London).
36. Păun G. (2002) *Membrane Computing. An Introduction*, Springer-Verlag (Berlin).
37. Prusinkiewicz P. (1998) Modeling of spatial structure and development of plants: a review, *Sci. Hortic.* **74**, 113–149.

38. Prusinkiewicz P. (1999) A look at the visual modeling of plants using L-systems, *Agronomie* **19**, 211–224.
39. Prusinkiewicz P. and Hanan J. (1990) Visualization of botanical structures and processes using parametric L-systems, in *Scientific visualization and graphics simulation*, edited by D. Thalmann, J. Wiley & Sons (Chichester), pp. 183–201.
40. Prusinkiewicz P., Lindenmayer A., Hanan J., *et al.* (1990) *The Algorithmic Beauty of Plants*, Springer-Verlag (Berlin).
41. Smith R.S., Guyomarc'h S., Mandel T., Reinhardt D., Kuhlemeier C., and Prusinkiewicz P. (2006) A plausible model of phyllotaxis, *PNAS* **103**, 1301–1306.
42. Takahashi K., Vel Arjunan S.N., and Tomita M. (2005) Space in systems biology of signaliting pathways — towards intracellular molecular crowding in silico, *Minireview. FEBS Letters* **579**, 1783–1788.
43. Varela F.J. (1979) *Principle of Biological Autonomy*, McGraw-Hill/Appleton & Lange (New York).
44. Wilcox M., Mitchison G.J., and Smith R.J. (1973) Pattern formation in the blue-green alga, *Anabaena*. I. Basic mechanisms, *J. Cell Sci.* **12**, 707–723.

Index

P. Bourgine, A. Lesne (eds.), *Morphogenesis*, DOI 10.1007/978-3-642-13174-5,